Human Settlements and Planning for Ecological Sustainability

Urban and Industrial Environments

Series Editor: Professor Robert Gottlieb, School of Public Policy and Social Research, University of California, Los Angeles

The U.S. Paper Industry and Sustainable Production: An Argument for Restructuring
Maureen Smith, 1997

Human Settlements and Planning for Ecological Sustainability: The Case of Mexico City
Keith Pezzoli, 1998

Human Settlements and Planning for Ecological Sustainability

The Case of Mexico City

Keith Pezzoli

The MIT Press
Cambridge, Massachusetts
London, England

This book was set in Sabon in Miles 33 by Crane Typesetting Services, Inc.

Printed and bound in the United States of America.

Library of Congress Cataloging-in-Publication Data

Pezzoli, Keith.
    Human settlements and planning for ecological sustainability : the
case of Mexico City / Keith Pezzoli.
        p.   cm. — (Urban and industrial environments)
    Includes bibliographical references and index.
    ISBN 0-262-16173-7 (hc)
    1. Mexico City—Environmental conditions. 2. Land use—Mexico—
Mexico City. 3. Ajusco (Mexico) 4. Sustainable development—
Mexico—Mexico City. 5. Land use—Mexico—Mexico City—Planning—
Citizen participation. I. Series.
HT243.M62M487   1998
363.7'07'097253—dc21   97-39497
                                                                        CIP

To the people of Bosques del Pedregal,
especially Miguel and Consuelo,
Polo and Maricela, Ernesto and Teri

Contents

Foreword

In her recent study on "human capabilities," Martha Nussbaum puts forward a strong argument for universal human values that transcend cultural specificities. Hers is a noble undertaking. But we don't need Aristotelian philosophy nor Kantian moral imperatives to demonstrate that there are two interconnected human struggles going on in the world today that, despite their local particularities, are fundamentally the same. I refer to the struggle for a subsistence livelihood and the struggle for the life space of land and housing. Focusing on urban population alone, we can make these rough numerical estimates: by the year 2000, the world's population will number about 6 billion people, of whom more than half will be living in urban environments. Of this number, from one-third to one-half are engaged in struggles for a basic livelihood and land, totaling between 1 and 1.5 billion people who have abandoned on a more or less permanent basis rural homes where they felt their life situation was even less promising than in the city. Here, then, we have a universal struggle about which we, the readers of this book, are only vaguely aware. We know about the curse of poverty, but most of us don't really know what it takes to put even the bare necessities of food and water on the table, each and every day, or gain a foothold in the city. Keith Pezzoli's book provides us with a window onto these struggles, with a special focus on the securing of life space in one community on the southwestern edge of the Mexico City metropolis.

And what struggles they are! Here, "struggle" is not just a metaphor, but a continuing, decade-long series of desperate efforts to occupy land; defend people's right to it against a state that is forever fearful of descending into what it chooses to call "anarchy"; self-build their homes and schools; provide for themselves the social infrastructure minimally necessary for a civilized life (water, electricity); and do all of this while those among them in the labor force commute enormous distances to find work in the city so that each household might replenish its energies from day to day. The tenacity of this struggle is heroic. Leaders are arrested. Some disappear. Others are killed in violent action. But this struggle for land is not a simple case of "the people vs. the state." Actors are numerous, pursuing very different objectives for different reasons. The people's struggle is a collective struggle, in the sense that it involves communities mobilized around specific issues that, depending on the moment, vary. Mostly, communities (based on territorial propinquity) act on their own; but sometimes they unite in more coordinated actions. And "the state" is, of course, not a monolith either but a set of different and partially autonomous agencies, from the president all the way down to the local ecoguard in Ajusco, involving different ministries, district governments, politicians, and the ever present PRI—the misnamed Institutional Revolutionary Party—which runs the government like a well-oiled machine, or at least used to. Its grasp on power is weakening. On the side of the people there are a number of "external" agents, ranging from nongovernmental agencies, foreign voluntary organizations, the Catholic Church, university students, and even an "appropriate technology" business enterprise. Aligned with the state (though one is tempted here to reverse the sequence of who is aligned with whom), the so-called private sector: business interests, both legal and illegal developers of new subdivisions, wealthy homeowners with friends in the government. All of them are sucked into the vortex of this intense struggle of hundreds of thousands of Mexican citizens to survive in a city that, to them, appears for the most part hostile and malignant. But of course cities are nothing of the sort. They are a space in which innumerable encounters take place between those who in various ways are powerful and those who are not.

The powerful usually win, but not always. Pezzoli describes a return visit to his beloved Ajusco region and is astonished at what he sees: paved roads, electric lights, houses built of brick, stores—in short, the beginnings of a working-class neighborhood, one of hundreds of such neighborhoods in this metropolis of more than 15 million. So, one might conclude that "all's well that ends well," and the struggle was worth it, and people can now get on with their lives. But this is not the story on which Pezzoli spends much time. Because there is another facet that has so far remained hidden from view but which now needs to be considered. Urban development, according to a frequently repeated refrain, must be sustainable. And Mexico City's development, we learn, is not. Aquifers are drying up; parts of the city are plagued with subsidence; the air is unhealthy; the city's "green belt," or ecological preserve, is being deforested. To make life in Mexico City possible, vast quantities of resources must be brought in from distant provinces at great cost to the public purse. Above all, water must be pumped up at enormous expense to the Valley of Mexico, whose altitude lies between 1,000 and 2,500 meters.

The story of the natural environment "coevolves" with the story of Mexico City's settlement. The human story intersects with that of Nature. The ideal would be a win/win outcome, as the current jargon has it, in which harmony between humans and their environment is reestablished. But this is not what happens. In reality, humans gain at the expense of Nature, which moves ineluctably toward a state of increasing entropy, or randomness. The so-called urban crisis is simply an outward manifestation of this rise in entropic energy levels.

Mexico City continues to add 2 percent annually to its population, that is, 300,000 residents or more. Thus, the people of Ajusco, in their search for land, comprise about a year's increment of the city's population. As it happened, they found a harsh, salubrious reserve of land in Mexico City's ecological preserve. Keith Pezzoli tells the story of this "preserve" and how it was set aside as a national park by President Lázaro Cárdenas in 1930 and subsequently turned to more profitable use on private account: by a pulp and paper mill, by upper-class villas and condominiums, and by the city, which in its wisdom decided to put one of its major garbage dumps into the ecological terrain! Part of this story is the illegal sale of *ejido* land, which was supposedly inalienable land turned over to indigenous communities of peasants by President Cárdenas in perpetuity. So, the invasion of this "sacred" natural terrain (and historical battleground) also reflects the dissolution of communal rights over land, yet another aspect of the rise in entropy.

Pezzoli makes the best of a failed story of "alternative development" in Ajusco that is centered on a phase in the community's struggle for land. Encouraged by various "external agents," the people of Ajusco declared themselves to be a *colonia ecológica productiva* (CEP)—a productive ecological popular settlement—that would plant trees, cultivate mushrooms and orchards, and use energy-friendly systems of waste disposal in an effort to rehabilitate the ecological preserve on which they had chosen to settle and whose "life space" they were now defending and so reverse the flow of entropy in their immediate environment toward lower levels. This story of a noble but ultimately failed experiment is many sided and complex. It has to do with the state's ability to satisfy the *colonos*'s hunger for land, which was their first concern, and thus reveal the alternative CEP approach to be what it had always been for the colonos, a strategic means toward an end. And the end was land, not ecology.

Pezzoli ends his study on a hopeful note, advocating a new slant on ecology, which he calls political ecology, and asking us to think holistically about the environment. I am less sanguine about our ability to "save the earth." Those in power are not inclined to curtail their rapacious drive for ever more power. The name of the game today is global competition in the interest of private accumulation. By the end of the next century, the world will be thoroughly urbanized and will have exhausted itself in the pervasive desire to enjoy the commodities that capitalism conjures up before eager masses of consumers. In the remaining enclaves of the "backward" parts of the world that have been unable to join this self-destructive march of the lemmings, warlordism of the sort we are witnessing in Somalia, Afghanistan, and Liberia today, along with periodic famines, is what the future probably holds in store.

But who knows? Perhaps Keith Pezzoli and many others like him can reverse this dismal prospect and get us back on track to a win/win situation. I would certainly hope so.

John Friedmann
Melbourne, Victoria
February 6, 1997

Preface

Mexico is mired in its deepest political and economic crisis since the Mexican revolution sixty-five years ago. Over the period of 1995 to 1996, the nation's gross domestic product (GDP) fell by 6.9 percent (more than any year in the past six decades), 2 million more people joined the ranks of the unemployed, inflation shot up 50 percent, and real wages dropped another 20 percent (Weintraub 1996). Although the economy began to strengthen during 1996, new armed uprisings heightened concern about Mexico's political stability.

During 1994, the insurrection led by the Zapatista National Liberation Army (Ejército Zapatista de Liberación Nacional EZLN) drew attention to extreme poverty in rural Mexico. In late 1996, another rebel force—the People's Revolutionary Army (Ejército Popular Revolucionario; EPR)—shook the nation, and financial markets, by carrying out a series of antigovernment attacks in several Mexican states (*Excelsior*, August 30, 1996). At a press conference held on the outskirts of Mexico City, two EPR *comandantes* claimed that their left-wing organization has 23,000 guerrillas across the country—including "fresh forces" in Mexico City "ready to be activated" (Guerrero Chiprés 1996). President Ernesto Zedillo vowed "to act with the full force of the state" against these rebels (Second State of the Union Address, September 1, 1996). The turmoil has heightened concern about the impacts and fragility of Mexico's socioeconomic and political restructuring.

Throughout Mexico and the rest of Latin America there is a huge and widening gap between people's needs for livelihood, housing, electricity, transport, education, water, and sanitation and meeting those needs. Politicians, corporate elite, and government functionaries continue to express faith in the capacity of free trade, privatization, and export-led growth to ultimately resolve these problems. But not everyone is so sure. A rising number of critics—including those who have picked up arms—argue that neoliberal policies have failed to foster sustainable, poverty-reducing development.

Barry (1995) argues that the neoliberal agenda is actually leading to reduced food security, environmental destruction, increased rural-urban polarization, depopulation of peasant communities, and social and political instability. Pradilla Cobos (1995a, 1995b) advances a similar argument and focuses on Mexico City to support it. He argues that Mexico's National Development Plan (Plan Nacional de Desarrollo), the Emergency Adjustment Program (Plan Emergente de Ajuste), and the Economic Reactivation Program for the Federal District (Programa de Reactivación Económica para el D.F.), do not address the root causes of the enduring crisis. Whether one agrees with such critics or not, one thing is clear: Mexico City is an increasingly troubled place. During 1995, more than 800 crimes were reported daily in the capital—a doubling of the rate reported in 1994 (Weintraub 1996, 6). The Bank of Mexico has been tracking the rise in consumer prices in forty-six cities across the nation. Out of all these urban areas, Mexico City has registered the highest price increases (González Amador 1996). To get access to the low-income housing provided with government support, a head of household must now earn 4.5 times the minimum salary—up from 2.5 times the minimum a year ago (Hidalgo 1996). This puts formal sector housing beyond the reach of nearly three-quarters of the families in Mexico City who need it. Newspaper headlines flash the warning that land invasions will be the only recourse left to the urban poor.

Mexico City's problems epitomize many of the Third World's worst urban-ecological ills. As one author aptly puts it, "Mexico City is virtually a metaphor for Third World urban woes. With nearly 20 million people, 3 million vehicles, and 35 thousand industries, Mexico City is a megacity with megapollution problems" (Mumme 1991, 11). At the same time, though, Mexico City is one of the world's most exciting places for the development of new approaches to urban problems. It is a dynamic setting ripe with lessons about opportunities as well as constraints.

According to Habitat—the United Nations Centre for Human Settlements with a mandate to help governments improve housing conditions and manage urbanization—an estimated 500 million urban dwellers in cities worldwide are homeless or live in inadequate housing (Habitat 1996a). Housing production has not kept pace with demand in fast-growing cities. The world's urban population—expected to double from 2.4 billion in 1995 to 5 billion in 2025—is growing 2.5 times faster than the rural population (Habitat 1996a). In many cities, housing shortages and poor housing conditions have become life threatening. Substandard housing, unsafe water, and poor sanitation in densely populated cities are responsible for 10 million deaths worldwide every year (Habitat Press Release, May 31, 1996). Habitat projects that "more than one and a half billion people in the world's cities will face life and health threatening environments by the year 2025, unless a revolution in urban problem-solving takes place" (Habitat media report, November 20, 1995).

In documenting the profound challenges facing Mexico City, this book weaves together local, metropolitan, regional, and global levels of analysis. The story centers on a case study about community mobilization and the interactive politics of environmental regulation and popular resistance around ecological and land use issues. The case documents the conflict-driven process whereby urban sprawl has penetrated deeper and deeper into Mexico City's so-called ecological reserve. Urban encroachment into the greenbelt zone has been fueled by the rich and poor alike. The government's approach to this "irregular" expansion of human settlements has been contradictory and reactive. The environmental politics involved are inimical to securing ecological sustainability. The consequences make it less and less likely that the more ambitious agenda calling for *sustainable development* will ever be realized.

The book links the urban land question and the concept of sustainability to questions about livelihood opportunities and political mediation in the production of human settlements. Looking at the heart of the matter, I describe the creative proposals of several grassroots movements. My analysis of these movements does not pretend to offer a panacea. But it does offer noteworthy lessons about urban development and opportunities for community groups to deal with aspects of the city's urban-ecological problematic. The bottom-up initiatives of such grassroots movements are promising for their capacity to promote much needed social experimentation and social learning.

Most of the fieldwork I did for this book focused on a mountainous and forested area in Mexico City called Ajusco. Ajusco is a miniregion corresponding to the *Sierra del Ajusco*—a volcanic mountain range with forested foothills in the southwestern part of the metropolitan area. Most of Ajusco is designated for rural use and as an ecological reserve intended to function as a greenbelt. The greenbelt is supposed to contain urban growth while protecting forest cover, open space, biodiversity, and the regional watershed. But the area is subject to enormous pressures of urbanization. The legally designated urban limits have been extended outward four times since 1980 to accommodate urban encroachment inside Ajusco. Expansion of these limits has involved social unrest, ecological disruption, and violence. In short, Ajusco is a hotly contested terrain where interests within and among a host of actors clash. The cast of actors involved includes the state, popular groups, *ejidatarios* (farmers), developers, and upper-class elite.

My experience in Ajusco over the past decade has enabled me to compile a comprehensive record of the zone's development and to witness the process firsthand. For months at a time, I lived with people in the illegal settlements I studied. This enabled me to develop close and lasting relations with many families and to follow the careers of local grassroots activists. At the same time, I observed the emergence, and in some cases the demise, of cooperatives, popular movements, and government programs. I also witnessed settlements expand and in some cases be eradicated.

The origins of this study go back to 1983. At that time, I began work on a thesis about Mexico City for a master's degree in urban planning from the University of California, Los Angeles. The thesis focused on social and political aspects of low-income settlements and housing provision in the zone of Ajusco. The early eighties were a time when the vast literature singing praise for self-help housing and "freedom to build" began to be critiqued. New research situated the housing problem and housing solutions in a political economy perspective. Attention was drawn to critical social movements and to the importance of collective consumption as a rallying point for social change. It was from this perspective that I completed my master's thesis, titled "Mexico City: Autonomous Urban Movements and the Production of Residential Space." The focus was on the social movements of Ajusco and their relation to the state.

I completed the master's thesis in 1985. My doctoral dissertation expanded the scope of this earlier study and examined the urban poor's access to land. Around this time, a great deal of literature became available on squatter property markets. Two factors spurred interest in the area. First, the rise of neo-Marxist critiques attracted attention to ownership relations and production processes. Second, the mid-1980s saw a change in government policies. Greater emphasis was placed on consolidating self-help settlements than on eradicating them. This shift called for increased insight into the dynamics of irregular settlement.

I thus designed my dissertation, titled "The Politics of Land Allocation in Mexico City's Ecological Zone," to answer two broad organizing questions. First, what makes the practices of irregular settlement in Ajusco possible? More specifically—focusing on the expansion of irregular settlements that took place in Ajusco Medio from the late 1970s to 1988—who were the key actors and what were the socioeconomic structures involved? Second, what were the mechanisms and strategies brought into play in the power relations among the collective actors in the zone? My central concern was to explain how urban planning initiatives and environmental policies impacted the organization and strategies of popular groups (and vice versa).

Upon completing the dissertation in 1990, it was clear to me that the land question is an increasingly pivotal one in urban studies. It has technical, economic, social, political and ecological aspects. The story I tell in this book builds upon that perspective and includes the results of additional research I completed from 1993 to 1996. I am now compelled to adopt an increasingly holistic view of the dynamics surrounding the urban poor's access to land for human settlements. To promote the viable development of human settlements in the 1990s, it is imperative to take into account the interplay of biogeophysical and ethicosocial concerns with concerns about livelihood opportunities and sustainability.

Standing back from the urban drama unfolding in the Valley of Mexico, I can best sum it up as *creative destruction*. Tremendous ingenuity, hard work, and sustained dedicated effort on the part of whole communities of people go into the making of Mexico City's human settlements. Much gets done. Until recent history, the destructive aspects of this process (e.g., ecological disturbances, forced evictions) were significantly buffered. But said buffer capacity—both in its ecological and political dimensions—has diminished considerably. More will have to be done with less. Building healthy human settlements at the dawn of the twenty-first century calls for a mobilized citizenry willing and able to engage in social experimentation. This achievement calls for political change as well as social, technical, and environmental innovation. There are no easy answers. But there certainly have been lessons learned—some outlined in the story told here—that suggest what does and does not appear to be working.

Acknowledgments

The research and writing of this book were made possible thanks to generous financial, intellectual, and moral support from many sources. First of all, I want to acknowledge the warm support I got from many families in Bosques del Pedregal, one of the low-income settlements in my zone of study. These families encouraged me to live and work in their community. They taught me the meaning of popular struggle and the importance of critical social movements. Other people that were a great help during my stays in Mexico City include Professor Martha Schteingart and Miguel Díaz Barriga. Miguel continues to be a good friend. He gave me detailed comments on an early draft of this manuscript that I found very helpful.

Much of the Ajusco case study discussed in this book was completed during my tenure as a master's and then doctoral student in the Graduate School of Architecture and Urban Planning at the University of California, Los Angeles (UCLA). While at UCLA, I worked closely with Professor John Friedmann. Friedmann created a research environment at UCLA that drew my attention to grassroots movements and alternative development. His dedication and insights concerning social change have provided me with a deep source of inspiration. Along these lines, I also owe a great debt to Professors Peter Marris and Allan Heskin.

In the early stages of my research, I was able to cover fieldwork expenses thanks to grants from UCLA's Program on Mexico and from UCLA's Urban Planning Department. Later, I received a doctoral dissertation fellowship from the Inter-American Foundation (IAF). The IAF, under the direction of Charles Reiley, provided much more than money; it tied me into a network of scholars and gave me a sense of community while I was in the field. I wrote my dissertation during my stay as a visiting research fellow (from August 1988 to August 1989) at the Center for U.S.-Mexican Studies in La Jolla, California. My fellowship at the center proved to be an invaluable experience for which I owe special thanks to Professor Wayne Cornelius. The center provided me a place within a vibrant research community where spirited conversations and debates played a major part in motivating and shaping this study. I especially benefited from the company of two other visiting research fellows: Neil Harvey and Maria Lorena Cook.

The final form of this book derives from my eight years of experience teaching and doing research in the Urban Studies and Planning (USP) Program at the University of California, San Diego. I owe special thanks to the coordinator of the USP program, Professor Amy Bridges. She skillfully edited an earlier draft of the book. She also created a supportive environment in which I was free to develop new USP courses, three of which helped me define key points of this book.

Small portions of the book have been published elsewhere. Specifically, part of chapter 10 appeared in *Environment and Development in Latin America,* edited by David Goodman and Michael Redclift (Manchester University Press). Part of chapter 9 appeared in *In Defense of Livelihood: Comparative Studies on Environmental Action,* edited by John Friedmann and Haripriya Rangan (UNRISD and Kumarian Press). It should be noted that throughout this volume translations of quotes in Spanish are my own unless noted otherwise.

Getting the book to press would not have been possible without the helpful guidance of Professor Robert Gottlieb, editor of the MIT Press Urban and Industrial Environments series. Nor would it have made it to press without the steadfast support of Madeline Sunley and Enza Vescera, two of the MIT Press' tireless acquisitions editors. I also owe a great deal to the four individuals who anonymously reviewed the manuscript at various stages. The peer reviews they provided were a great help.

When it came time to hurriedly prepare "camera ready" copy of all the figures, photos, and maps, one man saved the day: Mark Waggoner. Mark runs the outstanding GIS lab in the library at the University of California, San Diego (UCSD). He was there when I needed him, and I will always be grateful. I also want to acknowledge Virginia Steel, the head of UCSD's Social Science and Humanities Library, where the GIS lab is located. She obviously understands the importance of providing researchers with access to cutting-edge technology. The one piece of artwork commissioned for the book—a black-and-white sketch of the Valley of Mexico (figure 4.1)—was skillfully drawn by Belinda di Leo. Thank you Belinda.

Most of all, I am indebted to my wife, Janice. She is my best friend and closest colleague—always ready to share a laugh, challenge an assumption, and contribute fresh ideas about people, history, and ecology. This book has been a long time in the making; no one knows that better than Janice. Her special moral and intellectual support spans a full decade. This includes time living with me in the field (Ajusco, Mexico City) while helping gather participant observation data. Other pillars of support include Virginia McGarry; my parents, Dorothy and Dick Pezzoli; and my siblings, especially my philosopher brother, Richard Pezzoli.

Finally, I would like to acknowledge my ten-year-old son, Christopher. I witnessed his birth. That event and now his growing up in the world have distilled in me a certain awe. How vulnerable we are, how simple yet complex, how noble even. Children make this so clear. We are indeed a species with great promise as well as foibles. If we are to have a fruitful future, it will do us all well to embrace in our thinking and actions the great Kenyan proverb: We do not inherit the earth from our parents; we borrow it from our children.

I

Human Settlements and the Question of Sustainability

1

Introduction

Since the early 1980s, "irregular" (i.e., illegal) human settlements have encroached deeper and deeper into Mexico City's so-called ecological reserve. Social unrest, environmental degradation, and violence are all parts of the ongoing crisis. The government's land use intervention has been contradictory and reactive. The associated *politics of containment*—including official plans, programs, laws, and regulations—have aimed at enforcing a sharp divide between urban and environmental space. But such an objective has proved illusive. From a critical perspective, this book examines the state's problematic containment strategies in light of counter-vailing grassroots movements and proposals. The book's analysis of these countervailing forces does not offer a panacea. The intent is to cull lessons from social struggle over sustainable land use (i.e., land use incorporating ecological and conservationist principles) and to provide insight into the sociopolitical dimensions of sustainable development in an urban context.

Mexico City is an exemplar of promising as well as troubling features of urbanization in today's world system. In making this point, the book provides an historical-comparative perspective. The story told weaves together local, metropolitan, and global levels of analysis while integrating insights from planning theory, human geography, the environmental sciences, the social movement and empowerment literature, and postmodernist discourse. What emerges is a political ecology of human settlements. That is to say, the study provides a critical coevolutionary account of both state-society relations and environment-development relations in the dynamic growth of human settlements.

The book centers on a case study of Ajusco. In the Ajusco miniregion, located along Mexico City's southwestern urban edge, the politics of planning for ecological sustainability find their most serious testing ground. The case is a useful analogue for understanding other situations in urban Mexico, and in its more general aspects, the political ecology of urban conservationism elsewhere in the developing world.

The Ajusco case, based on twenty months of fieldwork completed over the course of a decade (1983–1993), documents the dynamics of community mobilization and the interactive politics of environmental regulation and popular resistance around ecological and land use issues. The case links the urban land question and the concept of sustainability to questions about livelihood opportunities and political mediation. The study advances one of the key arguments stressed during the 1992 World Conference on Environment and Development (Earth Summit) and the 1996 City Summit (Habitat II): rising demands for healthy human settlements—including adequate housing, jobs, food, water, social justice, and equity—must be reconciled with the necessity to sustain the ecological foundations of life and industry. The Mexico City case clearly shows that efforts to meet such a far-reaching challenge must go beyond issues of urban management. Bottom-up pressure must be brought against the state so that planning for ecological sustainability can take into account the fact that human settlements are increasingly places for production and income generation as well as a places for consumption and reproduction. This challenge calls for community empowerment and, ultimately, for integrated approaches to sustainable development in urban and rural areas simultaneously—not just in the Valley of Mexico, but regionally and around the world.

Management by Crisis

Concern about the development of human settlements and prospects for urban-ecological sustainability is rising around the world. This subject gained unprecedented international attention during June 1996 when the United Nations hosted Habitat II in Istanbul, Turkey. Habitat II—also referred to as the City Summit—was the last major UN conference to be held this century. Representatives of 171 countries and a total of 16,400 people addressed two major themes: (1) adequate shelter for all, and (2) sustainable human settlements development in an urbanizing world. The City Summit concluded with publication of the *Habitat Agenda,* "a global call to action at the international, national and local level and a guide for the development of sustainable human settlements in the world's cities, towns and villages into the first two decades of the next century" (Habitat II press release, June 3, 1996).

The decision by the General Assembly of the United Nations to organize the City Summit "was motivated by the continued deterioration of human settlements worldwide and the need to find new ways to address the human settlements challenge, particularly in the rapidly urbanizing countries of Africa, Asia and Latin America."[1] Habitat II's official objective was two-fold: "(1) In the long term, to arrest the deterioration of global human settlement conditions and ultimately create conditions for achieving improvements in the living environment of all people on a sustainable basis, [and] (2) To adopt a general statement of principles and commitments and formulate a related global plan of action suitable for guiding national and international efforts through the first two decades of the next century" (*Earth Negotiations Bulletin* 1996).

During one of the international preparatory meetings leading up to Habitat II, a group of development experts underscored the City Summit's historic significance in the following terms: "The urgent need for pragmatic programs to encourage sustainable development of human settlements by and on behalf of all stakeholders—including governments, urban managers, residents, businesses, producers and consumers of services—must be recognized as vital to global well-being. The world cannot afford to approach human settlements development through management by crisis responses."[2] No one can reasonably deny the merits of this argument. Why then does *management by crisis* continue to be so pervasive? Are worthwhile alternatives so difficult to locate, define, or implement? If there are alternatives within reach, what can local actors do to promote them? These are the broad organizing questions that motivated this book.

I approach these questions through a case study of Mexico City—one of the world's most noted scenarios of management by crisis. The study examines Mexico City's development and its prospects for the future. At the heart of the case is a story about community mobilization and the interactive politics of environmental regulation and popular resistance around ecological and land use issues. The story focuses on the growth of irregular human settlements in the greenbelt zone of Ajusco. The book's overall objective is to cull lessons for theory and practice by critically examining the interplay of state-society relations and environment-development relations in the dynamic production of human settlements.

The Ultimate World City

The drama unfolding in Mexico's capital has earned the metropolis a reputation as the ultimate world city. It contains nearly one-quarter of Mexico's population and has long been the nation's industrial (iron, steel, chemicals, textiles) and cultural center (*Business Monitor International* 1992, 28). Peter Hall (1984, 214), a noted planning theorist and historian of cities, writes that Mexico City is "ultimate in size, ultimate in threat of paralysis and disintegration, ultimate in the problems it presents for its politicians and planners. And these problems are of interest not merely to the people who live and work in the Valley of Mexico, but also to the whole of the world." In this respect the metropolis is exemplary: "if by some miracle Mexico City's growth can be controlled and serviced and planned, then perhaps any other city on earth has hope too."

The metropolitan area of Mexico City (MAMC) includes the Federal District (Distrito Federal; DF with 8.2 million people) and seventeen municipalities of the state of Mexico (with 7 million people) (Negrete, Graizbord, and Ruiz, 1993, 19). The Federal District is divided into sixteen political-administrative wards called *delegaciones*. Throughout this book I use the terms *Mexico City* and *MAMC* interchangeably as referents to the contiguous built-up area shown in figure 1.1. Mexico City is the nucleus of an urban subsystem—composed of Puebla, Querétaro, Toluca, Cuernevaca, and Pachuca—located within the west-central region of Mexico (see figure 1.2). Within this subsystem, the metropolitan zones of Mexico City and Toluca have already fused, thereby forming Latin America's first megalopolis (Garza and Schteingart 1986). The remaining cities of the subsystem may well join the megalopolitan belt over the next few decades. In view of this prospect, Garza and Schteingart (1986, 4) sound an alarm: "this enormous urbanized area, the life space of more than 30 million people, will accelerate the demands for public investment and create problems whose very contours can scarcely be imagined yet."

Scholars writing about Mexico City and its environs are understandably impressed—but they often exaggerate the case. By world standards, Mexico City is not ultimate in threat of paralysis and disintegration. To realize this is so, one only has to glance at the urban violence and disintegration that have taken place in the former republic of Yugoslavia or in parts of what used to be the USSR. Nonetheless, Mexico and its capital city certainly do have a special place on the world stage.

Mexico provides a crucial context for studies about the social causes and consequences of environmental action and change. Globally, Mexico may rank as high as ninth place with respect to the international production of greenhouse gasses (Liverman 1992, 72). Ecologically, Mexico is a major source of biological diversity. Together with Brazil, Colombia, Madagascar, Indonesia, and Zaire, Mexico encompasses 50 percent of all the planet's biotic wealth (Quadri de la Torre 1992, 5). Economically, Mexico owes a huge international debt at the same time that it is the source of some of the world's most essential raw materials. Politically, Mexico plays a leading role among nonaligned nations on environmental issues. Mexico was the first country in the world to ratify the Vienna Convention and Montreal Protocol agreements for the protection of the ozone layer. President Salinas was selected by the international agency United Earth (an organization sponsored by the UN Program for the Environment) to receive the award Earth Prize for Leadership. He was given the award, also known as the Green Nobel, "[i]n recognition for his concern as the Chief of State that has made the greatest efforts in environmental protection" (México 1991b, 127). In his fourth State of the Union Address (November 1992), Salinas announced his accomplishments along these lines: "In June, I attended the Summit on the Environment and Development held in Rio de Janeiro. There and in other forums, the environmental work being done in Mexico has been recognized. In the UN, and at the World Conferences on Children and on the Environment, we were initiators and active participants in topics of worldwide interest" (México 1992, 9).

Mexico's public investment for environmental concerns during 1993 totaled approximately US$2.5 billion, 39 percent over 1991's public investment of $1.8 billion and more than 2,000 percent over the $95 million invested in 1988 (Koloditch 1993). Mexico has been one of the world's developing nations with the highest per capita spending on envi-

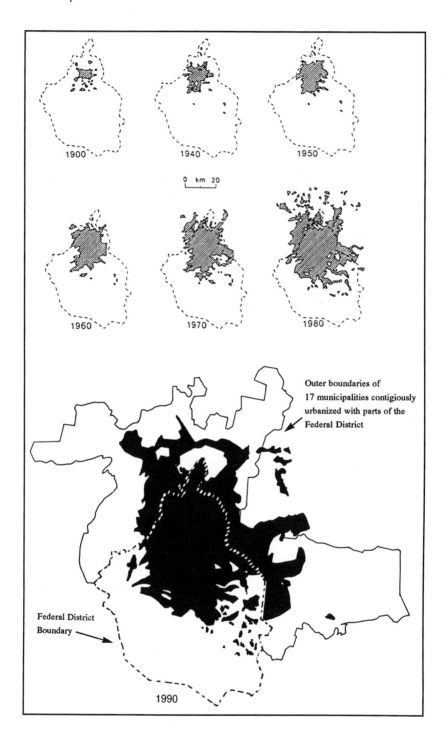

1900

1940

1950

0   km   20

1960

1970

1980

Outer boundaries of
17 municipalities contigiously
urbanized with parts of the
Federal District

Federal District
Boundary

1990

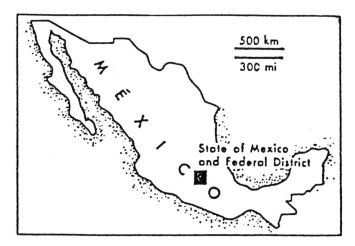

Figure 1.2
Location of the state of Mexico and the Federal District. Source: Dirección General de Reordinación Urbana y Protección Ecológica (1987).

ronmental protection and improvement. Mexico's National Council of Industrial Environmentalists (Consejo Nacional de Industrias Ecologistas; CONIECO) reported that Mexican industry invested a total of US$2.5 billion on environmental equipment and services during 1994. About US$1 billion of this was spent on pollution control equipment and cleaner production equipment (*Environment Watch: Latin America* January 1995, 5).

December 28, 1994, under the direction of President Ernesto Zedillo, the Mexican Congress elevated environmental concerns to the level of a new secretariat. Zedillo's rationale for creating the new secretariat was expressed in his formal proposal to the Mexican Congress: "The purpose of this initiative is to promote rigorous care of the environment by means of policies for economic development and the sustainable use of our natural resources and the prevention of contamination in all its manifestations. With the new Secretariat of Environment, Natural Resources, and Fisheries, we propose an efficient, participatory, and *productive environmental*

Figure 1.1
The expansion of the Metropolitan Area of Mexico City, 1900–1990. Source: Adapted from Cremoux (1991) and Ward (1990).

strategy" (my emphasis, cited in *Environment Watch: Latin America* January 1995, 2).

On the surface, this heightened attention to environmental concerns appears to be promising. Yet Mexico continues to be a nation where urban and demographic growth are rapid. Economic, political, and environmental strategies are conflict ridden and difficult to coordinate—a point readily acknowledged in the government's General Program for the Urban Development of the Federal District (Programa General de Desarrollo Urbano del Distrito Federal) (1995–2000). The program (called a "plan" in earlier versions through 1986) document notes how accelerated growth has "surpassed the administrative capacity of both the public and private sectors to resolve demands," and that "the precariousness of many sectors of the city has spurred the proliferation of irregular settlements" (cited in Urrutia 1995, 20).

Fragmented Development

The dramatic decline in the value of Mexico's currency (from 3.45 pesos to the dollar in November 1994 to 6.30 pesos to the dollar eight weeks later) precipitated a "national economic emergency."[3] Currency devaluation profoundly impacted the federal budget. For instance, the Undersecretariat for Environmental Policy suffered a 120 percent reduction in its budget compared to what had originally been appropriated for it in 1995. This translated into a reduction from US$49 million to US$27 million (Kiy 1995, 3). Mexico's economic crisis of the 1990s salts the wounds of the 1980s—widely considered a "lost decade." The 1980s was a period hammered by debt burdens, negative growth rates, austerity policies, net outflows of capital, deepening environmental problems, declining standards of living for the urban poor and middle classes, and a worsening of uneven development within cities and across regions (Barkin 1990; Walton 1991). While Mexico's total population grew by 2.1 percent per year during the 1980s, the growth rate of Mexicans living in "extreme poverty" increased 4.4 percent annually (Heredia and Prucell 1996, 7). Saxe-Fernández (1994a, 333) characterizes Mexico's development during this period as "fragmented" as opposed to "integral." Fragmented development concentrates or polarizes the distribution of wealth whereas integral development tends to promote economic equity. Despite optimism earlier in the decade, the 1990s may not turn out much better. From December 1987 to May 1994, the minimum wage in Mexico grew 136 percent. But, over the same period, the cost of living shot up 371 percent (Heredia and Prucell 1996, 7).

With respect to the nation's urban-ecological problems, Barkin (1990) persuasively argues that Mexico City is "Mexico's single most serious problem, the one that continues to be the acid test of the official determination to resolve environmental problems" (43). The rising incidence of cholera—1,062 cases in the Federal District during 1995—is one indicator of the seriousness of the challenge. Concern about the cholera outbreak has prompted the government to launch a prevention program (Bordon and Páramo 1996).

The Urban Environment and Human Settlements

It has been impossible for most of Mexico's urban families to purchase a housing unit in the formal (i.e., legally sanctioned) real estate market. Reasons for this include Mexico's widely skewed distribution of income, the rapid rate of urban and demographic growth, the lack of mechanisms for low-income housing finance, and the inability of the economy to generate enough well-paying jobs. Only the middle and the upper class have been able to significantly access formal sector housing, either in subsidized middle-class public housing or in private developments. At the national level, the unmet need for housing is estimated to be 7 million units (Contreras Salcedo 1996). The Mexican government's Federal Housing Program (Programa Nacional de Vivienda; 1995–2000) reports that one of every four housing units in Mexico are substandard (i.e., overcrowded, precariously constructed, and inadequately linked to public services) (Davelos 1996).

Actual and Projected Housing Deficits in Mexico City
A study conducted in 1994 by the Colegio de México estimates that there are 3,168,000 housing units in the MAMC (Méndez 1994). Of these, 10 percent have poor quality roofs made of bituminized cardboard or other makeshift material. Nearly 20 percent of the population live in units that

are seriously overcrowded (more than four people share a single room as sleeping quarters). Another 27 percent share a single room as sleeping quarters—sometimes the kitchen—with anywhere from two to four other people (see table 1.1).

Mexico's Secretariat of Urban Development and Housing (Secretaría de la Desarrollo Urbano y Vivienda) estimates that in the Federal District alone there is a need to construct at least 65,000 housing units every year into the foreseeable future (Méndez 1995). Yet, during 1995 only 45,000 housing units were produced. The backlog, estimated to be 400,000 units, has continued to grow. To begin closing the gap the government promises to promote sites and service projects. Yet, there is precious little land reserve available and "it is running out at an impressionable speed" (Juan Gil Elizondro, cited in Méndez 1995, 5B).

Access to Land and Irregular Settlements

Many of Mexico's urban poor live in overcrowded and substandard rental units (Gilbert and Varley 1991). Many others have constructed their own dwelling units through self-help. Settlements built up of such housing typically start out with an "irregular" (i.e., illegal/semilegal) status; the residents do not possess legal title to the land they occupy. "Irregular settlements" of this sort (also known as *colonias populares*) cover roughly 50 percent of the surface area of Mexico's large cities (Selby et al. 1990, 16). As Castells (1983) points out, this type of "popular sector" settlement is based on three interlinking elements: "first, a considerable amount of work provided by the housing occupants themselves; second, the state's tolerance of the illegal status of most housing settlements; and third, investment by speculative private capital operating outside the legal limits through a variety of intermediaries" (188).

Many researchers define irregular settlements as those that develop *outside of the law*. But such a notion can be misleading. As Antonio Azuela de la Cueva (1987) points out, despite the fact that such settlements are in some way illegal, or rather because of this, "the law becomes a real issue which influences the strategies of the social agents involved, thus shaping social relations and in some cases, the very structure of urban space" (523). Also, it is not just the illegal occupation of land that makes a settlement "irregular." Sometimes it is due to the fact that individual houses or infra-

Table 1.1
Housing conditions in the Valley of Mexico, 1995

Level of crowding and size of units	No. of units	% of total
Total number of housing units	3,168,000	100
Overoccupied units	988,000	31
One room only	507,000	51
Two rooms only	322,000	33
More than two rooms	158,000	16
Type of construction		
Solid roof of slab, concrete, brick, or tile	2,420,000	76
Roof of sheet metal, asbestos, or wood	434,000	14
Roof of cardboard or other material	313,000	10
Distribution of occupants per room		
From two to four people	858,000	27
More than four people	605,000	19
Water		
Running water (RW) inside dwelling unit	1,964,000	62
RW inside lot, but outside dwelling	950,000	30
No access to RW in immediate vicinity	253,000	8
Hooked up to sewer or septic tank	1,964,000	62
Has toilet with connection to RW	2,090,000	66
Has toilet without connection to RW	760,000	24
No toilet	316,000	10

Source: Méndez (1995).
Note: Totals vary slightly due to rounding.

structure do not meet standards set by official building codes (Burgess 1985).

Jorge Legorreta (1992), an investigator working out of Mexico City's Center of Ecodevelopment (Centro de Ecodesarrollo), estimates that 29 percent of Mexico City's total urbanized area currently has an illegal land tenure status. This amounts to 348 square kilometers, roughly the same area that makes up the city of Guadalajara. According to Legorreta's figures, this 348-square-kilometer area contains approximately 1,252,000 irregular lots inhabited by 6,640,000 low-income people—the equivalent of 2.2 times the entire population of Monterrey (128). The people in these settlements often lack adequate water, sewage, and electrical services

(128). Most of the Third World's urban poor face similar deficits (Gilbert 1992).

Irregular land tenure results from three basic forms of illegal/semilegal land alienation. One is the illicit sale or cession of communal lands including *ejidos* and *tierras comunales*. *Ejidatarios* are peasants who occupy ejidos—land holdings distributed to landless peasants through Agrarian Reform (Reforma Agraria) after the Mexican Revolution. The *ejidal* form of land tenure predates the Mexican Revolution. It has origins in both Spanish and Mesoamerican tradition. The name comes from the Spanish word *ejido*, which was the preindustrial term for commonly used woodlands and pasture at the outskirts of a pueblo (Silva Herzog 1964). Markiewicz (1980, 7) points out that communal forms of land use were also present in Aztec society: "Kinship groups (calpulli) held the land in common, though this land was not farmed as a unit, but was divided into family plots; land of a lesser quality, not specifically assigned or claimed by the village could be used at will by members of the group." The word *ejido* thus came to mean land held in common by the village for the use of its members.[4]

The term *ejido* now refers to the land itself, and to the legal corporation, composed of ejidatarios, that owns the land under important limitations. Until recently, ejido land could not be sold, rented, or mortgaged. The 1992 reform of Article 27 of Mexico's Constitution gave ejidatarios the right to legally sell, rent, sharecrop, or mortgage their land parcels. The reform also made it possible for private capital (including foreign) to purchase former ejido holdings. There are 28,000 ejidos in Mexico, occupying more than one-half of Mexico's arable land and employing over 3 million ejidatarios (Dewalt, Rees, and Murphy 1994). Given the institutional and traditional importance of the ejido in Mexico, the reform has been characterized as comparable in scope with the Dawes Act in the United States and the Eighteenth–Nineteenth-Century English Enclosure Movement (Cornelius 1992).

In the majority of Mexico's cities, ejido land constitutes the greatest source of urbanizable land. Given that ejido land has almost always been developed illegally, a large state and federal bureaucracy has grown up with the task of carrying out post hoc land tenure regularization. Researchers studying the dynamics whereby ejido land has been converted to urban use have tended to blame the ejidatarios as illicit profiteers. But it is more complicated than this. Some ejidatarios do profit, but many, if not most, do not. Cymet's (1992) detailed study, *From Ejido to Metropolis*, shows how most ejidatarios lose out in the process of urbanization. Specifically, they lose their livelihoods.

Recently, a group of scholars got together to consider how the reform of Article 27 may impact the urbanization of ejido land in Mexico. This group published a report referred to as the "Austin Memorandum" (Austin 1994). The report notes that more than 50 percent of the land development that has taken place in Mexico's rapidly growing urban areas has occurred on ejido land. Since the 1980s, this proportion has tended to increase as other land development modes have been foreclosed. In view of these trends, Austin argues that it is likely that the illegal sales of ejido land in the urban periphery will persist. The main problem, the report notes, "will continue to be the apparent inability of policy to control and/or direct growth towards certain parts of the city" (329). Moreover, Austin suggests that illegal land development may even speed up as "a result of the uncertainties introduced by the Reform and the fear that the state may intensify its 'land reserves' program through the mechanisms provided by deregulation" (330). The case of Ajusco examined in this book supports these observations. What is not clear is whether the popular sector's preferential access to ejidal land may decline. If it does, the result could be a deepening of social and spatial segregation (330).

The second basic form of illegal-semilegal land alienation is the unauthorized subdivision of private lands, which violates urban-planning laws and land use regulations. The third, and least common, is invasion—the occupation of land without or against the landowner's will. Legal forms of access to land for self-help construction have included the purchase of a plot in a sites-and-services scheme or the purchase of a plot in any kind of existing legal settlement.

From the 1940s to the 1970s, the production of irregular settlements in Mexico City alleviated an otherwise critical housing shortage. More recently, however, this traditional mode of access to land has been constrained by two sets of forces. First, after a generation of hyper urbangrowth, the available supply of unbuilt land in Mexico City has become more scarce and, as a consequence, more actively contested among potential users. Second, property development has become both subject to greater state control and more commercialized. In 1991, the government launched an "Urban Saturation Program" (Programa de Saturarcíon Urbana) that aims to account for and utilize every nook and cranny within the Federal District (*UnoMásUno,* April 3, 1991).

Ward and Macoloo (1992) have found that "access to land through invasion has declined as governments and landlords become more wary about the need to protect private property, and more efficient at doing so. Moreover, the *laissez-faire* attitude of many public authorities which prevailed in the past has been replaced by a greater willingness to intervene to control at least the excesses of illegal developers" (67). Compounding these constraints are the impact of the economic crisis and the politics of austerity.

Lacking viable alternatives, millions of Mexico City's inhabitants now live in low-income settlements occupying land that is unsuitable for urban development. Such settlements occupy the barren, desiccated lake bed of Texcoco; hills that are unstable from mining; sites next to railway lines or factories that emit toxic waste; or, as in the case of Ajusco, zones designated for ecological conservation. As the supply of land becomes increasingly scarce and subject to more rigid controls, the contest for the land heats up. This is especially evident in Ajusco. Characterizing the results of urban expansion under such conditions, Cymet (1992, 3) notes:

Most often this expansion is accomplished with little regard for the land's fitness for urban use, its degree of risk to human life and property, or its ecological role. The expansion has resulted in the destruction of environmentally sensitive areas and in a great increase in the city's vulnerability to natural and technological disasters. The occupation of the highly vulnerable, inhospitable, dry lake bed of Texcoco by one and a half million people and the contamination of the Ajusco aquifer recharge area by irregular settlements are just two examples of this destruction.

Such trends are troublesome in many Third World cities (Dunkerley and Whitehead 1983; Angel et al. 1983; Burnstein 1994). Among academics, there is a convergence of opinion that "The problem of adequate land for the urban poor in developing countries is bleaker today than it was 25 years ago, and almost surely will become more bleak in the future" (Doebele 1987, 7). Mitlin and Bicknell (1992a, 3) point out that an estimated 600 million urban dwellers in Africa, Asia, and Latin America currently "live in life and health threatening environments because of unsafe and insufficient water, poor quality and often overcrowded shelters, inadequate provision for sanitation, garbage and drainage, unsafe housing sites and a lack of health care."

The situation just described has led some researchers and policymakers to argue that any consideration of sustainable development in these regions must focus on improving the housing, living, and working environment of the urban poor. For instance, in the foreword to a paper prepared for the Human Resources Program of the International Institute for Environment and Development, A. Ramachandran (the executive director of Habitat) makes the following observation:

The condition of human settlements—how they develop, use natural resources and interact with the natural environment—will be central to any successful transition to sustainable development. This applies particularly to the settlements of the developing countries not only because an increasing share of the world's population lives there, as population growth and urbanization concentrate in the developing countries, but also because these settlements play a key role in social and economic development. They not only are centers for new jobs, for innovation and for expanding economic opportunities; settlements also play a crucial role in support of agricultural development as well as in the provision of social and other basic amenities. (Mitlin and Satterthwaite 1990)

The Case of Ajusco

The zone of Ajusco gets its name from one of the mountainous peaks within its bounds, the snowcapped Ajusco volcano. It is a verdant mountainous and wooded area covering roughly eighty square kilometers on the southwestern fringe of Mexico City. The rugged terrain—rising from 2,000 to 4,000 meters—is part of a volcanic mountain range called the Sierra del Ajusco. This range forms the southern limit of the Central Mexican Basin.

Ajusco is considered ecologically significant for a number of reasons. Its permeable volcanic subsoil absorbs the region's abundant winter rainfall; thereby recharging the city's underground aquifers (Contreras Domínguez 1988; México 1991b; Quadri de la Torre 1993a). Ajusco is valued for its rural uses, including farming, forestry, and ranching (Rapoport and López-Moreno 1987; México DDF 1989a, 1989b). Ajusco is also valued for its majestic open space and wealth of biodiversity, including a cover of trees and vegetation believed to be significant for sustaining favorable microclimatic and hydrological conditions (Wilk 1991a, 1991c; Soberón 1990).

The area is actually considered the richest biotic zone in the Valley of Mexico (Quadri de la Torre 1991a, 44). More than 1,000 species of plants have been counted there (México 1989, 5).

Most of Ajusco is designated as an ecological conservation area, natural reserve, or park. Yet it is a highly contested terrain subject to intense pressures of urban expansion, where interests of the state, popular groups, ejidatarios, and developers clash. The urban poor are not the only ones who have staked out claims for land in Ajusco. The zone's abundant green space, clean air, and panoramic vistas have attracted real estate developers and higher-income groups interested in upper-class development. Historically, economically better-off groups have concentrated in the southwestern part of the city and in the case of Ajusco the competition for land is especially intense (Schteingart 1987; Pezzoli 1991). Within the same ecological reserve, one finds well-protected enclaves of luxury villas not far from sprawling squatter settlements (Serrania Alvarez 1984; Schteingart 1989a; Figuereo Osnaya 1992).

Ajusco is not a political entity (i.e., it does not have officially set boundaries). However, it is generally referred to as the area shown in figure 1.3. In view of its great expanse and urban-rural mix, Ajusco is subdivided into a lower, middle, and upper section. Ajusco Bajo (lower Ajusco) was initially settled and to a considerable extent consolidated prior to 1980. Ajusco Medio (middle Ajusco) lies at the outermost edge of the officially designated urban area; land use there has been rapidly changing from rural to urban. This transitional part of Ajusco is where settlements, including Los Belvederes, took hold and began to consolidate during the 1980s. Los Belvederes is composed of thirteen contiguous, low-income settlements. The history of these particular settlements lies at the heart of this book. Ajusco Alto refers to the southernmost rural area and undeveloped open spaces of the zone. It contains the Xitle volcano and other peaks of the Sierra del Ajusco. It also contains three of the thirty-six rural settlements located in the Federal District's ecological conservation area.

Over the past decade, noted biologists, ecologists, and social scientists have drawn attention to the zone of Ajusco. Although these researchers have diverse views, there exists a certain level of consensus among them. As one environmental scientist puts it: "Ajusco painfully illustrates how the ecological crisis in Mexico involves interlocking relations among the

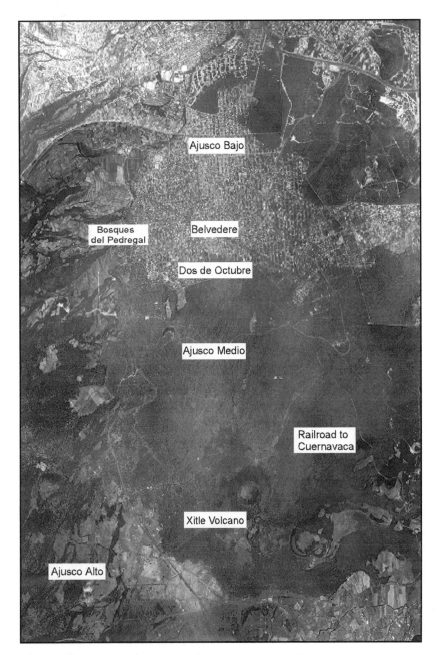

Figure 1.3
The Xitle Volcano shown within the miniregion of Ajusco. Source: Detenal Aerial Photography, Flight number R313, Scale 1:40,000, Photo number L-8, Mexico City, June 1979.

economy, ecology, and politics, and that our life style over the medium and long term is not viable."[5] Critics and supporters alike refer to the Ajusco greenbelt as the government's environmental litmus test.

The government claims that Ajusco is a proving ground—a place to demonstrate how growth can be controlled, social protest can be mollified, and the ecology can be protected and sustained. The first set of plans to limit urban growth in Ajusco were drawn up in 1976. On paper, Ajusco was designated as a buffer zone intended as a natural barrier against urban expansion. In practice, it has not worked that way. Between 1980 and 1986, the government extended the legally designated urban limits three times to accommodate urban encroachment inside Ajusco (see figure 1.4). During this same period, the population increased from around 20,000 to 200,000—the bulk of whom are low-income families living in self-built housing and lacking adequate public infrastructure and services. Over the ten-year period of 1986 to 1996, the urban-ecological boundary was moved again several times to accommodate urban growth. Urban expansion into Ajusco has involved fraudulent schemes by real estate developers, illegal sales of communal and ejidal property, land invasions and violence, mass eradication of incipient low-income settlements, popular resistance and opposition movements, widespread corruption, and deepening ecological disruption.

Grassroots Environmental Action
Planners and officials that I interviewed at the highest levels of the Federal District Department (Departamento del Distrito Federal; DDF) held the view that the intensity of social and political turmoil in the illegally urbanizing parts of Ajusco, coupled with the area's increasingly serious ecological problems, has confronted the city administration with its most serious and difficult challenges regarding land use and environmental planning.[6] The director responsible for setting environmental standards and criteria *(normas oficiales)* at the government's National Ecology Institute noted that "the fraudulent purchase-sale of land in the zone of Ajusco is one of the greatest political, environmental, social and juridical problems in Mexico City" (Quadri de la Torre 1993b, 37). The National Institute of Statistics, Geography, and Information (Instituto Nacional de Estadística, Geografía e Informática; INEGI) generated a list of the Federal District's most impov-

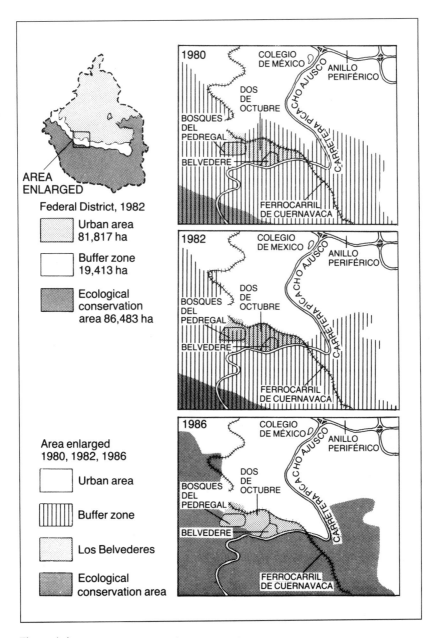

Figure 1.4
Limits to the legally designated urban area illustrated as shifting south in 1980, 1982, and 1986. Note: In 1986, the buffer zone was eliminated. It was added into the ecological conservation area. Source: Pezzoli (1991).

erished human settlements in 1996. One of the forty-four settlements on the list is Belvedere, a colonia popular in Ajusco (Joyner et al. 1996).

One need not rely on the authorities to gauge the gravity of the situation in Ajusco. Independent popular groups I worked with in the zone were quick to acknowledge the notoriety arising from their well-organized grassroots political opposition and their capacity to mobilize people. Furthermore, events in Ajusco have consistently made headlines in Mexico City's newspapers. From my reading of several thousand newspaper articles concerning urban land and related problems in Mexico City, I collected over 600 articles directly concerning social, political, and ecological problems in Ajusco.[7] Over the period of 1976 to 1988, the zone of Ajusco was the subject of more articles concerning irregular settlement and land use conflict than any other zone in the entire metropolitan area. Since the late 1980s, Ajusco has been one of two primary targets selected for special attention in the Federal District.[8] A series of ecological restoration initiatives, including the expropriation of land for the establishment of an ecological park, have been undertaken.

Ajusco's majestic landscapes, together with its social and political turmoil, has inspired a significant amount of academic research (including master's and doctoral theses, books, and journal articles) as well as poetry and songs. The case of Ajusco has even captured attention from around the world. At certain junctures, external agents from foreign countries (e.g., the Canadian International Development and Research Center, a German religious organization, and the U.S. government's Inter-American Foundation) have played a part in the zone's development. With technical and financial support from external agents outside of Mexico, as well as within Mexico, the zone of Ajusco has proved to be a fertile field for grassroots mobilization and social experimentation.

During the mid-1980s, the government announced plans to eradicate all the families illegally settled in the Ajusco zone. The area threatened was made up of the thirteen contiguous settlements called Los Belvederes. At the time, around 20,000 people occupied these settlements, which covered 171 hectares (431 acres) of land. Divided into 219 blocks, the area contained 5,000 lots of approximately 200 square meters each. Community groups in Los Belvederes mobilized to resist the forced relocation. Herein lies the most instructive part of the story: the innovative organized resistance. Besides well-worn strategies of marches and demonstrations, the grassroots movement generated bottom-up, proactive environmental action. Friedmann and Rangan (1993, 4) define environmental action as "the efforts and struggles by rural and urban communities to gain access to and to gain control over the natural resources upon which their lives and livelihoods depend." Aided by technical assistance form external agents, grassroots activists in Los Belvederes began to promote innovative productive ecology projects as a countervailing strategy to secure the presence of the families in the zone.

Productive Ecology
Community activists in Los Belvederes argued that they could point the way to a new development for Mexico City: far from being a blight on nature—which was the general view—each settlement would be transformed into a *colonia ecológica productiva* (CEP), suggesting a productive, sustainable future. This book provides a close analysis of the CEP initiative. Participants in the productive ecology movement argued that the demand for land in the area makes it impossible to maintain the entire zone as a greenbelt for consumption (in the form of a national park or ecological reserve). Yet it is possible to develop a greenbelt for production, a place wherein grassroots social experimentation may be promoted to generate new forms of development that are resource sustainable and minimize ecological disruption.

The grassroots activists argued that their call for an experimental greenbelt for production sharply contrasts with the government's approach to planning, which treats housing, economic, and ecological problems as separate issues. They argued that housing, economic, and ecological problems cannot be solved separately. Faced with chronic scarcities of income and resources, participants in the grassroots ecology movement saw no alternatives outside of promoting self-reliant forms of urbanization. Their proposals centered on productive ecology, a concept that calls for the development of appropriate technology that can reduce the cost of generating employment and recover capital investment through the rational use and reuse of resources. In short, they sought to promote a praxis of productive ecology; their objectives are listed in box 1.1.

The CEP movement called for adequate livelihood opportunities and income generation, the integration of urban-rural and society-nature relations, and collective strategies for community empowerment. This book examines each of these objectives in the context of the broad socioeconomic, political, and environmental changes that are sweeping the region and the world.

At its zenith, the CEP movement had hundreds of families rallying under its "productive ecology" banner. At the same time, the movement had networked across the urban-rural divide and had successfully articulated multiclass alliances with independent researchers, university students, newspaper columnists, international agencies, and campesinos both locally and from the northern part of the country. This environmental action proved to be an effective counteractive force.

In the case of Los Belvederes, when the legal status of the settlements changed, so did the terms of collective social struggle. The government promised to include all of Los Belvederes in the legally designated urban area. The government also promised that every family would eventually be given security of tenure on an individual basis. "Winning" the right to have their settlements incorporated into the legally designated urban area

Box 1.1
Objectives of the Colonia Ecológica Productiva

---

1. To promote an integral, barrio-based model of urban development for generating jobs and resources and fostering the production of goods in a way that is socially necessary, ecologically valid, and economically viable
2. To intervene positively in the transformation of daily life, generating new relations between humankind and the surrounding natural environment, combining production with an ethic of noncontamination and sustainability
3. To employ alternative technologies as part of a collective strategy of self-reliance
4. Over the medium and long term, to implement a conceptual transformation of residential space from a place of consumption to a place of production
5. To put into operation an urban buffer zone (a greenbelt for production) to mediate the contradictions between the city and the outlying rural land, recovering or reincorporating intrinsic rural values into urban lifestyles

---

Source: This list of objectives appears in a document titled "Colonia Ecológica Productiva," which was written by community activists in Ajusco in 1984 together with an external agent—the Grupo de Technología Alternativa.

reduced the settlers' sense of insecurity and urgency. The "victory" was the result of political mediation and tended to remove the impetus for further mobilization. From the standpoint of popular mobilization, this pattern of development is typical and presents a major dilemma. As one grassroots organization describes it: "Periods of strong mobilization correspond to the struggle to satisfy permanent needs (access to legally secure land for housing, piped water supply, a school) or to unexpected crises such as transportation fare increases or compensation for grave accidents, but as soon as the critical juncture is passed, demands resume on general terms and organization disappears" (MRP 1983, 6).

The environmental movement in Los Belvederes declined after mid-1985, although it did not subside altogether. Winning the right to have their settlements incorporated into the legally designated urban area highlighted problems outside the realm of ecological discourse, thereby prioritizing other strategies. New battles had to be waged to secure the best terms for regularization and to resolve litigation regarding fraudulent land transactions and boundary disputes. This tended to fragment the movement and sap its momentum. The waning of collective action has to be understood within the national and political context, where the provision of urban services by the state in Mexico "has been one of the most powerful and subtle mechanisms of social and institutional control of everyday life" (Manuel Castells, cited in Ramírez Saiz, 1990b, 235). Herein lies the most significant lesson brought to light in this book. To be effective, community-based environmental action like that promoted by the CEP activists requires continuity and cohesion in terms of social organization, and, ultimately, it requires the support of the state. Because the terrain to be navigated is beset by contradictions and dilemmas, the challenge eludes a simple characterization of the situation as good versus evil, or bottom up versus top down.

Not all of those who advance the CEP argument actually seek to create an alternative type of ecological human settlement. Quadri de la Torre (1993b) goes as far as to argue that colonias ecológicas populares have become the big ruse of the 1990s. Quadri blasts those who organize invasions in Ajusco's ecological reserve under the false pretext of establishing CEPs, and he argues that clandestine developers are getting rich by marketing the productive ecology argument in the sphere of irregular human settlement expansion. He says they are selling "the idea of colonias ecológicas populares to justify the permanence of irregular settlements inside the Federal District's ecological conservation area, in opposition to the forces of relocation" (37). Indeed, there are clandestine developers employing sophisticated strategies *(mercadotécnica)* whereby services of representation and political mediation are offered for a price—a high price that new *colonos* (settlers) pay. But their behavior does not warrant the wholesale rejection of CEP initiatives and arguments. Rather, it highlights the need for careful analysis distinguishing how ecological "models" get applied by market forces or the state from how they take shape through grassroots action.

Careful analysis shows that top-down or official efforts to conserve the environment are driven by multiple agendas including some unrelated to formal policy functions. In effect, planning and land use regulation distribute and redistribute public values, values that are socially constituted in ways often independent of the policy process itself. The Ajusco case shows how shifts in the legal-institutional terrain alter the rules of land settlement, thereby creating new opportunities as well as obstacles for all the players concerned. Policy development is best understood as a multifunctional and highly dynamic process. It is often suboptimal in terms of actual policy outcomes as well as fraught with unintended consequences. Such an understanding underscores the importance of social learning and social experimentation as core principles for engendering a more effective, strategic approach to planning for sustainable development. Along these lines, political ecology provides a useful analytic approach; it provides the governing conceptual basis for this book.

Political Ecology as an Analytic Framework

The term *political ecology* has been in circulation for at least twenty years (Atkinson 1991). However, in his review essay, "Political Ecology: An Emerging Research Agenda in Third-World Studies," Bryant (1992) points out that the concept has become a more frequent target for inquiry since the late 1980s. There is now a significant body of work from which to draw in articulating a political ecology of human settlements. Conceptually, such work builds on the ideas of internationally known thinkers such as Ivan Illich, Andre Gorz, Rudolf Barro, Ignacio Sachs, and especially

Enrique Leff (1993, 1994, 1995). Bryant (1992) does a good job conveying the promise of political ecology research: "By critically focusing on the relationship between environmental change, socio-economic impact and political process, such research addresses neglected issues. It rejects facile assumptions about environmental change and human welfare—for example, that ecological degradation is a universal evil befalling rich and poor alike. Rather, it explores how such change is incorporated into concrete political and economic relationships, and the ways that it may then be used to reinforce or challenge those relationships" (p. 27).

Mexican-based scholars have made major theoretical contributions to political ecology. In fact, this book builds on an excellent body of research about Mexico City that documents the linkages among urban processes, poverty, and the environment.[9] Political ecology links ecological themes with social struggles (Leff 1994; Martínez-Alier and Thrupp 1992). In Atkinson's (1991, 2) terms: "political ecology is about building a radically new social, political and cultural world out of the ruins of the old, one which will obviate the environmental catastrophe ahead and establish an economic and social system which incorporates a sustainable relationship between society and nature." At the heart of the matter lies the empowerment of the urban and rural poor. In this book, I embrace a working definition of empowerment as an enhancement of the capacity of individuals, households, and collective actors to engage in and learn from an ecosocial praxis that lessens the uncertainty and risk they face while increasing their collective security and prospects for sustainable development. A group of scholars and activists—recently brought together for a workshop sponsored by the International Institute for Sustainable Development (IISD)—also define empowerment in proactive and affirmative terms, stating that it "affirms the need to build the capacity of communities to respond to a changing environment by inducing appropriate change internally as well as externally through creativity, innovation and commitment to sustainable development goals. The principles embodied in the concept of empowerment include inclusiveness, transparency and accountability" (Singh 1995, 7).

Social learning is key to empowerment and begins and ends with action, that is, with purposeful activity. In addition to the action itself, social learning involves, "political strategy and tactics (which tell us how to overcome resistance), theories of reality (which tell us what the world is like), and the values that inspire and direct the action. Taken together, these four elements constitute a form of social practice" (Friedmann 1987, 181). To be effective in the mediation and advocacy of social practices for sustainable development, political ecology must take into account multiple levels of analysis. It must cull for insights from across disciplinary boundaries and theoretical frameworks. This calls for cross-fertilizing the fruits of a *sociological* and *biogeophysical imagination* and for conceptualizing a new *ecological production épistémé* (EPE) as an alternative to the dominant *industrial production épistémé* (IPE).

Political ecology conceptualizes as problematic several controllability assumptions underpinning the IPE and its theories of production and growth, including (1) controllability of a production process, (2) dominance over the environment, and (3) the independence of production processes from other processes (M. O'Connor 1994, 63–67). These assumptions underpin a reductionist view that fails to adequately take into account economic and biogeophysical interdependencies. In contradistinction to the IPE, those attempting to define an EPE do so in terms of an open system that involves a turbulent coevolution of interacting "entities." The emphasis here is humbling; it "connotes the indeterminacy, ambivalence, and mutability of historical trajectories" (M. O'Connor 1994, 66). This realist approach to epistemology and social analysis suggests that "it may be wise to avoid thinking of knowledge as attempting to "represent" or "mirror" the world like a photograph. A better analogy may be that of a map or recipe or instruction manual, which provide means by which we can do things in the world or "cope" with events" (Sayer 1992, 59).

What does all this mean with respect to human settlements and planning for ecological sustainability? Its significance has to do with control. Controllability assumptions, like those underpinning the IPE, also underpin the mainstream approach to urban planning. In a Habitat publication dedicated to the City Summit, Raquel Rolnik—former director of planning in the city of São Paulo—outlined how urban planning has historically been informed by certain post–Carta de Athenas conceptions of city planning. The fundamentals of this paradigm are "(a) urban management based on "rational" decisions taken by the State to creating a city which is a model of urban order; (b) an ideal model of the city as part of a modernizing and integrated project; and (c) guarantee of continuous national and international investment flows towards provision of basic infrastructure which, in turn, creates conditions for the growth of capital" (Rolnik 1995, 6).

As Rolnik and others point out, the assumptions underpinning mainstream approaches to planning are problematic. Their limitations are evident in the international context, where control has been elusive—whether one looks at planning efforts within Soviet-style socialism or post–New Deal capitalism. Acknowledging the limitations of planning underscores the importance of social experimentation and social learning as means for engendering sustainable development. But it does not necessarily suggest a diminishing role for the state. Rather, it calls for changing the approach to planning and for a redefinition of urban order and the state. Along these lines, Rolnik (1995) provides a useful insight: such changes "would involve a redefinition of the role of the State and, specifically, of local government which, far from becoming less important as market-oriented neo-liberals would wish, should act as a diffuser of negative influences on solidarity, self-reliance and distribution of income and power" (6). Douglass (1996) makes a similar argument: "Community mobilization is by no means a substitute for government intervention; government action is essential in tackling the interconnected problems of poverty and environmental degradation. But the potential for communities to help themselves can be a major force for change. Indeed, over the past three decades, most urban "success" stories have involved projects that have incorporated community action, from the Orangi Pilot Project in Karachi, Pakistan, to the Zabbaleen in Cairo, Egypt" (125).

Community mobilization is a potentially innovative and countervailing force for social change. Without such mobilization, no challenge will emerge from civil society able to shake the institutions of the state through which urban-ecological problems are (re)produced. In their write-up of a major transdisciplinary research project on this subject, Leff, Carabias, and Batis (1990, 10) argue that mainstream development policies in Mexico, as in other countries of the region, have failed to stop ecological degradation and the deterioration in the quality of life of millions of people, especially the poor. On a more upbeat note, Leff and his colleagues also argue that the problems stemming from this failure have driven the search for new survival strategies in urban and rural communities, thus giving rise to the proliferation of new productive experiences. These are candles lit in the darkness.

Although there may be no boilerplate answers to redefining planning and the role of the state in the face of the sustainability challenge, one thing is clear: "Whatever the case(s), from the standpoints of progressives, red or left greens, and feminists, the last thing in the world we need is factionalism, sectarianism, and 'correct lineism,'—instead, we need to scrutinize critically all time-worn political formulae and develop an ecumenical spirit, and to celebrate our commonalities or 'new commons' as well as our differences" (M. O'Connor 1994, 173).

The Book's Key Questions and Conceptual Approach

Based on principles of political ecology, this book advances what Friedmann and Rangan (1993) call a middle-range explanatory framework. Such a framework "integrates global processes with local environmental action and reveals the particular outcomes experienced by peoples and communities living within localities and regions" (11). Based on such a framework, this book aims to address three sets of questions:

1. *Coevolution and the urban prospect*   What is the likelihood that the so-called ecological reserves and other ecological resources in the metropolitan area of Mexico City will be effectively safeguarded given the coevolutionary, sociopolitical, and economic dynamics now in place? Are there innovative approaches to the development of human settlements, now or in the past, that can help bring about equitable and sustainable uses of space and resources?

2. *The political ecology of human settlements*   What is it about land use management and planning in Mexico City that makes the development of human settlements increasingly problematic? What are the constraints and opportunities posed by the shifting socioeconomic as well as legal and institutional terrain?

3. *Empowerment*   Are there initiatives spearheaded by popular groups in Mexico City that have potential to ameliorate urban-ecological problems? If so, how might such grassroots environmental action be improved?

In addressing these questions, this book provides an analysis of political strategies as well as social, ecological, and technical issues surrounding the policies, programs, and laws involved in the demarcation and enforcement of the area for ecological conservation in Ajusco. The focus is on strategies and actions, forces and countervailing forces. The analysis thus provides critical insight into the dynamics of community mobilization and the interactive politics of environmental regulation and popular resistance with respect to ecological and land use issues.

An analysis of power relations in interaction is central to my approach to the state's politics of urban containment in the greenbelt zone of Ajusco. The analysis aims to weigh contradictory facts with a view to the resolution of their contradictions. Of particular concern is the fact that government agencies dealing with human settlements and the environment often contradict one another. And what is publicly recorded as the purpose of a law or institution is often contradicted by what the entity actually does in practice. For instance, the legal and institutional terrain aimed at regularizing (i.e., legalizing) land tenure in low-income settlement is riddled with contradictions. It is supposed to prevent the growth of irregular settlements in the city's ecological reserve, but instead it helps perpetuate the process.

Ajusco is at the cutting edge of the city—a place where urban expansion is driven by intense conflict, where legal meets illegal, where creative and destructive social forces clash, and where the contradictions of state action are clearly visible in the discretionary enforcement of the law. With this case in view, my concern is "with the point where power surmounts the rules of law, which organize and delimit it, and extends itself beyond them, invests itself in institutions, becomes embodied in techniques, and equips itself with instruments and eventually even violent means of material intervention" (Foucault 1983, 307). From this angle, I view popular struggle and state-society relations concerning the ecological limits of Mexico City's legally designated urban area as constituting what Foucault (1983) calls the *frontier for the relationship of power*: "the line at which, instead of manipulating and inducing actions in a calculated manner, one must be content with reacting to them after the event" (225). Understanding the dynamics involved here helps to explain why management by crisis is so entrenched.

Careful analysis calls for charting out the field of relevant power relations or, to use Jessop's terms, to discern the overall balance of forces. From this perspective, "the analysis of power is closely connected with the analysis of the organization, modes of calculation, resources, strategies, tactics, etc. of different agents (unions, parties, departments of state, pressure groups, police, etc.) and the relations among these agents (including the differential composition of the structural constraints and conjunctional opportunities that they confront) which determines the overall balance of forces" (Jessop 1982, 256). My inquiry thus delves into the exercise of power where, to use Foucault's (1983, 220) terms, the exercise of power "is a total structure of actions brought to bear upon possible actions; it incites, it induces, it seduces, it makes easier or more difficult; in the extreme it constrains or forbids absolutely; it is nevertheless always a way of acting upon an acting subject or acting subjects by virtue of their acting or being capable of action." Hence the exercise of power puts into motion a set of actions upon other actions. This consists in *"guiding the possibility of conduct and putting in order the possible outcome"* (my emphasis, 221).[10]

By conceptualizing the exercise of power as guiding the possibility of conduct and putting in order possible outcomes, Foucault provides us with another useful insight: "Basically, power is less a confrontation between two adversaries or a linking of one to the other, than a question of government" (221). To govern, in the broad sense, is to structure the possible field of action of others. Foucault (1983) uses the term *government* here in the same broad way it was used during the sixteenth century when the term referred to more than political structures or the management of states: The term "designated the way in which the conduct of individuals or of groups might be directed: the government of children, of souls, of communities, of families, of the sick. It did not only cover the legitimately constituted forms of political or economic subjection, but also the modes of action, more or less considered and calculated, which were destined to act upon the possibilities of action of other people" (221). This insight helps explain the interactive politics of environmental regulation and popular resistance around ecological and land use issues in Mexico City's greenbelt.

Power is not a commodity or a hidden rationality to history orchestrated by the controlling layers of society. To characterize the state's technocentrism as entirely oppressive or rooted in exploitation and domination would be a mistake. It would be equally mistaken to indiscriminately characterize grassroots groups and movements as the unequivocal carriers of a progressive ecocentrism. Grassroots mobilization and self-help should not be romanticized nor characterized as the panacea for all of society's ills. As Rakodi (1990, 121) points out, "Local organizations are easily co-opted, and rarely provide the impetus for structural changes to the existing economic and political power structure where this is biased against poor people." Along similar lines, Marcuse (1992) argues that "pure" self-help projects cannot be a substitute for resources indispensable for housing provision. Nor can they provide major items of collective consumption such as hospitals and universities. Self-help, Marcuse maintains, often violates good planning. It can be politically reactionary and socially divisive as well. Thus it is not the task of political ecology to glorify the grassroots. Rather it is to pursue the benefits of careful study—historically grounded in biogeophysical and ethicosocial dimensions—that looks for the opportunities as well as constraints in human interaction.

The challenges posed are complicated by the far-reaching and highly interactive nature of the dynamics involved. With respect to city-hinterland and economy-ecology linkages and interaction, Carley and Christie (1993, 19) note:

> The urbanization process itself generates economic activities which raise income levels, draw in resources from the countryside and even from faraway savannas and rainforests, and generate enormous amounts of waste which end up as pollution of air, water and land. The increased income generates more consumption, more industry, more pollution, more automobility, urban sprawl, endemic traffic congestion, and so on. Urban sprawl results in loss of prime farmland which, when combined with rural population growth, contributes to lowland forest loss due to agricultural expansion in the countryside well away from the city, thus completing a cycle of interaction.

In view of such dynamics, I take as a starting point the fundamental argument of Leff and other political ecologists that new concepts of "the city" are needed. It is fruitful to think in terms of *greening the city*: "Greening the city implies the articulation of urban functions in an overall sustainable development process. It implies new functions for the city and its reintegration into the overall production process through a more balanced spatial distribution of agri-ecological, industrial and urban activities" (Leff 1990a, 56). In short, this book examines what Leff calls the social and natural processes that provide the *ecological conditions of production*. The state provides such conditions through urban and regional planning, including the establishment of ecological reserves, as well as through other juridical-administrative means. Civil society and mobilized communities provide ecological conditions of production through local resource management. They have the potential to foster productive and consumer practices that are environmentally sound and economically viable, many of them located outside the realm of capital and the market (Leff 1993, 54).

The book is arranged in four parts. The first part situates Mexico City and the question of urban-ecological sustainability in a global and comparative context. Specifically, part 1: (1) introduces Mexico City's development in relation to what is currently being said about human settlements problems and prospects in international policy circles, (2) lays the theoretical foundation for the book's critical approach in terms of political ecology, and (3) compares urban-ecological conditions and land use policies in Mexico City to other cities in Mexico and to other rapidly growing cities around the world—including Seoul, Jakarta, São Paulo, and Bangkok, among others.

Part 2 spells out the biogeophysical and historical factors underpinning the dramatic rise of Mexico City. As Kandell (1988) points out, Mexico City's tumultuous past "encompasses all the stages of urban development more clearly than any other community on earth. For Mexico City can trace its existence continuously from Paleolithic site to cradle of ancient civilization, and from colonial stronghold to contemporary megalopolis" (8). Mexico City is the only one of the great pre-Colombian urban centers that has maintained its political and administrative importance into the contemporary era (Meyer 1987). Part 2 advances the argument that the intersection of history, power, and environmental change in the Valley of Mexico can best be understood from an interactive, coevolutionary perspective (Norgaard 1994). This perspective helps define the urban land question and the challenge to achieve sustainability in the (re)production of human settlements.

Parts 3 and 4 of the book provide a detailed analysis of the urban-ecological problematic in Mexico City's miniregion of Ajusco. The main focus is on the growth of thirteen contiguous low-income settlements—Los Belvederes—which claimed a large section of the city's so-called ecological reserve. But I also draw attention to the most recent incursions of irregular settlements into the forbidden greenbelt. Over the past fifteen years, urban encroachment into Mexico City's greenbelt has been fueled by the rich and poor alike. The government's approach to this "irregular" expansion of human settlements has been contradictory and reactive. Social unrest, ecological disruption, and violence are all parts of the continuing crisis. Parts 3 and 4 of the book critically document the interplay of top-down and bottom-up initiatives—including the interactive politics of community development by the urban poor, growth management strategies, and grassroots environmental action. With this critical account of state-society–environment-development relations in view, the book concludes on a forward-looking, prescriptive note. Lessons from the Ajusco drama are spelled out, and an agenda for research and action is suggested.

2

Mexico City's Current Status: A Global and Comparative Perspective

Our species was on earth a million years before it numbered 1 billion people. That was at the turn of the nineteenth century. It took only 130 years to reach the second billion, 30 years to reach the third, 15 the forth, and 12 the fifth (MacNeill, Winsemius, and Yakushiji 1991, 3). This tidal swell of humanity has followed an equally impressive growth in the global economy and demands on the earth's resources. From 1900 to 1990, as the world's population multiplied more than three times, global economic output grew twentyfold, the consumption of fossil fuels grew by a factor of thirty, and industrial production by a factor of fifty. The great bulk of this growth, about four-fifths of it, has occurred since 1950 (MacNeill, Winsemius, and Yakushiji 1991, 3). Sachs (1993, 6) estimates that the world economy now increases every two years by about the size (US$60 billion) it had reached by 1900, after centuries of growth. In view of such growth, the World Bank calculates that total world output by 2030 may be as high as 3.5 times what it is today (World Bank 1992, 9).

Global economic growth—including fundamental shifts in the world economy's composition, geography, and institutional framework—has influenced the dynamics of urbanization in both First and Third World countries (Sassen 1994; Friedmann 1986; Johnston 1994). Sassen (1994) characterizes economic globalization in terms of two geographies: (1) *centrality* (defined as the concentration of capital investment and global control capability in particular urban centers), and (2) *marginality* (defined as the peripheralization, either in whole or in part, of cities and regions as a result of their exclusion from the major economic processes fueling global economic growth). Sassen (1994) also documents ways in which new dynamics of centralization are linking the First World's major international financial and business centers (e.g., New York, London, Tokyo,

Paris, Frankfurt, Zurich, Amsterdam, Sydney, Hong Kong) with each other as well as with some of the Third World's major cities (e.g., São Paulo and Mexico City). Concurrent with dynamics of centralization and the establishment of transnational linkages, there is a deepening of marginality taking place in both less developed and highly developed countries.

In view of the new geographies of centrality and marginalization, some authors argue that the First World–Third World typology has lost its meaning. Castells (1993) argues that what is commonly understood to be the Third World is no more; it is "rendered meaningless by the ascendance of the newly industrialized countries (mainly in East Asia), by the development process of large continental economies on their way toward integration in the world economy (such as China and, to a lesser extent, India), and by the rise of a *Fourth World,* made up of marginalized economies in the retarded rural areas of three continents and in the sprawling shanty-towns of African, Asian, and Latin American cities" (37).

Concern about the rise of a Fourth World can be found in Habitat II's 1996 *Global Plan of Action* (GPA). Authors of the GPA argue that the advancing degradation and the disorganization of cities in both the developed and developing worlds suggest the existence of a new *un*developing world (i.e., Fourth World):

Countries in this new world—countries in the North as well as the South—exhibit cities that are becoming less efficient as markets and producers; cities that are becoming less humane and attractive; faction ridden cities that disrupt national cohesion and solidarity; cities that drain national resources just to cope with chronic problems of health and safety; cities where people are mistrustful of one another and insecure; and cities that are marketed as commodities not communities. If such a new third world is to be averted, if gains in the quality of life for all are not to be lost, settlements—and especially cities—must be made to work. (Habitat 1996c, 3)

This brings to light a paradox. On the one hand, cities of the world are increasingly interdependent in the new global division of labor. Yet, it is a very uneven process. Whole cities, and more often, parts of cities, are marginalized in the process.[1] Saxe-Fernández (1994b) argues that economic globalization involves dynamics of disintegration, fragmentation, and de-linking as well as integration. There are winners and losers. As opportunities are created in certain localities and regions, they are dashed in others (Cook, Middlebrook, and Molinar Horcasitas 1994). This point was made painfully obvious on January 1, 1994, when a rural rebellion in Southern Mexico (Chiapas) seized seven towns to protest Mexico's entry into the North American Free Trade Agreement (NAFTA) (Hernández 1994; Heridia 1994). One of the first communiqués from the rebellious forces stated that NAFTA "is a death certificate for the Indian peoples of Mexico, who are dispensable for the government of Carlos Salinas de Gortari."[2]

Criticism of the neoliberal agenda has also been violently expressed by the newly formed guerrilla organization, the People's Revolutionary Army (Ejército Popular Revolucionario; EPR). In late August 1996, the EPR mounted a series of attacks aimed at army and police posts in several states in south and central Mexico. The attacks left at least sixteen dead and twenty-eight wounded (*La Jornada,* September 3, 1996). Contrary to the government's characterization of the EPR as an isolated fringe group confined to rural parts of the state of Guerrero, leaders of the left-wing force claim to have a contingency of armed rebels ready for action in the heart of the nation (i.e., in the Federal District and state municipalities that make up Mexico City) (Chiprés 1996). During a press conference held in a safe house on the outskirts of the capital, two EPR *comandantes* described how the worsening conditions in Mexico City helped motivate their rural-based insurrection. They argued that Mexico City is so racked by "unemployment, misery, lack of education, and the high cost of living" that it is no longer a viable receiving area for rural migrants; a historic escape channel from rural poverty has been cut off (Chiprés 1996). Concern about livelihood opportunities has become pivotal.

One of the trends cutting across national boundaries is the rise of open unemployment. Portes (1989) points out that "it was thought that mass unemployment in Third World cities was impossible, high underemployment, yes—but unemployment, no. The indisputable fact is recent years have witnessed a rise in the number of city dwellers deprived of any opportunities to earn income" (39). Portes argues that the rise of unemployment can be traced directly to structural adjustment policies (SAPs) intended to lessen foreign indebtedness. Cheru (1995) presents evidence that SAPs have "aggravated urban poverty and weakened the capacity of both national and municipal governments to manage more complex issues of the day such as environmental protection, poverty alleviation and the provision of basic services" (5). In many parts of the Third World, urban poverty has grown faster than rural poverty as a result of macroeconomic adjustments (Bartone et al. 1994, 10).

Livelihood Opportunities and the Impact of Economic Crisis

There is mounting evidence that the urban economy does not generate enough jobs for the fast-growing economically active population (EAP) in many Third World cities. Along these lines, Rondinelli and Kasarda (1993, 103) argue that

The enormous number of people added to the urban labor force in developing countries is placing severe strains on the ability of private and public organizations to increase production and expand employment opportunities sufficiently to absorb the growing numbers of people looking for work. Unless productive capacity can be expanded rapidly enough to create about 1 billion new jobs in urban areas over the next 35 years, widespread urban unemployment and underemployment will undermine progress toward economic growth and improved living conditions.

Rondinelli and Kasarda (1993, 101) estimate that the urban labor force growth in developing countries will more than double from 409 million in 1980 to 839 million in the year 2000. They project it will double again to 1.7 billion by the year 2025. More than 228 million new jobs will have to be created in Third World cities during the 1990s. Beyond this, "[a]n additional 875 million urban jobs will be needed in the following 25 years. To absorb the enormous growth of the urban labor force, at least 280 million new urban jobs will be needed in Africa, 180 million in Latin America and the Caribbean, and almost 672 million in Asia" (101).

In the case of Mexico, Cordera Campos and González Tiburcio (1991, 28) note that throughout the 1980s, the nation's EAP grew at a faster rate than the general population (2.6 percent versus 2.2 percent). Mexico's EAP numbered 25 million people in 1991. The relatively small size of the labor force compared to the size of the general population—of more than 80 million—is due to the country's young demographic profile. In 1990, over 80 percent of the population was under forty years of age; 38 percent were under fifteen.

Over the period of 1980 to 1990, the number of people working in the primary sector of Mexico's economy remained stable at 5 million.

Employment in the secondary sector rose from 2.9 million to 6 million. The most rapid growth occurred in the service sector, where the number of jobs increased from 4 million in 1980 to 10.8 million in 1990 (Business Monitor International 1992). Rondinelli and Kasarda (1993, 92) estimate that roughly 1 million new jobs will be needed every year up through the year 2010 to keep pace with the growth in Mexico's EAP. By no means is it preordained that this demand will be met. From a historical perspective, there is room for doubt. Few developing countries have been able to create enough formal sector livelihood opportunities in their fast-growing cities (Rodgers 1989b). Even developing countries that have had high rates of economic growth face serious problems generating enough adequate livelihoods.

To stimulate economic growth and job creation, the emphasis in Third World economic development policy shifted from import substitution industrialization (ISI) to export led industrialization (ELI). Mexico is a classic illustration. Between 1940 and 1976, the dramatic expansion of Mexico's economy was based on ISI. During this period, most Mexicans experienced improved socioeconomic conditions (Barry 1992, 94). Living standards rose as a large middle-class of bureaucrats, professionals, and industrial workers grew to comprise nearly 30 percent of Mexican society. But as millions of people swelled the ranks of the nation's burgeoning cities—especially Mexico City, Guadalajara, and Monterrey—not everyone enjoyed the same benefits.

In 1977, the year after Mexico's first major economic crisis, the lowest 40 percent of Mexico's population had less than 12 percent of the country's personal income (Barry 1992, 94). By 1980, according to a World Bank report, Mexico had "one of the worst profiles of income distribution of any nation on earth" (94). Notwithstanding the grossly unequal distribution of income, Mexico's sustained economic growth from the 1940s to the 1970s made it one of the richest nations in Latin America. Presently, when ranked according to the value of manufacturing exports, Mexico is listed among the world's top-ten developing countries (United Nations Development Program 1992). Its industrial base is larger than that of Belgium, Spain, or Sweden (Mexico Business Monthly 1992, 12).

For a whole generation, Mexico based its development policy on ISI. This urban-biased, resource-intensive model of development eventually produced many of the same problems that characterized the process elsewhere in Latin America: technological dependence, a balance of payments deficit, inefficient industries, hyperurbanization, environmental degradation, and the impoverishment of rural communities. By the end of the 1970s, it was obvious that formal sector job growth as a result of ISI did not keep pace with the growth in Mexico's economically active population (Ward 1990; Cockcroft 1983). ISI as a development strategy came to be viewed as part of the problem; it was thus abandoned in favor of ELI (González de la Rocha and Escobar Latapí 1991; Nuccio 1991).

In response to the debt burden, soaring inflation, and economic recession that took hold of the country in the early 1980s, President de la Madrid (1982–1988) initiated a far-reaching process of economic restructuring. Subsequent administrations have continued this process with an increasing emphasis on (1) the privatization of the economy, and (2) trade liberalization in an attempt to strengthen Mexico's export industries. As a result of restructuring, Mexico's economy began to experience economic growth in 1991, and the nation's international image improved. But the benefits were uneven. According to a 1992 report in *El Financiero Internacional* (April 20), the wealth created during the so-called economic recovery of the early 1990s was not evenly distributed; it deepened the inequalities in Mexican society. This report cites official data showing that the principal indicators of social welfare—real salaries, nutritional indexes, food consumption, and labor's share of the GDP—declined (3). Silva (1994, 9) reports similar results. Although Mexico's GDP grew 18 percent over the ten-year period of 1984 to1993, the gains achieved per capita declined 3 percent for the same period.

Policies of macroeconomic restructuring implemented in Mexico since the early 1980s have had a major impact on the urban poor. Prices of food, water, energy, and housing have increased, while real wages have fallen (Gilbert 1992, 4; Heredia and Purcell 1995). Between 1982 and 1992 the purchasing power of the minimum wage fell 72 percent (Business Monitor International 1992).[3] From January 1994 to January 1996 real wages dropped 20 percent more (Banamex-Accival, Grupo Financiero 1996). Currency devaluation and reduced government backing for the demands of organized labor have contributed to the dramatic fall in real wages. The result has been a decline in the living standards of Mexico's poorest citizens. This group includes rural people, who as a result of the recession have increased the flow of rural to urban migration (Gilbert 1992, 438). The government has justified the decline by arguing that a reduced wage bill for the private sector will translate into greater competitiveness, higher employment, and greater production (Business Monitor International, 1992, 29). At the same time, wages will keep consumption pared to the bone, thus reducing both inflationary pressures and imports (29). President Zedillo presented an emergency plan, including wage and price controls and significant cuts in government expenditures, during a special address to the nation. The speech rang with words such as "drastic," "grave," and "urgent."[4]

The negative impact on the urban poor's capacity to access land and materials for housing has been significant (Ward and Macoloo 1992). This trend illustrates well what Brecher and Costello (1994) have characterized as the "downward leveling" associated with economic globalization. Based on an analysis of the cost of land, materials, and infrastructure, Glibert (1992) documents how the recession has damaged housing standards. He cites evidence that the recession has had a negative effect on housing tenure, housing densities, plot sizes, service and infrastructure provision, levels of rent, and frequency of eviction. Along similar lines, González de la Rocha and Escobar Latapí (1991, 11) point out, "the expansion and improvement of the stock of popular housing has slowed as a direct consequence of the need to devote a larger part of household income to the satisfaction of more basic needs." For instance, in 1981 one day's minimum wage could buy about 18 pounds of beans; by 1991 it could buy only 5.5 pounds (Barry 1992, 98). The magnitude of the decline in purchasing power has translated into a significantly diminished capacity to purchase building materials (Schteingart 1993).

Economic liberalization has become the target of rising criticism in Mexico and throughout Latin America. It is on increasingly shaky ground as public support for it weakens. The violent protests in Argentina and Southern Mexico clearly show the strain (Barry 1995; Lowenthal 1994). Critics point out that the macroeconomic bias associated with economic liberalization and ELI misses important opportunities. For instance, a widely noted scholar of Mexican studies argues that "a government dominated by macroeconomically oriented technocrats, international big business, and the upper-middle class in large cities risks overlooking the opportunities for socially beneficial micro-level intervention by the state" (Cornelius 1994). Critics also express concern about the widening gap between the haves and have-nots. In 1984, the top 10 percent of the Mexican population earned 33 percent of the national income. By 1989, this figure had risen to 38 percent. In contrast, the poorest 30 percent of the population received only 8.5 percent of the national wealth in 1989, down from 8.8 percent in 1984 (INEGI 1992a). Mexico is one of the world's top nations in terms of billionaires, only the United States, Germany, and Japan have more billionaires. Reportedly, the billionaire roster in Mexico rose from one in 1988, to thirteen in 1993, to twenty in 1994.[5] At the other extreme, a study released by the prestigious research university El Colegio de México reports that 54 percent of those living in the MAMC are in extreme poverty; 1.7 million (1990 figure) people live in housing considered indigent.[6]

Urbanization under conditions of resource and income scarcity demands the integration of concerns about the environment and development. Douglas and Zoghlin (1994) capture this point well: "While much of the concern over the environment has disregarded or even opposed questions of income and livelihood, the concern over the economics of livelihood in the city has tended to dwell on employment and income generation over environmental resource conservation and renewal. Unless both questions are tackled simultaneously, sustainable cities as a concept that entails human-environmental relations has little utility" (171).

The Growth of Large Cities

The *World Development Report 1992* contains the following warning: "Between 1990 and 2030 the world's population will almost double, and industrial output and energy use will probably triple worldwide and increase sixfold in developing countries. Under current practices, the result could be appalling environmental conditions in cities and countryside alike" (World Bank 1992, 310).

According to UN projections for the period from 1990 to the year 2000, 80 percent of the growth in the world's population will be in urban areas. At the current growth rate of 2.8 percent yearly, the number of people living in cities throughout the world is projected to double in about twenty-five years. Nearly nine-tenths of this expansion is expected to occur in the Third World, where the annual urban growth rate is 4 percent—more than triple that of the industrial world (UNDP 1992, 169). Starting with a baseline of 2.4 billion people in 1990, the United Nations Development Programme (UNDP) forecasts that the world's total urban population will rise to 3.2 billion in 2000 and 5.5 billion in 2025. The developing countries' share in these totals is expected to rise from 63 percent in 1990, to 71 percent in 2000, and to 80 percent in 2025 (UNDP 1992, 212–213). Megacity growth is expected to be a significant part of this scenario. By the end of the 1990s, Mexico City and São Paulo may top the 20 million mark; Calcutta, Shanghai, and Bombay are expected to swell to at least 15 million; and thirteen other cities in developing countries will each be the life space of at least 10 million residents: Seoul, Cairo, Dacca, Delhi, Lagos, Beijing, Manila, Jakarta, Karachi, Tianjin, Buenos Aires, and Rio de Janeiro.

Currently, two-thirds of the total annual population increases in the developing world accumulates in cities. An estimated 60 million people join the Third World's urban masses every year (Sitarz 1993, 170–171). If this rate holds, nearly 2 billion people will populate the urban areas of developing countries by the year 2000. By the year 2020, some estimate that the urban population of Africa, Asia, and Latin America will have increased from its 1980 total of 1 billion to reach more that 3 billion people (Douglass 1992, 32). Rural-to-urban migration has historically accounted for most of the increase in urban populations. Such is still the case for many African cities. But in urban areas of Asia and Latin America, most of the population growth is due to births among city dwellers already in place (Hesselberg 1995, 8).

Satterthwaite (1991, 1996) cautions that, in many cases, population projections for Third World cities may be significantly overestimated, as they have proved to be in the past. He argues that it is unlikely that the rapid urban growth rates of the past will continue up to the year 2025. Mexico, presently the world's eleventh most populated country, is a case in point. Projections made in 1970 suggested that Mexico's total population would reach 130 million by the year 2000. It is now obvious that such an estimate was too high; Mexico's total population will probably not reach 130 million until 2025. The nation's total population growth rate actually decreased from 3.2 percent per year in 1970 to 1.8 percent in 1990. And it is expected to continue dropping to 1.2 percent per year by the year 2000 (*Union Tribune* (San Diego), August 29, 1994).

Although growth projections do vary, there is a general consensus that the historic shift toward an increasingly urban global population will continue into the near future. In 1940, only one person in eight lived in one of the world's urban centers, and only one in one hundred lived in a city with a million or more inhabitants. By 1980, nearly one in three persons was an urbanite, and one in ten lived in a city with a million or more (WCED 1987, 236). In 1990, most of the world's population still lived in rural areas. But by 2030, according to World Bank (1992, 27) projections, the opposite will be true: urban populations may be two times the size of rural populations.

In Mexico, nearly one out of every two urban dwellers (45 percent of the total urban population) currently lives in a city of 1 million or more (UNDP 1992, 168). The annual growth rate of the urban population in Mexico, from 1960 to 1990, was a remarkable 4.1 percent. Over the same period, Mexico's urban population—as a percentage of the total population—shot up from 51 to 73 percent. Most of the increase registered in the nation's three largest cities: Guadalajara, Monterrey, and especially Mexico City.

Mexico City's Primacy in the National Economy

If the Metropolitan Area of Mexico City (MAMC) were a separate nation, it would have the world's thirty-fifth largest economy (Cremoux 1991, 7). Approximately 30,000 industrial entities produce roughly half the nation's nonoil manufactured output (Business Monitor International 1992). In services too, the MAMC has exercised a position of paramount importance in the national economy: about one-half of the total national production generated by this sector is created in the metropolitan area (Ward 1990, 19).

Eighty percent of Mexico City's inhabitants are less than forty-four years old (López and Hidalgo 1996). Over 7 million economically active people in the city make it the single-largest labor market in the world (Jusidman 1988, 225). Mexico City's economy, with a GDP of US$63.6 billion in

1990, is nearly equal to that of Argentina's national economy and is larger than those of Panama, Belize, Costa Rica, Nicaragua, El Salvador, and Honduras combined (Cremoux 1991, 7). The MAMC generates 36 percent of Mexico's total GDP.

Sprawled out over 950 square miles, Mexico City receives 25,000 tons of food daily, delivered by 50,000 trucks. This amount would suffice to feed for one day the entire combined populations of Costa Rica, Nicaragua, Belize, Panama, and El Salvador—or, Copenhagen, Berlin, Madrid, Rome, Athens, and Budapest (Cremoux 1991, 19). On any given weekend, 279 cultural events take place, including 22 lectures, 68 films, 59 exhibits, 4 book presentations, 105 plays, and 21 concerts. And, over the course of a year, the city is host to at least 1,000 major public demonstrations (19–22). On an ongoing basis, the city's historic central plaza, the Zócalo, is occupied with protesters flying banners of all sorts. One report notes that in the first seven months of 1992, there were 2,412 marches and demonstrations that ended up in the Zócalo—that's an average of eleven per day.[7] Such events make the city one of the world's most active centers of popular mobilization.

Mexico City's urban primacy in not uncommon. Urban growth and distribution in Latin America, the Caribbean, large parts of Asia, and to some extent Africa has long been characterized by the macrocephalic domination of the country by its national capital and sometimes by one or two secondary cities (see table 2.1). The metropolitan area of São Paulo, for example, generates 36 percent of Brazil's national domestic product and 48 percent of its net industrial product (Sassen 1994, 30). In many countries, urban primacy appears to be intensifying. Sassen (1994) notes that the "disintegration of rural economies, including the displacement of small holders because of the expansion of large-scale commercial agriculture, and the continuing inequalities in the spatial distribution of institutional resources are recognized as key factors strengthening primacy" (34). But Sassen also notes that certain countries in Latin America experienced a deceleration of primacy in the 1980s. Mexico is one such country.

Mexico City's growth rate declined from 5.1, to 3.2, to 2.0 percent over the period from 1950 to 1980 to 1990 (Negrete et al. 1993). The city's growth reached a threshold in 1970, after which massive immigration ceased and growth slowed to a 3 percent annual increment over the extent

Table 2.1
Urban indicators and growth rates in select countries, 1970–1992

| Nation | Urban population | | | | Population in capital city as % of | |
| | As % of total population | | Average annual growth rate (%) | | Urban | Total |
	1970	1992	1970–1980	1980–1992	1990	1990
Latin America						
Bolivia	41	52	3.4	4.0	34	17
Chile	75	85	2.4	2.1	42	36
Colombia	57	71	3.3	2.9	21	15
Ecuador	40	58	4.8	4.4	21	12
Mexico	59	74	4.1	2.9	34	25
Peru	57	71	4.0	2.9	42	29
Uruguay	82	89	0.7	1.0	44	39
Venezuela	72	91	5.0	3.4	23	21
Asia						
Bangladesh	8	18	6.8	6.2	37	6
China	18	27	2.7	4.3	4	1
India	20	26	3.9	3.1	4	1
Indonesia	17	32	5.1	5.1	17	5
Pakistan	25	33	4.4	4.5	1	0
Philippines	33	44	3.8	3.8	32	14
Thailand	13	23	5.3	4.5	57	13
Africa						
Ethiopia	9	13	4.8	4.8	30	4
Kenya	10	25	8.5	7.7	26	6
Mozambique	6	30	11.5	9.9	38	10
Nigeria	20	37	6.1	5.7	23	8

Source: World Bank and International Bank for Reconstruction and Development (1994, table 31).

	Pop. in urban agglom. of 1 million or more as % of			
	Urban		Total	
Nation	1970	1992	1970	1992
Latin America				
Bolivia	29	29	12	15
Chile	40	44	30	38
Colombia	40	41	23	29
Ecuador	50	55	20	31
Mexico	43	41	25	30
Peru	39	45	22	31
Uruguay	51	47	42	42
Venezuela	28	30	20	27
Asia				
Bangladesh	47	56	4	9
China	48	35	8	9
India	32	34	6	9
Indonesia	42	36	7	11
Pakistan	49	53	12	17
Philippines	29	36	9	15
Thailand	65	60	9	13
Africa				
Ethiopia	29	30	3	4
Kenya	45	30	5	7
Mozambique	69	43	4	12
Nigeria	26	29	5	10

population. The city may have added only 800,000 people during the 1980s. It is also widely believed that the 1980 population level was overestimated, so the real increase for the 1980s may be somewhat higher (see Gilbert 1993, 726). Mexico City's standing is expected to drop (over the period of 1995 to 2015) from fourth to tenth place on the list of the world's most populated cities (López and Hidalgo 1996).

The MAMC's declining overall growth rate is due in part to the slackening of Mexico City's primacy in the national economy (de la Pena et al., 1990). Among other things, the creation of export processing zones and expanding opportunities in cities near Mexico's border with the United States have redirected some investment away from the capital (Portes 1989, 34). In addition to Mexico City, Guadalajara, and Monterrey, there are now fifteen medium-size cities of 500,000 inhabitants or more that show significant increases in population, income, employment, and economic activity in general.[8]

Over the period of 1980 to 1988, the number of industrial establishments in Mexico City fell from 35,400 to 29,400 (a reduction of 750 per year) (Negrete et al. 1993, 18–24). During the same period, the total number of formal sector employees dropped from 981,000 to 732,000, and the city's contribution to the GNP decreased from 43 percent of the nation's total to 32 percent (18–24). Gilbert (1993, 727) describes this trend in terms of "polarization reversal." He cites evidence that ELI in Korea, China, and Mexico has encouraged a more deconcentrated pattern of industrial growth (730). Lustig (1992) also documents the decline in Mexico City's share of the country's GDP. Lustig concludes that "the concentration of activity and income in a region that is far from maritime outlets and the border areas might be working against the expansion of exports" (237).

It is important to note that aggregate figures about Mexico City's overall growth rate and declining primacy can be misleading. Parts of the metropolitan area are currently growing at rates of 8 percent or more, while some of the inner-city wards have actually been loosing population. Despite the slowing of the overall growth rate, urban sprawl has continued due to the ongoing redistribution of the metropolitan area's population. A number of factors have contributed to this expansion at the metropolitan level, including the relatively low cost to consumers of (subsidized) public transportation, the decentralization of sources of employment out of the Federal District to peripheral municipalities of the state of Mexico, and the decline in environmental quality of life in certain parts of the city to the extent that said parts have become "population repulsers" (Negrete, Graizbord, and Ruiz 1993, 24).

With respect to the demographic shifts taking place in the Federal District, Quadri de la Torre (1993a, 39) calls the situation perverse. He notes that the Federal District's central wards, including Cuauhtémoc, Miguel Hidalgo, Benito Juárez, and Coyoacán, are losing population while the peripheral, semirural, ecologically vital wards—including Tlalpan—are experiencing exponential growth. Cymet (1992, 2) reports that during the 1980s more than 500,000 people migrated out of the Federal District's inner wards only to resettle in places at the city's outer periphery. Most of the peripheral expansion is taking place in the state of Mexico, especially an area called Chalco.[9] But certain wards within the Federal District are also experiencing a rapid expansion of irregular settlements. Specifically, rapid growth is taking place in Xochimilco, Tláhuac, Ixtapalapa, and most of all in Tlalpan—especially the part designated as the Ajusco greenbelt zone.

A burgeoning literature now draws attention to the serious problems of environment and development facing fast-growing cities around the world. The ongoing concentration of urban populations in large cities raises serious concerns about the future of such metroregions and the hinterlands supporting them. The preamble to the *Habitat Agenda* signed during the 1996 City Summit specifies a long list of what are labeled as the most serious problems confronting cities and their inhabitants. The list includes inadequate financial resources; lack of employment opportunities; spreading homelessness and expansion of squatter settlements; increased poverty and a widening gap between rich and poor; growing insecurity and rising crime rates; inadequate and deteriorating building stock, services, and infrastructure; lack of health and educational facilities; improper land use; insecure land tenure; rising traffic congestion; increasing pollution; lack of green spaces; inadequate water supply and sanitation; uncoordinated urban development; and an increasing vulnerability to disaster. The City Summit document goes on to acknowledge that all of these problems have "seriously challenged the capacities of Governments, particularly those of developing countries, at all levels to realize economic development, social development and environmental protection, which are interdependent and mutually reinforcing components of sustainable development" (Habitat 1996b, preamble).

In his article about megacity growth in Latin America, Enrique Leff (1990a, 56) points out that "the centripetal forces of urban concentration have already trespassed the physical and social absorbing capacities of the megacities." In some cases, such as Mexico City, this process has begun to undermine the city's original advantages and "has externalized social and ecological costs in the form of saturation levels of air, water and noise pollution. It has also generated acute regional imbalances and social inequalities. Ultimately, it has degraded the basic ecological mechanisms that assure sustained productivity of natural resources and the social basis for a democratic management of the productive processes by communities. The metropolitanization process has generated a growing deficit in public services and the reversal of *economies of agglomeration*" (Leff 1990a, 50). Leff views the consequences of this type of urban growth in bleak terms. He argues that it is "leading to the degradation of the quality of life for people, to pressing social inequalities and increasing ecological costs, and to the rising prices of inputs from the regional environment" (50).

Insofar as efforts have been made to better manage urban development, the World Bank highlights the following concern: "While urban settlement patterns have varied across countries, in no countries have efforts to restrain migration or urban growth been successful" (World Bank 1991, 200). The UNDP has highlighted similar concerns. It describes the growth of the world's cities as relentless, inevitable and irreversible. Critically reflecting on the type of urban demographic trends noted above, the UNDP notes that

Despite the obvious efficiency advantages of cities, the negative consequences of urbanization for low-income groups are overwhelming. Simply, many city dwellers in developing countries live in crushing poverty—more than 300 million, or a quarter of all those in urban areas. The number promises to swell. By 2000 more than half the developing countries' poor will be in cities and towns: 90 per cent in Latin America, 45 per cent in Asia and 40 per cent in Africa. Their living conditions are alarming, for their numbers far outstrip the supplies of water, waste removal, transport and clinics. (UNDP 1992, 213)

When such daunting figures are documented, the Mexico City case is often presented as an exemplar—the black hole of megacity doomsday theory. The "population explosion" in the Mexican capital and in other fast-growing Third World cities is sometimes characterized as an out-of-control cancerous growth that will ultimately destroy the whole earth. But careful analysis gets beyond the human numbers game. Cities of the North and South are complex arenas of social and economic cleavages (Douglas and Zoghlin 1994). The relationship between urban size, prosperity, and environmental conditions is not clear cut (Drakakis Smith 1995; Kasarda and Parnell 1993b). Sensitive analysis must take into account the dynamics of uneven development in a historical and an international context while also keeping in view a particular city region's biogeophysical dimension.

The Valley of Mexico: An Increasingly Poor Host to Megacity Sprawl

Mexico City harbors a wide range of sources of contamination: erosion; exposed trash and feces; entry into the subsoil of untreated waste water and polluted runoff; emissions from factories, workshops, thermoelectric plants, refineries, petrochemical plants, iron and steel foundries, and a large quantity of industrial and domestic incinerators; and millions of internal combustion engines (Nuccio 1991; Herrera Legarreta 1990; Ibarra, Puente, and Saavedra 1987; Puente and Legorreta 1988). On a wider scale, the Central Mexican Basin has become seriously degraded by the toxic output from industry and vehicles, extensive deforestation (1,000 hectares per year), the nearly total desiccation of its lakes, and the exploitation of materials for construction (Rivera 1987; Velázquez Zárate 1992). And the particular geographic, hydrological, and subsoil conditions that constitute the valley make Mexico City especially vulnerable to the region's powerful earthquakes.

Out of all the city's environmental problems, air pollution gets the most attention. Barry (1992, 269) notes that breathing the city's air is unhealthy during eight of every ten days. During 1991, air pollution levels exceeded health standards on 350 of 365 days (*Mexico Business Monthly* 1992). The situation is aggravated by the fact that the city lies within the Central Mexican Basin—a valley nearly 6,800 feet above sea level and enclosed by mountains. Given its high altitude, Mexico City's air contains 23 percent less oxygen than what you would find at sea level. Consequently, the operation of internal combustion engines is dirtier and less efficient

(Velázquez Zárate 1992, 36). The valley is subject to thermal inversions that trap a noxious mixture of industrial emissions, vehicle exhaust, smoke, and human waste.

Bartone et al. (1994, 103) report that lead, ozone, carbon monoxide, and fine particulate matter are the MAMC's most threatening air pollutants. National and international norms of ozone, sulfur dioxide, nitrogen dioxide, and suspended particles are regularly surpassed (Barry 1992, 269). Indeed, Mexico City has the dubious distinction of being the world's metropolis with the highest urban concentration of ozone and suspended particulate matter. Carbon monoxide is by far the emission with the greatest mass. But atmospheric lead, from the combustion of lead antiknock compounds in gasoline, has been the most life damaging (Bartone et al 1994, 103).

A study conducted by a United Nations agency reports that the average lead content in the blood of a Mexico City urbanite is four times greater than one living in Tokyo and more than two times greater than one living in Baltimore, Stockholm, or Lima (Rohter 1989). In some areas of the city, breast-fed infants are ingesting toxic levels of once airborne lead that has become concentrated in their mother's milk. The high concentration of lead found in newborns has been shown to be related to low birth weights and developmental problems such as altered reflexes and learning disorders. It may be that as many as 29 percent of all children in the city have unhealthy levels of lead in their blood (Schteingart 1989a). Bartone et al. (1994) report that 140,000 children suffer a reduction in IQ and agility, requiring remedial education, due to excessive lead exposure. And roughly 46,000 adults suffer from hypertension for the same reason (103).

The quantitative estimates of the health and economic impacts of pollution in the MAMC are troubling. Hundreds of thousands, if not millions, of people suffer pollution-related illness in Mexico City (e.g., emphysema, pneumonia, bronchitis, asthma, cardiovascular complications, conjunctivitis, sinusitis, laryngitis, allergies, and bloody noses) (*Los Angeles Times*, April 21, 1991, sec. M, p. 1). A researcher at Mexico City's National Institute of Respiratory Infections cites evidence that 4 million people in the city suffer from respiratory infections and that pulmonary cancer and fibrosis have increased significantly during the past several years (*Mexico Business Monthly*, April 1991, p. 8). Every day 80 tons of fecal dust join the ranks of other airborne contaminants riding the winds in the Mexico

City. This contamination leads to some killer diseases including gastroen-teritis (the baby stealer), typhoid, and hepatitis among others (*La Jornada*, April 22, 1991).

In a speech made just before he became president, Ernesto Zedillo lamented the heavy toll pollution exacts on the labor force. He cited an estimate that pollution-induced illness among Mexico City's economically active population results in lost income and medical expenses totaling at least 4,500 million new pesos (approximately US$1.3 billion).[10] Some experts calculate that the number of pollution-triggered deaths in the MAMC may be 5,000 annually (*Los Angeles Times*, April 21, 1991, sec. M, p. 1). Bartone et al. (1994) report that "excess levels of fine particulate matter may be responsible for 12,500 extra deaths and 11.2 million lost workdays a year, while ozone may account for 9.6 million lost workdays per year, both due to respiratory illnesses" (103).

With respect to global warming, Liverman (1992, 72) estimates that Mexico City's monthly temperature may increase by five degrees Celsius (over the current norm) in both winter and summer. This would generate higher levels of heat stress and air pollution in an already frequently hot and polluted environment. Liverman's analysis suggests that the Valley of Mexico would become significantly drier as soil moisture and water availability decreased. Moreover, "this has serious implications for water planning in Mexico City, where water is already limited, groundwa-ter levels are declining, and considerable water is pumped in over moun-tains at great expense" (Liverman 1992, 73). Rising concerns about Mexico City's air pollution are not limited to the negative impacts it has on peoples' physical well-being. Air pollution is also blamed for causing acid rain, which damages the cultural heritage (e.g., buildings, monuments, artwork) of the city as well as its flora, fauna, and soil resources (Leff 1990a, 53).

Each of the listed problems is cause for concern. However, it should be noted that much of the reporting about Mexico City's air pollution is exaggerated and sensationalist. For instance, in one article the city is referred to as an "urban gas chamber" and as the "anteroom to an ecologi-cal Hiroshima" (*Time Magazine*, January 2, 1989). Such exaggerations are not confined to street talk, the media, and academic circles; they enter the business world as well. For example, a commercial vendor of oxygen in Mexico City uses fear to sell its product. Called Oxi Medic, the company markets portable oxygen tanks for $US90.00. The advertisement promoting the sale of the oxygen tank reads, "surely you've noticed at some time respiratory problems caused by the excessive pollution in the air we breath. . . . What would happen if this situation suddenly worsened and became an emergency? Only three minutes without oxygen can cause irreparable damage. . . . When you need it, it could be too late to buy it" (cited in Cleeland 1991).

Sensitive researchers are quick to note that such reporting and the marketing of fear are grossly irresponsible (Ward 1990, xvii). Castillejos (1988) explains that it is difficult to gather good data; there are shifting peaks and valleys in the content, levels, and spatial expression of environmental contamination. Within the city, certain sectors of the population suffer from air pollution more so than others. Hardoy, Mitlin, and Satterthwaite (1992, 103) note that the highest concentration of airborne particulate matter exists in the southeastern and northeastern areas of the capital, where the population is predominantly lower income.

The government has taken some significant steps to curb air pollution in the capital (Dirección General de Comunicación Social 1992). In October 1990, the Mexican government launched a multi-billion-dollar antipollution campaign in the capital, which included measures to convert some 460,000 public transportation and cargo vehicles from gasoline or diesel to cleaner-burning natural gas (*Mexico Business Monthly* 1992, 8). The government has also closed down certain heavily polluting industrial facilities—most notably, the "March 18" oil refinery at a cost of 5,000 jobs and at least $500 million dollars.

However, a recent report by *Environment Watch: Latin America*[11] notes that "Industrialists and residents alike are severely criticizing Mexico City's government, and especially the Metropolitan Commission for the Prevention and Control of Contamination (CPCC), for continued air pollution that has not been significantly alleviated by numerous antipollution contingency programs." The report cites data indicating that air pollution in 1993 remained largely unchanged, with only thirty-seven days considered to have "satisfactory" ozone levels compared to thirty-one days in 1992.[12]

Although the campaign to curb air pollution has not been very effective, it has nonetheless been expensive. During the past administration, President Salinas noted that the demand for resources to clean and protect the environment competes with other priorities: "To clean the air in Mexico City, we are investing $4 billion—resources we could have used to build schools, health centers or improve the quality of life in other regions of the country. But we are convinced that those are resources well invested."[13] Given the fact that air pollution control has been given high priority, other issues get less attention. In this respect, Hardoy et al. (1992, 204) argue that the focus on air pollution may be misguided. That is, it concentrates on eradicating chemical pollution and damage to ecosystems at the expense of underfunding efforts to address two other pressing environmental problems: (1) biological pathogens, and (2) people's access to natural resources (especially freshwater and safe land sites for housing). Damaging levels of biological pathogens come from, among other sources, impure water, inadequate disposal of solid and liquid wastes, inadequate health services, and poor-quality housing. Millions of the Third World's urban poor cannot afford safe and healthful housing. Hardoy, Mitlin, and Satterthwaite, (1992, 204) note that these factors contribute to the serious ill health and even death of millions of people in Third World cities.

In view of such factors in the Mexico City context, it stands to reason that significant benefits could be derived from focusing more attention on the environmental problems stemming from biological pathogens and people's inadequate access to natural resources. This argument appears to be gaining support in international development circles. For instance, Leitmann (1994, 117) describes a shift in World Bank policy toward a *brown agenda*. The brown agenda focuses on the most pressing environmental problems facing cities in developing countries. It includes, among other things, the lack of safe water supply, sanitation, and drainage and the inadequate solid and hazardous waste management (Leitmann 1994, 117).

The Water Crisis

According to data calculated by the Mexican government's Water Commission of the Valley of Mexico (Comisión de agua del valle de México), the Federal District supplies 35,000 liters of running water per second to its consumers. The distribution network for this supply includes 10,700 kilometers of underground pipes, a significant portion of which were first installed more than 100 years ago (Esquinca and Núñez 1995). Major leaks occur throughout the system. More than 30 percent of the total volume of piped water is lost as a result.

Of the total water supply that does arrive to its intended destination, 67 percent goes to housing units, 17 percent to industry, and 16 percent to businesses and services (Esquinca and Núñez 1995). Although 97 percent of the housing units in Mexico City's Federal District are reported to have some kind of access to piped water, only 74 percent have an in-house source. Within the seventeen municipalities of the state of Mexico that belong to the metropolitan area of Mexico City, only 52 percent of the housing units have an in-house source of potable water. Roughly 3 percent of the homes in the Federal District and 9 percent of the homes in the state of Mexico do not have access to the public water supply (they obtain their water from surface sources, illegal wells, or private vendors). The most poorly served areas of the MAMC include the southern part of the Federal District (including the zone of Ajusco Medio) and the eastern areas in the State of Mexico.[14]

An estimated 30 percent of Mexico City's families live without adequate sewage facilities. As a result, large quantities of untreated sewage regularly flow into area waterways. The Panuco River, considered the valley's most polluted waterway, takes in roughly 2,000 tons of untreated sewage every day (Barry 1992, 270). In terms of other types of hazardous solid waste, only 3 percent of all that is generated on a daily basis in Mexico City gets treated. The rest (97 percent) is disposed and mixed with solid and nonhazardous waste and is either deposited in a landfill (legal and illegal), a river, lake, or sewer (Koloditch 1993).

In an attempt to safeguard water quality in Mexico, the federal government passed the National Water Law (Ley National del aqua). The 1991 law mandates that all of the nation's municipalities install wastewater treatment plants. But analysts estimate that 85 percent of the total volume of municipal wastewater is still discharged without any treatment at all (*Environment Watch: Latin America* January 1995, 4). Water supply problems are chronic in Mexico City (National Research Council 1995). Water from the valley's underground aquifer is being sucked up two times faster than it gets recharged (Quadri de la Torre 1993a, 37). The water table has been declining approximately one meter per year (Herrera-Revilla et al.

1994), and this dramatic fall in the water table's level has caused buildings and whole sections of the city to sink at the rate of six to forty centimeters per year (Esquinca and Núñez 1995). The ground level in some places has sunk as much as six to ten meters—causing rifts that provoke severe damage to pavement, buildings, and underground networks of water and drainage (National Academy of Sciences 1995).[15] To a certain extent this preexisting condition explains why the 1985 earthquakes were so utterly devastating.

Mexico City draws 72 percent of its total supply from the aquifer underlying the Central Mexican Basin. Another 2 percent is drawn from surface supplies. The rest of Mexico City's water supply comes from sources outside the basin including the Cutzamala pipeline and the Lerma River. It has become increasingly cost prohibitive to pump water up into the valley from the lower-lying, outside sources (Brown and Jacobson 1987; Duhau 1991a). Likewise, it has become increasingly expensive to pump used waters (i.e., *aguas negras,* which include liquid industrial wastes and human sewage) back out of the city. The valley has no natural drainage (Kandell 1988). Habitat has examined these trends and concluded that, by the year 2010, Mexico City will suffer a serious crisis due to water problems (López 1996).

Mexico City depends on water from the Cutzamala Basin, a site 100 kilometers away and 1,000 meters lower than the city. Quadri de la Torre (1993a, 37) calculates that in 1992 it took 3.4 million barrels of oil to supply the energy needed to pump, lift, and transport the water from Cutzamala. This had an opportunity cost of $US60 million, representing 6 percent of the total energy budget in Mexico City, including the energy budgets of industry, services, residential, and transport (37).[16]

From a holistic perspective, it should be noted that the Lerma and Cutzmala basins have been impacted by high social, economic, and environmental costs as a consequence of being tapped by Mexico City. Quadri de la Torre (1993a, 37) describes the impact as twofold: (1) it deprives agricultural producers of a vital resource, which may stimulate rural-urban migration, and (2) it degrades the lacustrine and river ecosystems.

It has become especially difficult and expensive to supply water to parts of Mexico City that have expanded up rocky mountainsides. For this reason, the government enacted a ban against building higher than the 2,350-

meter contour. The ban effectively outlaws the establishment of new urbanization on practically the entire western side of the Federal District (Connolly 1982). In irregular settlements that have established themselves above this height, water gets delivered by truck and is stored in open fifty-gallon drums. Water delivered in this way is often heavily contaminated with parasites and other organisms that cause a high incidence of intestinal illness. Such is the case in the low-income Ajusco settlements of Los Belvederes, which are above 2,600 meters in altitude.

To confront Mexico City's water crisis, the government has stepped up measures to conserve water (Merino y Guevara 1991). There has been growing support for investment in facilities to treat water for reuse (Ramírez Saiz 1992). But this has raised the questions: What type of investment? By whom? For whom?

Presently, less than 2 percent of Mexico City's demand for water is supplied by wastewater treatment plants. In this light, the efforts made in Ajusco to build an innovative water reclamation facility are laudable. Yet, in the Ajusco case, the government was disinclined to have community groups set up and operate such facilities. The government's approach emphasizes competition in the free market. In the Federal District there are over 2 million households and firms using water, yet only 1.2 million pay monthly bills. The government has been conducting a water census in order to increase the number of paying accounts. Once the census is completed, the government will invite private firms to take over administration of separate water districts. It is hoped that the expected increases in the price of drinking water will encourage development of water recycling projects (Mexico *Business Monthly,* July 1991, p. 11). It is not clear that the urban poor will benefit from this strategy.

The price paid for water and the amounts consumed vary greatly. Mexico City's poorest residents use the smallest amount of water but typically pay the most for it. Schteingart (1993) reports that the urban poor in the city's irregular settlements use as little as twenty liters of water per person per day. Legorreta (1992) reports an even lower figure: twelve liters of water per person per day. So few liters may be enough for cooking and drinking, but more than double this amount is required for maintaining adequate sanitation and a healthy environment. World Bank data suggest that at least fifty liters per person per day are necessary to ward off health-related problems (Falkenmark and Suprapto 1992). In comparison, people living in middle-class housing developments consume on average 300 to 400 liters of water per person per day. The respective figure for those living in upper-class housing estates is between 800 and 1,000 (Schteingart 1993).

The question of who pays for how much water is likely to become an increasingly contentious issue. Historically, water supplies in Mexico have been provided with high public subsidies.[17] Water subsidies have long been used to encourage industrial development. And, in cases where water rations have been provided free of charge to the urban poor, public sector expenditures have been defended as antipoverty measures. However, Mexico's fiscal crisis has tightened the supply of public revenue. Existing funds are inadequate to pay for water system operation and maintenance. Historic levels of water sector subsidies have become difficult to maintain. This has prompted water authorities in the government to collect more for delivery and to privatize service wherever possible. Along these lines, Mexico has been characterized as the vanguard among developing countries (National Research Council 1995). Water is increasingly recognized as a private economic good. However, reversing past trends will be difficult, and the issue of equity looms large. Prospects for making the water delivery system more equitable is inextricably bound up with the urban land and housing crisis.

Access to Land for Human Settlements: A Comparative Perspective

Mexico's National Housing Plan (Plan Nacional de Vivienda), published by the government in 1990, estimated the total urban and rural deficit to be 6.1 million housing units (see table 2.2). To eradicate such a staggering deficit by the year 2000 would require an annual production of *610,000 housing units every year from 1990 to 2000* (assuming for the sake of illustration that the deficit does not get worse—i.e., that the annual increment in demand due to population growth will be met). Given the state of the economy and the austerity policies in place as a part of Mexico's economic restructuring, it is very unlikely that the deficit could be reduced so fast (Schteingart 1993). Worse still, it is unlikely that supplies will soon begin to meet the additional increments of demand, much less the backlog of un-met needs. Whereas the total population is expected to increase by

Table 2.2
Deficit in housing units in select areas, 1990

Total deficit in housing units	6,100,000
Mexico City	2,700,000
Monterrey	500,000
Guadalajara	500,000
Border states	1,200,000
Rural areas	1,200,000

Source: National Housing Plan, 1990, cited in *Mexico Business Monthly* (1992, 12).

30 percent, it is estimated that the population that will need housing (ages 20–49) will increase by 45 percent (*Mexico Business Monthly* 1992).

Between 1980 and 1990 roughly 400,000 housing units were added to Mexico's total (formal and informal) stock every year, yet the housing deficit worsened. On average, only 150,000 formal sector housing starts have been realized annually since 1988. In 1987, a peak year, 300,000 housing starts were made, but most of this (about one-third) represented earthquake reconstruction in Mexico City (Business Monitor International 1992). This deficit sets up an unprecedented challenge. But Mexico is not alone. Nearly all Third World countries face a similar situation.

In chapter 7 of *Agenda 21* ("Promoting Sustainable Human Settlement Development"), details are provided about the extent of the global housing deficit. The first program area outlined in the chapter is titled "Providing Adequate Shelter for All." It stipulates the basis for emphasizing this program as follows: "Access to safe and healthy shelter is essential to a person's physical, psychological, social and economic well-being and should be a fundamental part of national and international action. The right to adequate housing as a basic human right is enshrined in the Universal Declaration of Human Rights and the International Covenant on Economic, Social and Cultural Rights" (sec. 7.6 cited in Robinson 1993, 86).

The global housing objective spelled out in *Agenda 21* presents a mighty challenge indeed. At least 1 billion people do not have access to safe and healthy shelter (sec. 7.6). The World Health Organization (1992) estimates that only 70 percent of the people living in Third World cities are served by some form of sanitation. Only 40 percent are connected to sewers, and where sewer systems do exist at least 90 percent of the wastewater is discharged without treatment (Bartone et al. 1994, 11).

Given their lack of money to compete for serviced land and adequate housing, the urban poor have had little recourse other than to establish their household in substandard rental units or in illegal-semilegal settlements. In major African cities such as Cairo, Lagos, Kinshasa, and Nairobi there are large areas of informal settlements that contain a significant portion of the city's urban area and population (Mabogonje 1994, xxiv). Called by various names—*bidonville* in French-speaking Africa, *shanty town* in English-speaking Africa, *barriada* in Peru, *favela* in Brazil, *katchi abadi* in Pakistan, and *colonia popular* in Mexico—the growth of "irregular settlements" is widespread in Third World cities (see figure 2.1). Typically such settlements are in the most hazard-prone, ecologically fragile, or polluted parts of an urban area (Bernstein 1994; Bartone et al. 1994). In Mexico City, the urban poor have built on steeply rising slopes, thereby creating new spaces that Legorreta (1992) has termed the *Ciudad de las Alturas*. This "City of Heights" includes the mountainsides of Ajusco, Magdalena Contreras, Cuajimalpa, Huixquilucan, Ecatepec, and Tutitlán. Legorreta (1992) documents how inhabitants in these areas have had to originate new survival strategies, principally in the supply of food, water, gas, and electricity (128). In Rio de Janeiro roughly two-thirds of all *favela* dwellers occupy steep slopes on the city's outskirts. Mudslides are a common occurrence in these favelas; they claim hundreds of lives and leave thousands homeless during the annual rainy season (128).

Forms of access to land for housing the world's urban poor vary considerably from city to city. In Mexico City, Guadalajara, and Puebla, there are few land invasions (i.e., takeovers without purchase or official consent). Most new settlements are established either through *ejidal* subdivision or other kinds of semi-legal and illegal subdivision (COPEVI 1977; Azuela de la Cueva 1989a, 1989b; Gilbert and Ward 1985; Gilbert and Varley 1991; Mele 1986). In Monterrey, Tijuana, Chihuahua, and Durango, land invasions have traditionally been much more important (Gilbert 1989; Hoenderdos and Verbeek 1989; Pozas-Garza 1989).

Land that has unclear and unenforced land tenure or zoning arrangements is the terrain most frequently subject to irregular settlement. This is

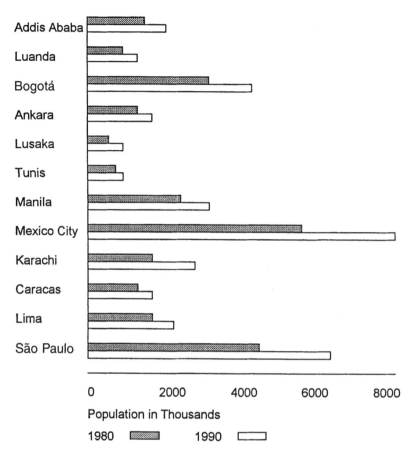

Figure 2.1
Estimated population of irregular settlements in selected cities, 1980 and 1990.
Source: CHF (1992).

true of cities throughout Mexico and the Third World more generally (Burnstein 1994). For instance, in and around Kuala Lumpur,[18] Malaysia—one of the fast-growing megacities of Southeast Asia—there exists Malay Reservation Land with similarities to the MAMC's ejido land. Malay land is subject to restrictions in the transfer of ownership. This has given rise to problems similar to those experienced in Ajusco—including irregular settlement by both low- and upper-income groups, urban strife, loss of resources, and obstacles to planning (Brookfield, Samad Hadi, and Mahmud 1991). As in the Mexico case, there are many constraints on the Malaysian government when it comes to dealing with irregular settlement problems on Malay reservation land: "Although the State has the power to sieze control of reservation land, just as it has to evict and re-house squatters, it is politically increasingly difficult for the authorities to exercise this power" (170).[19]

In Mexico City the politics of containment have slowed down urban encroachment into the so-called greenbelt and ecologically sensitive lands. But they have certainly not stopped it. Researchers at the World Bank argue that the ongoing degradation of resources, occupation of hazard-prone areas, and excessive urban sprawl taking place in cities throughout the Third World are caused by six primary factors: (1) inappropriate regulation, (2) lack of tenure security, (3) inadequate infrastructure capacity, (4) inadequate information, (5) inappropriate pricing and taxation, (6) weak institutions and poorly coordinated actors in the land market (Burnstein 1994, 1). This book draws attention to such factors in the context of uneven development, social control, and political mediation.

Urban-environmental problems are best characterized as embedded in broader uneven social and economic processes. This predicament raises some crucial questions: What can local actors do to gain the best advantage for the majority of its citizens? What contributions should one expect from the household, neighborhood, city, and state? Is it accurate to argue, as do Rabinovitch and Leitmann (1993, 49), that "the city administration should be in a position to determine structural guidelines for the city and its wider region, whereas citizens can better determine what is best for their own street or neighborhood"? To a certain extent the answer to this last question has to be yes. But leaving it at that misses something. To posit a division of labor between the city administration on the one hand and household and neighborhood level groups on the other underplays forms of domination, political mediation, and exploitative power relations that typically bind the two. Class struggle, not just for the means of production but also for the means of administration and for access to other bases of social power, is a deep part of the scenario. This is the blind side of mainstream policy analysis even when recommendations for productive ecology are advocated. But it is very much visible from the critical perspective of political ecology.

From the perspective of political ecology, the larger context within which localities are embedded is defined by uneven development and the new international division of labor. Economic globalization spurs resource-intensive industrialism and gives rise to biogeophysical as well as ethicosocial problems. Low-income households and neighborhood-level groups in Third World cities are at the front line. Women and children in these places typically endure the greatest hardship (Tuñón Pablos 1992; Bystydzienski 1992; Logan 1988; Joekes 1994). The deepening of urban poverty in Mexico and throughout Latin America has seriously impacted households. A household is defined here as "a nexus of social relationships that provide crucial linkages between producing and maintaining access to the material means of sustaining life, managing environmental resources, and passing on life chances to succeeding generations" (Douglass and Zoghlin 1994, 175). How have households responded to the crisis? Escobar Latapí and González de la Rocha (1995) document that

> To avoid a drastic reduction in food consumption, households have sent more members out to work. More youths, women and children have entered the workforce to earn the income needed for the survival of the household. Home-making duties have increased because many goods and services that were previously purchased in the market place must now be generated within the household (for example, mending and reconditioning domestic items). As a result, the domestic workload of women has increased, due both to the greater number of household members and to the greater dependence on at-home production. (74)

On a foreboding note, Escobar Latapí and González de la Rocha (1995) warn that households and microeconomic organizations may have exhausted their capacity to buffer hard times. This realization underscores the importance of promoting alternative development strategies that can generate new sources of income and sustainable livelihood opportunities. According to Chambers (1995), development policies can be "livelihood intensive without economic growth. To search for and implement livelihood-generating and supporting policies is a priority" (202). Scaling up the grassroots mobilization of civil society and political community is a crucial part of the struggle to search for and implement alternatives.

3

Urban-Environmental Policy and Planning: International Agendas/Local Realities

A new recognition of profound interconnections between social and natural systems is challenging conventional intellectual constructs as well as the policy predispositions informed by them. Our current intellectual challenge is to develop the analytical and theoretical underpinnings crucial to our understanding of the relationships between the two systems. Our policy challenge is to identify and implement effective decision-making approaches to managing the global environment.
—Nazli Choucri, *Global Accord: Environmental Challenges and International Responses*

Rethinking "Development"

Rising concerns about environment-development interrelations and prospects for "sustainablity" have prompted a rethinking of "development." The ninth session of the UN Conference on Trade and Development drew attention to this shift in thinking. "In the 50 years since the United Nations was established, the concept of development has evolved significantly. From a narrow focus on economic growth and capital accumulation, development has come to be widely understood as a multidimensional undertaking, a people-centered and equitable process in which the ultimate goal of economic and social policies must be to better the human condition, responding to the needs, and maximizing the potential, of all members of society." (United Nations Conference on Trade and Development 1996, 1)

Challenges to "conventional wisdom" are emerging from within as well as outside mainstream discourse. Yet, significant North-South differences in approaches exist to questions about the environment and development (MacNeill, Winsemius, and Yakushiji 1991; Sitarz 1993). Redclift (1987, 159) notes that environmental movements in the South are more likely to be motivated by livelihood struggles. Hardoy, Mitlin, and Satterthwaite (1992, 218) point out that "Most citizens in Africa, Asia and Latin America find it difficult to share the environmental concerns of the North. Questions of survival 20 or more years into the future have little relevance to those concerned with survival today." Environmental problems in Third World cities are typically more profound than those of cities in the North. Likewise, environmental problems in the Third World urban milieu are typically far more serious than in cities in Europe, Japan, or North America (218). This contrast does not mean that Third World cities make a larger contribution to global environmental problems than First World cities. On the contrary, the mass consumption maintained in the North places a much higher burden on global sources of environmental low entropy and sinks for waste. Ghandi was well aware of this fact a long time ago. In the 1920s, voicing his concern that the classical path of resource-intensive development engendered social alienation as well as resource overexploitation and exhaustion, he wrote: "God forbid that India should ever take to industrialism after the manner of the West. The economic imperialism of a single tiny island kingdom (England) is today keeping the world in chains. If an entire nation of 300 million took to similar economic exploitation, it would strip the world bare like locusts" (Ghandi, cited in Bandyopadhyay and Shiva 1988, 1224).

Sachs (1993) makes a similar point: "If all countries followed the industrial example, five or six planets would be needed to serve as "sources" for the inputs and "sinks" for the waste of economic progress" (6). Any attempt to discuss prospects for a city in the North or South demands much more than a calculation of population and carrying capacity within a select city's immediate hinterland. Such a task must also take into account the dynamics of world system linkages that connect cities as resource transformers and nodal points in the new international division of labor. The International Labour Office's (ILO) World Employment Program has published an edited volume that documents the linkages among urban processes, economic opportunities, and human settlements (Bhalla 1992). Essays in the volume draw attention to the informal sector's potential to create employment and to engender environmental action on a sustainable basis. Authors of one chapter argue that "Urbanization is not only a process of economic development, but a direct consequence of anthropogenic intervention in natural environments, following which physical space is converted into built environments. Thus, natural systems are turned into human settlements. Urbanization should therefore be seen to be as much a process of interaction between natural and built environments as an economic process" (S.V. Sethuraman and A. Ahmed, cited in Bhalla 1992, 122).

From this perspective, the authors raise a series of important questions linking the urban informal sector and the environment. For instance, "[I]s it possible to organize the collection and disposal of garbage in a more efficient way, which would also raise the incomes of the workers concerned?" (128). This perspective has gained currency in international policy circles. The agenda of the UNDP lists four major challenges: (1) urban poverty, (2) enabling and participative strategies for the provision of infrastructure and urban services, (3) improving the urban environment, and (4) strengthening the capacity of local government administration.[1] To combat urban poverty, the UNDP spells out strategies for organizing the urban poor at the community level and for increasing the poor's productive capacity by supporting informal sector activities (e.g., strategies call for public sector financial support).[2]

Although the informal sector does appear to be an increasingly important source of employment, it should be noted that the "informal sector" concept is problematic. Certainly it is misleading if it is understood to be a sector opposite to or outside of the "formal sector." These two sectors are inextricably linked (Portes, Castells, and Benton 1989; Portez and Walton 1981). Informal sector jobs include, among others, hawking and vending, cottage industry, small or microsized enterprises, and positions providing low-cost services (Rondinelli and Kasarda 1993, 106). Such informal sector enterprises are small-scale, mostly family-operated or individual activities. They are not legally registered and usually do not provide employees with social security or legal protection (106).

Economists estimate that approximately 30 percent of the economically active population (EAP) in Latin America currently hold informal sector jobs (Schteingart 1993, 228). More than 75 percent of urban employment in many countries of sub-Sarahan Africa can be characterized as informal (World Bank 1992). In the case of Mexico City, Jusidman (1988, 226) estimates that the portion of the EAP engaged solely in informal sector activity rose from 15.6 percent in 1975 to 19.2 percent in 1987. Palacios (1990) also cites evidence that informal sector activity has increased in Mexico City. A study by the Inter-American Development Bank (IDB) reports that 42 percent of the EAP in Mexico's three largest cities (Mexico City, Guadalajara, Monterrey) earn their livelihoods in the informal sector (cited in Palacios 1990, 123).

It is now understood that informal sector activity is not a marginal or vestigial component of the so-called modern industrial economy. Despite rapid increases in industrialization in many Third World countries, the informal sector portion of the total labor force has not declined since the 1950s (Rondinelli and Kasarda 1993). In many cases, such as in Mexico, it has grown. Mexico's informal sector activities multiplied rapidly during the seventies and eighties, reaching an unprecedented proportion within the productive apparatus of the country. This trend coincides with the most profound and generalized crisis that Mexico has suffered in more than half a century (Escobar Latapí and González de la Rocha 1995). Attention to poverty-environment linkages, green works programs, and community empowerment is driven by a kind of default scenario. The fate of cities in the Third World, including Mexico City, will depend in large part on the capacity of the poor majority to finance, build, and sustain the urban milieu. This acknowledgment underpins the shift that has begun to take hold in official circles with respect to urban land and human settlements policy.

Shifting Policies: 1980s–1990s

In 1983, Habitat convened a seminar of experts called "Land for Housing the Poor." The central argument that emerged is the need for government action to speed up and increase the supply of land for housing the poor and to ensure that the poor receive secure tenure. Seminar participants recommended that "the emphasis of public policy must shift from the housing construction process to the land delivery process, so that governments take responsibility for providing secure land and affordable infrastructure, while households or community groups take responsibility for building the shelter's structure" (Habitat 1983, 11). During the early 1980s it was argued that these are the most urgent and intractable elements of the urban land and housing problems (Hardoy and Satterthwaite 1981; Payne 1984).

Policy analysis concerning the urban poor's access to land has tended to focus on issues that can be dealt with at the municipal level. In some ways this has been a productive focus. It has helped draw attention to the advantages of alternative forms of land tenure and to innovative forms of government intervention in the land market. It is commonly argued that besides the many negative aspects, irregular settlement processes also have positive elements. Furthermore, benefits could be derived from attempts to "carefully study how illegal developers operate, to follow their lead wherever possible without compromising the law, and to generate appropriate responses to their actions in line with government objectives." (Angel, Archer, and Wegelin 1983, 11).

By the early 1980s, efforts were under way to assemble case studies of the informal land supply system, the actors in this system, and public intervention in this system. Research was conducted on different means of granting tenure and different means of resolving tenure disputes. There is now an extensive literature on strategies including land sharing, land pooling, divisions of tenure rights between development and use, and forms of collective or neighborhood ownership (Burnstein 1994; Linden 1994).

Today the UN and World Bank promote the concept of *enablement*.[3] Harris (1992, xix) characterizes the institutional stance this way: "Government should now supposedly seek to facilitate action by its citizens, private firms or non-governmental organizations, to provide for themselves such services and at such standards as people themselves might choose. The capital project now became replaced by the Programme of technical assistance and enabling."[4] With respect to human settlements, the argument for enabling people takes as its starting point the view that the problem of adequate land is not so much related to the rapid growth of the need for urban land as to the institutional and economic constraints of the supply side (Linn 1983; Burnstein 1994). This view is an important half-truth. It is only half-true because it is insufficient to argue that the scarcity of adequate land for housing low-income groups is engendered by constraints on supply or by inefficient administrative procedures. The concern to make land available is only partly a question of how to improve bureaucratic functions, open up markets, or facilitate access. The urban poor's access to land depends on their *capacities to pay* for the land and infrastructure. At the same time, the question of expanding access must take into account environmental concerns including the *regenerative capacities* of the land, watersheds, and other features necessary for ecological sustainability.

Careful analyses of *environment–poverty–land use linkages* have been slow in coming. This is not to say that there has been a shortage of critical studies. On the contrary, much ink has been spilt in efforts to debunk the "conventional wisdom" generated by international agencies such as the World Bank and Habitat. A number of Marxist and neo-Marxist studies began to address the urban land and housing problematic back in the late 1970s and early 1980s (Burgess 1982; Ward 1982; Castells 1983; Garza and Schteingart 1978). These studies have generated critical theories on illicit property markets by drawing attention to ownership relations and production processes. Castells, for instance, has provided insight into the politics of land allocation by examining popular mobilization and the popular sector's production of land and housing. Other Marxist and neo-Marxist studies have focused on ground rent and capital accumulation through land transactions. Some of these studies have greatly contributed to our capacity to understand the political economy of land and state-community relations. But they make little or no mention of the types of concrete changes (e.g., in practices, institutions, laws, policies, and regulations) that may be necessary to ameliorate urban land and housing problems. Nor did they address the question of sustainability. The earlier critical studies elaborated on political economy, not political ecology. As I noted in chapter 1, political ecology as a critical discourse has been around for more than twenty years. But not until recently has this perspective been applied to questions about human settlements and sustainability. The turning point for both critical studies and mainstream work occurred during the late 1980s. Around this time the theme of "sustainable development" entered the international spotlight.

The Rise of *Sustainability* Discourse

It is popular to refer to the "impact of human settlements on the environment," as though the environment were some independent entity, separate from human beings and their way of life. The fact, however, is that human beings live in human settlements which are the greatest part of the environment, and the real question to be faced is how the environment can support human settlements production and sustainably, to the greatest long-term benefit of the people who live in them, not just for now but for always. This is the intersecting point between human settlements, the environment and sustainable development.

The preceding quote is from A. Ramachandran, the executive director of the Habitat. It appeared in the foreword to Diana Mitlin and David Satterthwaite's (1990) report *Human Settlements and Sustainable Development: The Role of Human Settlements and of Human Settlements Policies in Meeting Development Goals and in Addressing the Issues of Sustainability at Global and Local Levels*. This form of discourse, a linking of sustainability and human settlements development, has taken center stage in both local and international policy arenas ever since the World Commission on Environment and Development published *Our Common Future* (WCED 1987). Why all the current interest in sustainability? There are at least two fundamental reasons: one rooted in epistemology, the other in the status of life systems underpinning production and reproduction. First, on the epistemological level, the term *sustainability* can be used to mean almost anything one wants it to mean:

> The earliest meaning of *sustain* is to "support," "uphold the course of" or "keep into being." What corporate chief, treasury minister, or international civil servant would not embrace this meaning? Another meaning is "to provide with food and drink, or the necessities of life." What underpaid urban worker or landless peasant would not accept this meaning? Still another definition is "to endure without giving way or yielding." What small farmer or entrepreneur does not resist "yielding" to the expansionary impulses of big capital and the state, and thereby take pride in "enduring"? (M. O'Connor 1994, 152)

The point to emphasize here is that the term *sustainability* has ideological and political content as well as ecological and economic content. There is a "struggle, worldwide, to determine how 'sustainable development' or 'sustainable capitalism' will be defined in the discourse on the wealth of nations" (M. O'Connor 1994, 153). The second reason that helps explain the great interest in sustainability concerns the status of life systems. Today's so-called modern development is characterized by a wide range of both domestic and international policies that seriously overload or deplete the earth's endowment of environmental low entropy—what some refer to as *ecological capital*—including the earth's rivers, lakes, and oceans; its soils and forests; its flora and fauna; and its ozone shield (Mac-Neill, Winsemius, and Yakushiji 1991). Part of the problem stems from the failure to factor the value of the environment into decision making. It is generally not known how much environmental quality is being given up in the name of development, nor how much development is being given up in the name of environmental protection (World Bank 1992, 34).

In an attempt to clarify such trade-offs at the national level, the World Bank points out that efforts are under way in a number of countries to adjust national accounting methods. In its *World Development Report 1992*, the World Bank cites a pilot study conducted in Mexico that illustrates the potential magnitude of the adjustments required: "When an adjustment was made for the depletion of oil, forests, and groundwater, Mexico's net national product was almost 7 percent lower. A further adjustment for the costs of avoiding environmental degradation, particularly air and water pollution and soil erosion, brought the national product down another 7 percent" (35–36).[5] The figures cited by the World Bank draw attention to the real and potential problems stemming from counting the depletion of geological capital and ecological life-support systems as net current income. In view of such dynamics, Mexico's president together with the director of Mexico's new environmental Secretariat argued before the nation's Congress: "if we ignore the costs of environmental deterioration, we overestimate the benefits of economic growth."[6] The Mexican government reportedly embraces sustainability as an organizing concept for development policy.

The most oft-cited definition of sustainable development was coined by the WCED: sustainable development is an environmentally sound development strategy that "*seeks to meet the needs and aspirations of the present without compromising the ability to meet those of the future*" (WCED 1987, 40). This definition embraces the Kenyan proverb "We do not inherit the earth from our parents; we borrow it from our children," and it raises sustainability to a global ethic by extending the United Nation's Universal Declaration of Human Rights (i.e., to meet each person's right to a standard of living adequate for health and well-being and including food, clothing, housing, medical care, and necessary social services) to future generations.

Most of the literature on sustainable development focuses on rural and natural resource management issues (e.g., social forestry, agroecology, fisheries). However, a rapidly growing body of work deals with urban issues. In both First and Third World contexts, ever more books, articles, conferences, symposia, policy forums, and even community group workshops and strategy sessions are being written or conducted on the subjects of "sustainable cities" and "planning for sustainable urban development."[7] The widespread adoption of sustainable development terminology within the international development community signals increasing concern about the adverse impact that standard prescriptions for development have had on the environment in both the First and Third World. In the words of Gro Harlem Brundtland, "As people continue their endless quest for new raw materials, new energy forms, and new processes, the constraints imposed by depletion of natural resources and the pollution caused by human activity have brought society to a crossroads" (Brundtland 1990, 147).

Until recently, most professionals tended to view the environmental costs of development as external to development, to be dealt with by measures external to development. In other words, the environment was viewed largely as an *add-on* to development, seldom as an integral *build-in*. As MacNeill, Winsemius, and Yakushiji (1991) point out, "this mind set was pervasive and became reflected in add-on institutions promoting add-on policies often requiring add-on technologies. The environment debate thus focused mainly on the adverse impacts of development on the environment; the impacts of a degraded environment on the prospects for development were largely ignored" (29). The concept of sustainable development has shifted attention to this latter concern. In the landmark publication on this subject, entitled *Our Common Future* (WCED 1987, 5), the authors write: "We have in the past been concerned about the impacts of economic growth upon the environment. We are now forced to concern ourselves with the impacts of ecological stress—degradation of soils, water regimes, atmosphere, and forests—upon our economic prospects." This shift in thinking has accompanied new efforts to reconceptualize the measurement of human development. Economic indicators alone are now viewed as inadequate: "GNP and GDP are inadequate measures of sustainability. They measure production but provide little information about people or the state of their living environment. If a deteriorating environment causes disease, resulting in increased health expenditures and thus increased GNP, the increased GNP would be interpreted as a higher level of development—even though the people and their environment are worse off. Similarly, current income measures do not factor in the inevitable future costs of current depletion of resources" (UNDP 1992, 24).

Despite all of its ambiguities (probably because of them) the term *sustainable development* has become a widely invoked trademark of local, regional, national, and especially international organizations dedicated to promoting environmentally sound approaches to economic development. The literature on the subject is burgeoning.[8] Hardoy, Mitlin, and Satterthwaite (1992) have helped clarify the components of sustainable development by distinguishing social, economic, and ecological dimensions (see figure 3.1). It is beyond the scope of this book to provide a comprehensive transdisciplinary review of all the insights, arguments, and perspectives in this vast literature. Suffice it to note here that Mexico has played a significant part in the internationalization of the discourse on sustainable development. As I will point out, Mexico played a significant role in events leading up to and during the 1992 Earth Summit.

Our Own Agenda

In preparation for the 1992 World Conference on Environment and Development held in Rio de Janerio, Brazil (the Earth Summit), heads of state from the Caribbean and Latin America, including Mexico's President Carlos Salinas de Gortari, formed the Latin American and Caribbean Commission on Development and the Environment. The commission's task was to develop a regional outlook on issues concerning the environment and sustainable development. One of the commission's products was a report titled *Our Own Agenda*. This report lists (1) land use and (2) environmental deterioration in human settlements as the top two priority issues (19). More specifically, the principal environmental problems targeted for urgent action include human settlements and sanitation; deficiencies in housing, basic services, and work environments; lack of land tenure security; industrial and domestic contamination; and vulnerability to natural disasters and accidents (23).

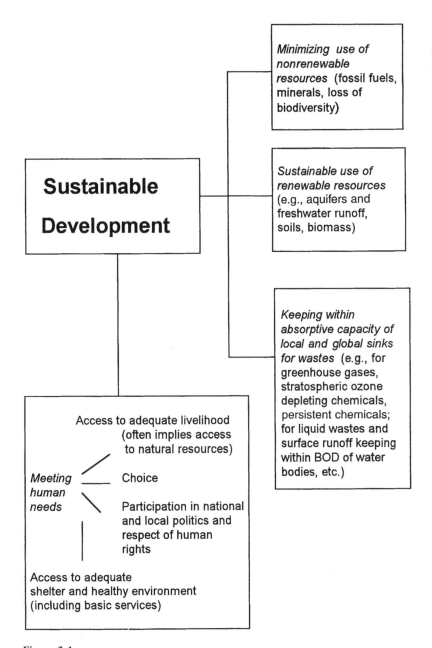

Figure 3.1
Components of sustainable development. Source: Hardoy, Mitlin, and Satterthwaite (1992, figure 6.1).

On a wider scale, *Our Own Agenda* places high on its priority list the uncontrolled physical expansion of the cities, the elimination of liquid and solid wastes, and contamination of the air. The Latin American and Caribbean Commission forecasts that if decisive steps are not soon taken to deal with these problems, a dismal future will result: "[F]or a high and growing percentage of the urban inhabitants of Latin America and the Caribbean, life will represent an experience in privation, without stable employment or an adequate income, without access to the most essential services and without insurance coverage—and with the fear caused by urban contamination and violence" (23).

The commission wrote *Our Own Agenda* as part of the preparatory process that led up to the 1992 Earth Summit. The Earth Summit was the largest conference ever hosted by the United Nations. It was attended by 116 heads of state or government, 172 states, 8,000 delegates, 9,000 members of the press, and 3,000 accredited representatives of nongovernmental organizations (Robinson 1993, xiii). The summit was the culmination of a series of parallel preparatory processes that spanned nearly five years— more than two of which involved formal negotiations (Grubb et al. 1993). *Agenda 21* was one of the key documents to come out of the process. Although *Agenda 21* contains contradictions and conceptual confusion, it nonetheless serves as a useful benchmark by which to elaborate a constructive critique of Mexico's, particularly Mexico City's, politics dealing with the environment and development.

Agenda 21

With more than 500 pages, divided into forty chapters, *Agenda 21* is designed to be an action plan for achieving sustainable development. The document's preamble is as dramatic as it is optimistic:

Humanity stands at a defining moment in its history. We are confronted with a perpetuation of disparities between and within nations, a worsening of poverty, hunger, ill health and illiteracy, and the continuing deterioration of the ecosystems on which we depend for our well-being. However, integration of environment and development concerns, and greater attention to them, will lead to the fulfillment of basic needs, improved living standards for all, better protected and managed ecosystems and a safer, more prosperous future. No nation can achieve this on its own; but together we can—in a global partnership for sustainable development. (Preamble to *Agenda 21*, section 1.1, cited in Robinson 1993, 1)

It has been more than three years since *Agenda 21* became established in international policy circles. In an effort to implement urban aspects of *Agenda 21*, the World Bank has focused its attention on the national and local management of cities and efforts to enhance productivity (Harris 1992, xix). The World Bank calls for action in two priority areas: (1) ameliorating poverty, and (2) protecting and improving the environment. The world's largest multilateral aid agency, the World Bank provides two-thirds of global-planning aid. The proportion of the bank's funding for urban projects is expected to rise substantially as a result of the joint World Bank, UNDP, and UNCHS Urban Management Programme (UMP).[9] Under the auspices of the World Bank, Burnstein (1994) has written a detailed policy paper for the UMP. She argues that in order to prevent further degradation of resources, "governments should exert some control over urban land use and development, but not necessarily constrain the supply of land for housing or discourage the private sector from providing affordable housing in safe locations. An important challenge is to achieve a balance between urban development and environmental protection, taking into account linkages among land use, poverty, and the environment" (i).

Agenda 21 and the changes in policy taking place at the World Bank are cited by many as a sure sign of new thinking in international development. But critics voice skepticism about such changes (Hildyard 1993; Finger 1993). In their analysis of recent World Bank activities, Jones and Ward (1994) ask themselves the question, Paradigm Shift or Policy Continuity? In sum, they answer that the changes are little more than "old wine in new bottles." While this may underestimate the significance, it does seem clear that the changes do not weigh in as a paradigm shift.[10] Simply put, the heart of the matter still beats with growth-oriented, market-driven solutions. André Gorz made a similar observation as witness to the first wave of environmentalism nearly two decades ago. In the opening to his book *Ecology as Politics,* he writes: "Ecology is like universal suffrage or the 40-hour week: at first, the ruling elite and the guardians of social order regard it as subversive, and proclaim that it will lead to the triumph of anarchy and irrationality. Then, when factual evidence and popular evidence can no longer be denied, the establishment suddenly gives way— what was unthinkable yesterday becomes taken for granted today, and *fundamentally nothing changes*" (my emphasis, cited in the edition translated from French to English by Vicderman and Cloud 1980, 1). To entirely dismiss the worth of *Agenda 21* and related international policy changes on the basis of such criticism would be a mistake. True, much of *Agenda 21*'s text appears to be statements of the obvious, and much of it can be criticized as a simplistic policy wish list. But it is also a great deal more. By endorsing the theme of sustainability as an organizing principle, it raises important considerations heretofore officially underplayed. Bartone et al. (1994) go as far as to say that "A milestone was reached at the UNCED Earth Summit (Rio de Janeiro, 1992) when cities were successful in broadening the environmental debate to focus attention on urban priorities" (1). Without my going into too much detail, two of the major subject areas covered in *Agenda 21* merit closer attention, namely, (1) combating poverty, and (2) promoting sustainable human settlements development.

Combating Poverty

One of the first chapters in *Agenda 21* is titled "Combating Poverty: Enabling the Poor to Achieve Sustainable Livelihoods." There is much to be lauded in this chapter. It underscores the importance of antipoverty strategies in the quest for sustainable development, emphasizing that "the long-term objective of enabling all people to achieve sustainable livelihoods should provide an integrating factor that allows policies to address issues of development, sustainable resource management and poverty eradication simultaneously" (sec. 3.4 cited in Robinson 1993, 25).

To actually integrate sustainable livelihoods and environmental protection will require initiatives from a wide range of actors across the local to global spectrum. *Agenda 21* places the greatest expectations on the community and local levels. Specifically, it calls for a community-driven approach to sustainability that promotes community empowerment and capacity-building. A selection of text under the heading "Activities" is reproduced in box 3.1.

The selection in box 3.1 illustrates several themes that run through *Agenda 21*. First, there is the advocacy of a bottom-up approach that assigns a vital role to civil society, particularly community groups and NGOs. Second, an emphasis is placed on the importance of adequate information and network building. And third, there is the call for open government. The emphasis on community participation—especially that of women

Box 3.1
Agenda 21: Activities for Sustainable Development

Sustainable development must be achieved at every level of society. Peoples' organizations, women's groups and non-governmental organizations are important sources of innovation and action at the local level and have a strong interest and proven ability to promote sustainable livelihoods. Governments, in cooperation with appropriate international and non-governmental organizations should support a community-driven approach to sustainability, which would include, *inter alia:*

(a) Empowering women through full participation in decision-making;
(b) Respecting the cultural integrity and the rights of indigenous people and their communities;
(c) Promoting or establishing grass-roots mechanisms to allow for the sharing of experience and knowledge between communities
(d) Giving communities a large measure of participation in the sustainable management and protection of the local natural resources in order to enhance their productive capacity;
(e) Establishing a network of community-based learning centres for capacity-building and sustainable development

Source: *Agenda 21*, sections 3.7, cited in Robinson (1993, 26).

in local decision making—has led some to observe that *Agenda 21* represents (at least on paper) a significant advance in promoting participatory democracy.

Promoting Sustainable Human Settlements Development

The overall human settlement objective is to improve the social, economic and environmental quality of human settlements and the living and working environments of all people, in particular the urban and rural poor. Such improvement should be based on technical cooperation activities, partnerships among the public, private and community sectors and participation in the decision-making process by community groups and special interest groups such as women, indigenous people, the elderly and the disabled. These approaches should form the core principles of national settlement strategies.
—*Agenda 21,* sec. 7.4 cited in Robinson 1993, 83

In documenting the *basis for action* in the sphere of human settlements development, *Agenda 21* notes that access to housing is considered a basic human right in international covenants (e.g., the Universal Declaration of

Human Rights). Moreover, access to safe and healthy shelter is recognized as essential to a person's physical, psychological, social, and economic well-being. Thus it should be a fundamental part of national and international action. At the same time, *Agenda 21* draws attention to the dire unmet need for shelter: at least 1 billion people do not have access to safe and healthy shelter. *Agenda 21* argues that "if appropriate action is not taken, this number will increase dramatically by the end of the century and beyond" (sec. 7.4 cited in Robinson 1993, 83).

Agenda 21 calls for a much greater commitment on the part of international support and finance organizations. It is instructive to note that the returns for investment in human settlements seem to be high. According to data presented in *Agenda 21,* every dollar of UNDP technical cooperation expenditure on human settlements in 1988 generated a follow-up investment of $122, the highest of all UNDP sectors of assistance (sec. 2.2 cited in Robinson 1993, 4). These data support the argument spelled out nearly a decade ago in Habitat's *Global Report on Human Settlement* (1986); that is, investments in human settlements are not welfare investments. They are investments in the productive capacity of the nation: its fixed capital asset (Habitat 1987, 6). Housing is now characterized as a tool of development policy (Habitat 1996a, 1996b).

To avoid the kind of urban encroachment taking place in Mexico City's ecological reserve, *Agenda 21* emphasizes the importance of sound urban management. And beyond this commitment, efforts are called for to relieve pressure on megacities by encouraging intermediate city development. In terms of improving shelter management, *Agenda 21* calls upon countries to tap into the framework offered by the joint World Bank, UNDP, and UNCHS (Habitat) UMP. Box 3.2 lists some of the recommended activities.

The second UN Conference on Human Settlements, Habitat II, 1996, built upon the recommendations put forth in *Agenda 21*.[11] The first Habitat Conference was held in 1976. Two years later Habitat (the United Nations Centre for Human Settlements; UNCHS) was established. Habitat is responsible for the formulation and implementation of the human settlements programs of the United Nations. Habitat describes itself as a think tank within the United Nations system. It conducts research and technical analysis to help governments improve the development and management

Box 3.2
Agenda 21: Activities for Urban Management

---

1. Adopt and apply urban management guidelines in the areas of land management, urban environmental management, infrastructure management and municipal finance, and administration.
2. accelerate efforts to reduce urban poverty through a number of actions:
a. Generate employment for the urban poor, particularly women, through the provision, improvement, and maintenance of urban infrastructure and services and the support of economic activities in the informal sector, such as repairs, recycling, services, and small commerce.
b. Encourage NGO initiatives.
3. Adopt innovative city-planning strategies; for instance,
a. Institutionalize a participatory approach to sustainable urban development
b. Start green works programs to create self-sustaining human development activities and both formal and informal employment opportunities for low-income residents.
c. Empower community groups, NGOs, and individuals to assume the authority and responsibility for managing and enhancing their immediate environment through participatory tools, techniques, and approaches embodied in the concept of environmental care.

Source: *Agenda 21*, sections 7.13–7.21, cited in Robinson (1993, 87–92).

of human settlements. The 1996 *Habitat Agenda,* including a declaration of principles and commitments and *Global Plan of Action* (GPA), was adopted at the Habitat II Conference. The purpose of the GPA is to "guide national and international action on human settlements for the coming two decades."[12]

Habitat II and the International Arena

We recognize the imperative need to improve the quality of human settlements, which profoundly affects the daily lives and well-being of our peoples. There is a sense of great opportunity and hope that a new world can be built, in which economic development, social development and environmental protection as interdependent and mutually reinforcing components of sustainable development can be realized through solidarity and cooperation within and between countries and through effective partnerships at all levels.

—*Habitat Agenda,* Preamble

The preparatory committee for Habitat II laid down three principles as the ethicoconceptual foundation for the *Habitat Agenda*'s GPA. They are the principles of civic engagement, sustainability, and equity. Based on these principles, the strategy of the GPA is that of *enablement*. The specific commitments agreed to under the rubric of enablement target all key actors in the public, private, and community sectors so that they may collectively "play an effective role at the national, state/provincial, metropolitan and local levels in human settlements and shelter development" (*Habitat* 1996b, chap. C, par. 28–29).

The enablement section of the *Habitat Agenda* spells out six commitments:

1. Exercising public authority and using public resources with transparency and accountability
2. Decentralizing authority and resources, as appropriate, and functions and responsibilities to the level most effective in addressing the needs of people in their settlements
3. Promoting institutional and legal frameworks and capacity building conducive to civic engagement and broad-based participation in human settlements development
4. Capacity building for human settlements management and development
5. Supporting enabling frameworks—both institutional and legal—for mobilizing financial resources for sustainable shelter and human settlements development
6. Promoting equal access to reliable information, utilizing, where appropriate, modern communications technology and networks

Once such commitments to enablement are put into practice, it is presumed that the three principles of civic engagement, sustainability, and equity can be translated into action in a variety of cultural and political contexts. As the following statement suggests, it is hoped that the synergy from combining these principles will provide a basis for "new thinking" about the urban prospect.

[T]he principles are neither abstract nor beyond reach. Civic engagement implies that living together is not just a passive exercise; in settlements, people must contribute actively to the common good, both directly and through their representatives. Sustainability is a measure of adaptability and depends upon people working together to modify their own behavior and their institutions as they face environmental, social and economic opportunities and constraints. Equity is the application of fairness in the allocation of resources to meet the basic needs of all people; applied equity helps to create solidarity, warding off the social unrest and entropy (degradation and disorganization) that is becoming the hallmark of many settlements in both the developed and developing worlds. This combination of principles provides a basis for new thinking about humanity's second habitat—the city.[13]

The commitments to "enablement" are intended to engender sustainable human settlements development and the provision of adequate shelter for all. What commitments will it take to promote "sustainable human settlements"? The nine-point list agreed to by the signatories of the *Habitat Agenda* is shown in box 3.3.

Habitat II was the culmination of a group of UN conferences referred to as the Rio Cluster (see box 3.4). Some observers note that Habitat II may best be remembered for the unprecedented participation of local authorities, global civil society parliamentarians, nongovernmental organizations, the private sector and private foundations, trade unions, the scientific community, professionals, and researchers, as well as Habitat's UN partners. Another notable outcome of the conference was the reaffirmation of the commitment to the "full and progressive realization to the right to adequate housing."

In addition to Habitat II and the other Rio Cluster conferences, a series of related conferences outside the UN system have addressed the theme of urban sustainability. For instance, the World Bank convened an "Environmentally Sustainable Development" conference (Washington, September 19–21, 1994). The theme was "The Human Face to the Urban Environment" with a focus on the consequences of environmental problems for the welfare and productivity of people.[14] Four main questions were addressed: (1) Are cities sustainable? (2) Have the megacities reached their environmental limits? (3) What policies and approaches have allowed some cites to be successful in managing the impact of environmental problems on people? and (4) Does the management of urban environmental resources require new approaches and forms of governance at the city level?

The City of Manchester (United Kingdom) and the Centre for Our Common Future hosted "Cities and Sustainable Development" in mid-1994. It brought together delegations from fifty cities around the world. The

Box 3.3
The *Habitat Agenda's* Commitments to Engender

We commit ourselves to the goal of sustainable human settlements in an urbanizing world by developing economies that will make efficient use of resources within the carrying capacity of ecosystems and by providing all people with equal opportunities for a healthy, safe, and productive life in harmony with nature and their cultural heritage and spiritual and cultural values, thereby ensuring social progress. We further commit ourselves to the objectives of

1. Promoting socially integrated human settlements, combating segregation and discriminatory and other exclusionary policies and practices, and recognizing and respecting the rights of all, especially women and the poor
2. Acknowledging and harnessing the potential of the informal sector, where appropriate, in providing housing and services for the poor
3. Promoting changes in production and consumption patterns and settlements structures that will protect natural resources, including water, air, biodiversity, energy, and land, thereby providing a healthy living environment for all
4. Promoting spatial development patterns that reduce transport demand, and creating efficient, effective, and environmentally sound transport systems that improve access to work, goods, services, and amenities
5. Preserving productive land in urban and rural areas and protecting fragile ecosystems from the negative impacts of human settlements
6. Protecting and maintaining the historic and cultural heritage, including traditional shelter and settlement patterns, as appropriate, as well as landscapes and urban flora and fauna in open and green spaces
7. Enabling competitive and sustainable economic development that will attract investments, generate employment, and provide revenues for human settlements development
8. Alleviating the undesired impacts of structural adjustment and economic transition on human settlements
9. Reducing the impact on human settlements of natural and human-made disasters

Source: *Habitat Agenda* (1996b, chap. B, par. 26–27, UN document A/Conf. 165/ 1j).

Box 3.4
The "Rio Cluster" of UN Proceedings

Habitat II (Istanbul, 1996)
UN Conference on Trade and Development (Midrand, 1996)
Fourth World Conference on Women (Beijing, 1995)
World Summit for Social Development (Copenhagen, 1995)
Migratory and Straddling Fish Stocks (New York, 1995)
Conference on Population and Development (Cairo, 1994)
Sustainable Development of Small Island Developing States (Barbados, 1994)
World Conference on Natural Disaster Reduction (Yokohama, 1994)
World Conference on Human Rights (Vienna, 1993)
Earth Summit (Rio de Janeiro, 1992)
International Conventions on Climate, Biodiversity, and Desertification

Note: Details about these conferences and conventions, including references to related literature, can be found on the Internet at http://www.igc.apc.org/habitat/un-proc/index.html.

conference was announced as a follow-up to the 1992 Earth Summit and was convened "in the spirit of creating solution-driven responses for the achievement of sustainable development in urban areas across the globe."[15]

Habitat II, the Rio Cluster conferences, and other major meetings in the international arena have helped fuel worldwide interest in sustainability. The rising interest and concern have been accompanied by a rising tide of initiatives as well as literature (Holmberg 1992; OECD 1995; Bartelmus 1994; Brenton 1994). Most of the initiatives and literature are preoccupied with redefining *managerialism, policy,* and *planning* (Pezzoli 1997). Along such lines, most of the work produced by the World Bank, United Nations, the Mexican government and those writing about sustainable cities, is *technocentric*. Adrian Atkinson (1992) notes the distinction between *technocentric* and *ecocentric* versions of environmentalism.

The technocentric approach "assumes that with sufficient political commitment and financial resources, we can tackle any environmental problem. All it takes is the right technical and administrative measures" (Atkinson 1992, 338). The assumption underpinning this version of environmentalism is that "we are not trying hard enough to implement poverty alleviation programs and environmental controls; and that if we can put more resources into these all will be well" (338). The technocentric view of the urban-environmental problematic does not go to the heart of the matter. Its conservative bias precludes addressing relevant issues of social structure, issues rooted in history and power relations that reproduce reckless exploitation and domination.

The ecocentric approach, on the other hand, does not preclude questioning social structure and power relations. It rejects *technological optimism*—defined by Eckersley (1992, 50) as the overly confident belief that with further scientific research, societies can rationally predict, manipulate, and control all the negative unintended consequences of our actions on earth. Likewise, ecocentrism rejects the epistemological foundation of *scientism;* that is, it rejects the conviction that empiric-analytic science is the only valid way of knowing (50). This framework opens the door for a deeper appreciation of indigenous practices, as well as social experimentation at the grassroots level of society and social learning.

In place of technological optimism and scientism, the ecocentric approach posits that "real solutions can only be achieved by changes in attitudes and lifestyles that substantially reduce our impact on the biosphere and, furthermore, which abandon notions of progress and back off the advance into an increasingly problematic unknown" (Atkinson 1992, 328).

Of course, one must be careful not to take this kind of academic labeling too far. Casting issues in either technocentric or ecocentric terms can be overly simplistic. The worth of the World Bank and UN initiatives, and for that matter the Mexican government's initiatives, should not be so easily dismissed—even if they are fundamentally technocentric. Government action in policy arenas is not unilateral. It involves a give and take. This is especially true where the implementation of policy takes place at one of society's *frontiers in relationships of power*—i.e., the urban-ecological boundary at the periphery of a rapidly growing city.

Greenbelts and Urban Containment Strategies in Mexico and around the World

In their thorough review essay about the role of human settlements in meeting development goals, Mitlin and Satterthwaite (1990) argue that "In regard to land use, sustainability goals emphasize the need to limit the encroachment of settlements on agricultural land and on those parts of the natural landscape with important cultural, ecological or recreational value. Very few cities in Asia, Latin America or Africa even begin to achieve this" (6).

Few would argue with the fundamental premise underlying Mitlin and Satterthwaite's observation. Certain terrains are better suited for urban development than others, and some land should be set aside for ecological purposes. But which land, by what criteria, according to who, and how? There is no universally applicable rationality to answer such questions. While useful as an organizing concept, "sustainability" is far too ambiguous a concept to provide operational criteria (Norgaard 1994, 17–20). Urban land use patterns result from a complex interplay among social and political relationships of power, history, and the environment. While some form of control over land use for human settlements is certainly necessary, it has proven to be elusive.

Strategies for gaining control over land use vary by country. In her review essay on this subject, Ellen Brennen (1993, 74) writes: "Cities like Bangkok, Dhaka, and Metro Manila have had virtually no effective measures to influence or control land development; Seoul represents the opposite extreme. Independent of the degree of intervention, there are similarities. Many cities have master plans prescribing directions of urban growth, but these plans rarely are realized and languish in metropolitan planning offices as irrelevant documents" (77). Again, Mexico City is a case in point. It's ecological buffer zone and greenbelt line have repeatedly been extended to include urban sprawl beyond the legal limits. Cymet (1992), who documents this expansion from a rural perspective, also notes that "chaotic urban expansion continues in spite of the government's declared policies in favor of planned urban growth" (8).

Ecological reserves and greenbelts around fast-growing cities tend to be highly problematic spaces. Most of the research that examines such spaces, in the North and South alike, documents a range of serious problems. Friedmann, Nield, and Weed (1994), in a case similar to that of Ajusco, document the case of the recently declared Antisana Ecological reserve in Ecuador. The Antisana miniregion (named after the 5,705-meter volcano within its bounds) was declared a 120,000-hectare ecological reserve because of its immense hydrological resources and its importance as a watershed for the rapidly expanding city of Quito (the nation's capital, less than sixty kilometers away). Yet, urban encroachment into the Antisana reserve continues. Douglass (1989a) documents similar dynamics within the designated ecological conservation area in Jakarta, Indonesia. Likewise, urban encroachment continues within a so-called ecological reserve in Acapulco, Mexico. In his review of eviction trends worldwide, Audefroy (1994) notes that 25,000 people have been evicted from Acapulco's ecological reserve.

Seoul, South Korea's capital city, is often cited as one of the world's strictest enforcers of a greenbelt policy. But even there, encroachment is a source of conflict (Joochul 1991). Seoul's population grew from 3 million in 1960 to more than 10 million by the start of the 1990s. In 1971, greenbelt areas in and around all of Korea's major cities were designated with the aim of controling urban sprawl. The Seoul greenbelt area is about 538 square kilometers. Joochul notes that more and more of the designated greenbelt areas have been released for urban development and that the criteria by which this has occurred have been very problematic and involve widespread corruption (29).

In the greater metropolitan area of São Paulo (MASP)—Brazil's most urbanized, industrial, and affluent region—a number of laws have been passed to protect green space, including legislation to create ecological reserves (Grena Kliass 1990). As in Mexico City, the rapid expansion of the MASP's urban periphery continues even though the MASP's overall population growth rate has slowed down. Of the more than 10 million people living within São Paulo city proper, as many as 1.5 million live in favelas—low-income squatter settlements located for the most part on floodplains and steep slopes (Grena Kliass 1990). Most of the floodplains and steep slopes (those slopes with inclines of 30 percent or greater) have been designated as ecological reserves in order to protect the urban bioregion's watersheds and to avoid costly outlays for infrastructure (Bartone and Rodríguez 1993). However, as in the case of Rio de Janeiro noted earlier, the majority of these areas have been settled by squatters who have formed favelas (Grena Kliass 1990).

It is not just cities in the South that have had problems with ecological reserves and greenbelts as urban containment devices. In his comparative study of physical planning in the United Kingdom and California, Simmie

(1993) comes to the conclusion that the greenbelts in both countries have failed. Simmie argues that the greenbelt policy in the United Kingdom and California should be replaced with something more like the regional planning policies for Paris, where large-scale growth is channeled in corridors (173).

Greenbelts in London and Jerusalem have also run into trouble. According to Cohen (1994):

> Common wisdom holds that greenbelts are an effective and popular planning strategy that allows authorities to restrain and shape urban growth and simultaneously provide for the welfare of residents. . . . In spite of the consistency and longevity of greenbelt planning—more than a century for London and seventy-five years for Jerusalem—the same problems and definitional issues still face the cities. Moreover, the greenbelts are confronted with ever greater challenges and new criticism concerning the balance of their benefits and costs. (86–87)

Cohen (1994) argues that the limits posed by the greenbelts have resulted in leapfrog development and expensive outlays for infrastructure, among other problems (87). Indeed, "despite the institutionalization and expansion of the greenbelts, and in some cases because of them, urban growth has spread over a far greater area than envisioned in the original plans for the two belts" (87).

One of the few places in the North that receives praise for setting and maintaining urban growth boundaries (UGB) is the state of Oregon. With reference to Oregon as a successful example, Calthorpe (1993) argues that "urban growth boundaries should be established at the edge of metropolitan areas to protect significant natural resources and provide separation between existing towns and cities" (73). The trick, Calthorpe argues, is to establish a regional agency responsible for the limits—otherwise the line can easily be moved at the change of the guard (i.e., next elections). Where urban growth takes place across multiple jurisdictions, Calthorpe argues that building in the need for regional consensus to shift the UGB would safeguard its integrity (73).

On the face of it, Calthorpe's argument makes a lot of sense. But again, it is a complicated issue. In the case of Mexico City, the Federal District shares boundaries with seventeen municipalities of the state of Mexico. Yet, there is very little cooperation across these jurisdictions. To establish the means to resolve issues of coordination is crucial—that much is certain. But the problems also run deeper than this; as I have been arguing, they go beyond the scope of securing good urban land management techniques. Even in parts of the Third World where the state has had complete control over the land market, establishing control over access to land for low-income human settlements has been elusive.

The Chinese government, for instance, has long had a national urban policy aimed at controlling the growth of large cities. However, Wei (1994) observes that China's large cities continue to grow rapidly and that urban land use controls have not been well implemented: "Since the early 1980s national urban policy aimed at controlling large city growth has been advocated and implemented. However, the growth of large cities remains appreciable and has not been much slower than the growth of small cities. The increasing percentage of population living in million-plus cities has been one of the marked characteristics of Chinese urbanization" (57).

In view of the urban land crisis facing Shanghai, Tang (1994) argues that socializing land is not necessarily the solution. Tang observes that even where land is nationalized, urban land development problems do not necessarily disappear. Kombe (1994) and Lusugga Kironde (1995) supports this argument with reference to Dar es Salaam—the biggest city and most important socioeconomic center in Tanzania. Although all of Dar es Salaam's land has been nationalized and put under total state control, most access to land for human settlements has taken place under illegal and semilegal (i.e., informal) conditions. Kombe (1994, 24) estimates that between 60 percent and 70 percent of Dar es Salaam's total urban area is made up of irregular settlements.

One city that offers lessons of a positive, rather than negative, type is Curitiba. Curitiba has a population of 1.6 million in the city proper and 2.3 million in the greater metropolitan area. It is Brazil's fastest-growing urban area (Rabinovitch and Leitmann 1993). Because of the city's innovative approach to urban-environmental problems, Curitiba has earned widespread recognition among agents of the international media, researchers and academics, government officials, planners, and development institutions.[16]

Perhaps the most noteworthy aspect of Curitiba's planning process is the way in which the city's master plan integrates traffic management with land use to limit urban sprawl. Through the Curitiba Integrated Transport Network the city's road network, public transportation and land management systems are each designed to reinforce one another. As opposed to the radial urban sprawl that typifies Mexico City and most other cities worldwide, Curitiba has managed to direct growth along preplanned linear corridors and structural axes. This is not to say that every city should aim to guide its development along linear corridors. Each city has a unique terrain that should be taken into account. However, as Rabinovitch and Leitmann (1993, 9) point out, the case of Curitiba does suggest "that cities should make a conscious decision concerning their spatial structure in relation to patterns of travel and land use at the outset of rapid growth." The environmental benefits of such an integrated approach in Curitiba are significant: "From a plane or from the State Telephone Company Observation Tower, it is possible to visually determine the borders between different residential zones, distinguish commercial collecting roads from neighborhood ones, see the built-up line of "structural sectors" defining urban growth, and the protected green areas redirecting it. These have resulted in a more energy-efficient, greener city" (Rabinovitch and Leitmann 1993, 29).

If Curitiba's innovations were limited to integrative spatial strategies, perhaps it would not be "the most lauded city in Latin America" (Gilbert 1994, 166). But there is more to the story. Curitiba's successful experimentation with other environmental innovations has earned it the title of "environmental capital of Brazil" (66). Rabinovitch and Leitmann (1993, x) identify the major reasons for the distinction: 70 percent of Curitiba's households participate in one of a series of programs set up to recycle the city's solid waste, green space per capita has increased, water and park development policies are integrated and innovative sewage treatment systems have been developed, an ecoindustrial park has been established that attracts low-polluting enterprises, and lessons about ecology are integrated into the educational systems for grade-school children as well as adults. These steps at least move in the direction of productive ecology and a more sustainable form of development.

One feature of the Curitiba case that Rabinovitch and Leitmann (1993) suggest may not be transferable is the "political commitment, leadership, and continuity that Curitiba has enjoyed over the last twenty years" (50). The city's charismatic mayor (Jaime Lernor, a talented architect, planner, and consummate politician) is credited for putting many of the innovations into practice during his three terms in office (1971–1975, 1979–1983, 1989–1992). It would be wrong, however, to dismiss the Curitiba case entirely in terms of top-down managerialism. One of its features has been the involvement of grassroots initiatives in efforts to create income and livelihood opportunities. Two notable examples are the Garbage That Is Not Garbage Program and the Green Exchange, both of which create income opportunities by enabling the recycling of solid waste in low-income communities (Rabinovitch 1992).

Rabinovitch and Leitmann's (1993) presentation of Curitiba does not pretend to offer a perfect model for all cities. But the authors do hope that their "presentation of the main difficulties, innovations, principles, and results achieved in Curitiba can serve as a basis for discussion and guidance for other cities" (vii). In their final analysis, Rabinovitch and Leitmann point out that Curitiba is not without its problems: only 55 percent of the city's dwellers are connected to the sewage system, roughly 8 percent of the population live in slumlike conditions, irregular settlement continues to take place along streams and water springs, and nearly one-half of the city's children do not complete grade school. The authors rightly note that "many of the ongoing problems stem from the fact that the cities cannot be managed in isolation from state and national governments. Curitiba is not an island within Brazil and presents structural characteristics that reflect a broader social and economic context" (51). The same can be said of the Mexico City case. To understand the constraints as well as opportunities it is necessary to take into account the *politics of urban containment* and the *dynamics of political mediation*.

The Politics of Containment and Political Mediation

Despite the gains made and all the rhetoric, there is still a deeply seated perception in Mexico, and elsewhere, that *cityscapes* (areas designated for mixed urban uses) and *landscapes* stand in opposition to one another. There is a long history of critics arguing that such a separation of cityscape from ecology involves misconceptions and associated practices that bode disaster. Ebenezer Howard made such an argument almost a hundred years ago in his classic work about garden cities. Recently, some excellent literature, concerning both First and Third World contexts, has been generated along these lines. Although the call for an integrated view of cityscape and ecology has become increasingly popular in the 1990s, there is a long way to go.

That the cityscape is widely conceptualized as something categorically distinct from ecological space has implications for policy and planning. Cityscape and ecology are viewed as mutually exclusive; indeed one is viewed as the negation of the other. Ecology becomes something "out there" to be protected and preserved. This perspective is a central feature in the Mexican administration's approach to land use and the question of sustainability. It underpins the government's failed *politics of containment.*

The politics of containment can be defined as the emergent legal and institutional means to control the horizontal expansion of irregular settlements on the basis of ecological arguments. The politics of containment is an ensemble of numerous—and often contradictory—plans, programs, laws, regulations, and institutions.[17] It is both the form and action of a growing ecobureaucracy.[18] Two strategic dimensions are involved: (1) efforts to legalize (regularize) land tenure and consolidate irregular settlements that have already become established, and (2) efforts to set limits to the legally designated urban area by designating ecological reserves at the city's rural periphery (see figures 3.2 and 3.3). An excerpt from the 1986 General Program for the Urban Development of the Federal District epitomizes the latter strategy:

The demarcation of the area for ecological conservation is one of the most urgent measures needed to contain horizontal urban growth. . . . In order to keep the ecological conservation area free from the pressures of urban expansion, its limits will be determined and marked by physical barriers. . . . The barriers may be natural, such as ravines or rivers, or man-made such as stone walls and concrete boundary markers. . . . Supporting the limits will be large signs, watchguard stations, and control offices—particularly at roadway access points where the pressures of urban expansion are very strong. . . . Also the special police force of ecology guards *(ecoguardias)* will protect the land. (DDF 1986, 9, 67)

As the government has stiffened its policy prohibiting land invasions and the formation of incipient colonias populares, evictions have increased (Gilbert and Varley 1991, 49). There is now a streamlined, usually well-orchestrated procedure for evictions. Waste and destruction are familiar parts of the process (see figures 3.4 and 3.5) . I refer to the mechanism to

Figure 3.2
A sign posted in Ajusco marking the "limit of the urban area."

conduct such evictions as the *desalojo machine*.[19] The desalojo machine includes the tools of eradication (e.g., sanitation trucks, walkie-talkies, horses, hammers, and guns), as well as the labor (e.g., police on foot and on horseback, demolition, and clean-up crews, officials) (see figures 3.6 and 3.7).

The politics of containment, enforced by the desalojo machine, have given rise to an *industry of destruction;* that is, a cycle of irregular settlement being formed, then eradicated, then formed again, in which certain public and private individuals reap huge profits through illicit land transactions. This cyclical process results in violent confrontations and loss. Simply put, the politics of containment have not worked. This book notes the failure in the context of Mexico City's ecological reserves. Quadri de la Torre (1994b, 38–39) notes the failure of the politics of containment throughout Mexico's sixty odd national parks: "Practically all the national parks and biosphere reserves present to a greater or lesser extent phenome-

Figure 3.3
A sign posted in Ajusco reading "refuse illegal lots."

nons of irregular settlements, fires and deforestation for agriculture and ranching, invasions, and illegal forestry exploitation." Quadri de la Torre argues that the only parks that are in good shape are those that are inaccessible.[20] This is due to the fact that Mexico's national parks, including those in the Valley of Mexico, exist in name only. Usually there is no mandated change in land tenure arrangements, no compensation to the owners of the land, and no clear delineation of park boundaries. In the case of Mexico City, urbanization continues to press further and further into the supposed ecological reserve. Clearly, an alternative approach is needed, and this is where lessons may be culled from the social experimentation at the grass-roots level in Ajusco.

Ecological arguments articulated by state functionaries are not simply cogent critiques of urban development advanced in the public's interest. It is more accurate to view them as elements of an ecological strategy in a context of *political mediation*. My use here of the term *political mediation* draws from the work of Joseph Foweraker (1982, 98), who argues that

Figure 3.4
Results of an eviction from Lomas del Seminario, Ajusco Medio, 1988.

mediation "constitutes the social forces in struggle in so far as it fixes the institutional terrain where struggles between them take place." In Mexico City, the state's prioritization of urban planning and its emphasis on ecological conservation are important aspects of political mediation.[21]

A central component of political mediation is the purposeful activity of state functionaries. The state mediates among conflicting individuals and collective actors with the intent to effect an agreement or reconciliation, although the deliberations that result in "solutions" may or may not be accepted by all contending parties. Political mediation also involves the production and reproduction of structures and power relations that actually shape the players' actions, individually and collectively, on the stage. This always involves a "mix" of acknowledged and unacknowledged conditions and of intended and unintended consequences. Political mediation thus implies more than political intervention. It involves the institutionalization or repression of political struggle through a wide range of forms,

Figure 3.5
Evicted children waiting in the rain for their parents, who were not home when the eviction was executed.

including bureaucratic procedures, legal maneuvering, co-optation of grassroots leaders by the state, accommodation by the state to certain claims or demands of popular groups, and violence.

The connotation of violence as an unruly or chaotic eruption of extreme force does not apply here. Rather, I use the term *violence* in the sense that Max Weber employed it when he discussed it in relation to the state: "The state is considered the sole source of the 'right' to use violence" (Max Weber, cited in Held et al. 1983, 112), where violence is understood as the "legitimate" use of physical force. It may seem contradictory to consider violence as a form of mediation. But, as I will point out further on in the book, I have observed that violence—specifically, the destruction of housing units during *desalojos* (evictions)—has come to be an expected part of the irregular settlements process in the ecological zone of Ajusco. Recently, a woman (re)staking out her claim to a plot of land in a part of Ajusco the day after her house was eradicated told me that she expected

Figure 3.6
Frontline demolition crew ready to eradicate housing in Lomas del Seminario, Ajusco Medio, 1988.

to suffer three, maybe four, desalojos before her claim would stick. Indeed, the desalojo machine is institutionalized violence and is an often used form of mediation employed in the contest for land. Enduring the desalojos appears to be a way for low-income and disadvantaged groups to strengthen their hand in their protests and negotiations for land.

This critique of the politics of containment—including its desalojo machine and the industry of destruction—certainly does not mean that any effort to set aside land for an ecological reserve in inherently misguided, corrupt, or doomed to failure. Nor does it suggest that all land should be opened up to provide land for housing. Rather, as Mitlin and Satterthwaite (1990) point out with respect to land use, "sustainability goals emphasize the need to limit the encroachment of settlements on agricultural land and on those parts of the natural landscape with important cultural, ecological or recreational value." This point is unassailable. The same can be said about the logic underpinning Mitlin and Satterthwaite's more general argument:

Figure 3.7
Female foot police on eviction day in Lomas del Seminario.

The effectiveness of physical planning and land-use management as instruments to optimize the use of land and other natural resources depends on the quality and nature of the institutional structure for implementation. They are most likely to be effective at balancing societal needs against individual interests where implemented by competent, effective and representative local government. This implies locally generated plans; it also implies the need for democratic processes if societal interests are to be protected. It also implies regional authorities which work with local governments to ensure a compatibility between local and regional plans and which work to ensure the protection of regional resources such as watershed protection and water-basin management. (1990, 43)

Yet, the problem remains: how can the balance, integration, and much vaunted ideals of democratic participation noted above be realized? The *urban challenge* is much more than a question of capacity to produce and manage some additional quotient of urban space and services. The challenge calls for the redefinition, in theory and practice, of the very form and meaning of urbanization. It calls for *empowerment* and *ecologically integrated proactive planning* that can effect social transformation. It calls for critical insight into the way history, power, and the environment intersect in the development of human settlements.

II

The Valley of Mexico: History, Power, and the Environment

4

Mexico City's Urban-Ecological Coevolution

A common feature of human populations over the past few thousand years is their dramatic fluctuation. As Richerson (1993, 127) describes it, societies rise to prominence, then decline, often ending in both political and demographic collapse. The rise and fall of so many civilizations indicate the elusiveness of sustainability. Lewis Mumford (1972), a social critic and historian of cities, blames what he saw as the urban displacement of nature:

[T]he displacement of nature in the city rested, in part, upon an illusion—or, indeed, a series of illusions—as to the nature of man and his institutions: the illusions of self-sufficiency and independence and the possibility of physical continuity without conscious renewal. Under the protective mantle of the city, seemingly so permanent, these illusions encouraged habits of predation or parasitism that eventually undermined the whole social and economic structure, after having worked ruin in the surrounding landscape and even in far-distant regions. (144)

Mumford's sweeping indictment is an elegant oversimplification that continues to find support in the literature. For instance, Ponting (1991, 434) argues that "the story is repeated throughout human history and all over the globe, from Sumeria to ancient Egypt to pre-Columbian North America to tiny Easter Island: Human beings prosper by exploiting the earth's resources until those resources can no longer sustain the society's population, which leads to the decline and eventual collapse of that society." Despite the appeal of this line of argument, it is not at all certain that endogenous environmental changes due to human impact (such as aquifer depletion and soil salination and erosion) comprise the main causal factor or driving force in the collapse of past societies. In addition to the factor just mentioned, Richerson (1993) offers three other plausible hypotheses:

(1) exogenous environmental shocks or changes such as a series of dry years, long term climate deterioration, or the introduction of new diseases; (2) exogenous

political or economic changes, such as the rise of pastoral nomad confederations or the shifting of critical trade systems; and (3) endogenous political or economic changes, such as a stress on ideological and political legitimacy when elite manipulation of the economy fails to keep up with population growth, or when the limits to imperial conquest lead to an inability to reward the military establishment. (127)

In view of the complexity of real cases, simple monocausal explanations are clearly inadequate. It is necessary to examine interrelations among environmental problems, security, and the political system (Aguayo Quezada and Bagley 1990; Liverman 1990a; Taylor and Garcia Barrios 1995). Zimmer (1994) expresses enthusiasm that the integration of "new ecology" and human geography is inspiring a deep rethinking of human-environment interrelations. Along such lines, Turner and Meyer (1993, 44) provide a useful review titled, "Environmental Change: The Human Factor." They spell out a range of driving forces in history and argue that historical shifts in production and consumption, including rises and falls in societal resource demand and waste emission, can best be explained in terms of a series of interacting factors: population growth, state of economic development, technological change, social organizations, and beliefs and attitudes.

Turner and Meyer (1993, 42) outline two major categories of ecological research that focus on the impact human beings have on the environment. On the one hand, there are studies that focus on the past 300 years and emphasize the Industrial Revolution as the fundamental turning point in global nature-society relations. On the other hand, there are studies covering the past 500 years and emphasizing the landscape-type changes typified in the Columbian encounter (Crosby 1972). As Turner and Meyer (1993) point out, "Whichever of the two one chooses to stress—and it is clear that they were closely interrelated—their results today are evident in the rise of mass-consuming societies whose reach in affecting the global environment in general and specific ecosystems in particular is enormous" (42).

On a millennial time scale, Whitmore et al. (1990) did an interesting historical-comparative study of society-nature relations in the Central Mexican Basin, the central Mayan lowlands of Mexico and Guatemala, the Tigris-Euphrates lowlands, and the Egyptian Nile Valley. Turner and Meyer (1993, 43) provide a useful synopsis of the insights that have emerged from the Whitmore et al. (1990) and related studies (see box 4.1).

Based on the findings outlined, Turner and Meyer (1993) argue that the doomsday implications of contemporary population growth are not necessarily supported by historical research on population-environment relations. They argue that "Despite the antiquity and increasing ubiquity of human use, the capacity of the environment to recover—albeit altered—or to sustain human occupancy, even in larger numbers at higher standards of living, has been the norm" (Turner and Meyer 1993, 49). Be this as it may, these authors are also quick to emphasize that such a generalization does not necessarily uphold a "cornucopian" view of today's nature-society relations (49).

Contemporary environmental change at the global level and in most regional and local contexts is novel in pace, magnitude, and kind. Humani-

Box 4.1
Synopsis of Whitmore et al.'s (1990) and Related Findings on Society-Nature Relations

---

1) Dense settlement and long-term high rates of population growth are not exclusively recent phenomena, at least at the regional level, and the human transformation of regional environments is of equal antiquity;
2) For the most part, premodern changes involved the landscape elements basic to the agricultural resource base (soil, vegetation, surface water, and so forth), and they often substantially altered these elements;
3) Nevertheless, over millennia of intensive use of the regions, sustained occupation did not necessarily lead to persistent degradation of the environment to the point of land abandonment, and the sustained use and/or recovery of land long cultivated and highly altered was not uncommon;
4) Human populations have experienced periods of sharp decline as well as sustained growth in all of the regions. In the extreme case of the central Maya lowlands, the highest regional population was attained by A.D. 800, a figure several dozen times that of today;
5) Recovery of the environment following population loss rarely involved a return to the system that had existed prior to human use. For example, though the deciduous tropical forests of the central Maya lowlands returned upon abandonment after years of clearance and intensive use, the relative abundance of species (e.g., *Brosimum alicastrum*) apparently changed; and
6) While population increase through time, regardless of the technological base and social organization, invariably led to major ecosystemic changes, it is not at all clear that the transformation and biotic simplification of the environment translated into socioeconomic deterioration.

Source: Turner and Meyer (1993, 43).

ty's ecological footprints now substantially impact basic biogeochemical cycles that sustain the biosphere (Kates, Turner and Clark 1990).[1] Pickett and McDonald (1993, 311) document ways human activities now significantly impact materials, energy, organisms, and water within the earth's ecological systems.[2] They note that the cumulative impact of human activities currently rivals, in both magnitude and scale, natural disturbances such as floods and hurricanes.[3] With respect to the world's cities, Hardoy, Mitlin, and Satterthwaite (1992) note that large areas of the earth's surface are transformed:

Hillsides may be cut or bulldozed into new shapes, valleys and swamps may be filled with rocks and waste materials, water and minerals may be extracted from beneath the city, and soil and groundwater regimes modified in many ways. The construction of buildings, roads and other components of the urban fabric modifies the energy, water and chemical budgets of the affected portions of the earth's surface. As cities expand, they not only alter the earth's surface but create new landforms. . . . Urban developments greatly affect the operation of the hydrological cycle, including changes in total runoff, alteration in peak-flow characteristics and a decline in water quality.

The point of these observations is not to suggest that the scale or magnitude of human activities is so great that the quest for sustainable development is an impossible one. Rather, the point is that the earth's buffer capacity, with respect to our activities, can no longer be taken for granted. The quest for sustainability should be seen as an ethicosocial-biogeophysical imperative; the continuance of (un)sustainable development is fast narrowing options for today's and tomorrow's generations. Framed in such a way the tenets of modernism loose their transhistorical validity: "knowledge, technologies, and social organization merely change, rather than advance, and the 'betterness' of each is only relative to how well it fits with the others and values. Historical change in the coevolutionary explanation, rather than a process of rational design and improvement, is a process of experimentation, partly conscious, and selection by whether things work or not" (Norgaard 1994, 37). Enrique Leff's work complements this argument. He emphasizes the need for researchers and policymakers to take into account the importance of history and the interaction of biogeophysical and social forces in a way that is linked with social movements of the poor. Leff (1994, 1995) employs a historical-ecological and materialist perspective, arguing that the environmental costs of development are not taken into account (i.e., internalized) by Mexico's market economy. From this angle, claims about the growth machine become suspect. Conditions that are vital for the sustainability of production, but that are difficult to value in capital terms, tend to be excluded from economic rationality and measurement.

Environmental Low Entropy and Closed Loop Development

Life can be sustained only as long as it can continuously feed itself on environmental low entropy (Georgescu-Roegen 1971). Entropy is a qualitative measure of the degree to which matter-energy is potentially accessible for human use. Something with "low entropy," such as a lump of coal, contains a concentrated source of energy; it embodies greater potential usefulness than high entropy sources (e.g., the diffuse heat stored in the earth's oceans). As Daly and Cobb (1994) point out, "It is the quality of low entropy that makes matter-energy receptive to the imprint of human knowledge and purpose. High-entropy matter-energy displays resistance and implasticity" (195).

Environmental low entropy comes not only from terrestrial stocks of stored fuel energy such as plant, fossil, nuclear, and geothermal heat sources. It is also available in flows from the sun and water. Furthermore, it includes useful (nonfuel) energy embodied in, for example, fertile land, germ plasm, and ecosystems. That we can view land, germ plasm, and ecosystems as embodied energy was a point well noted by Aldo Leopold. He characterized the economy of nature as "a fountain of energy flowing through a circuit of soils, plants, and animals" (Leopold 1966, 253).

The economic process (production followed by consumption) is entropic. In other words, as work is done, entropy increases. The decrease in entropy in one place occurs at the expense of increasing entropy (disorder) in another (Chiras 1991). Whether the unit in question is a person, a city, or the larger economy, each exists by sucking low entropy (raw materials) from the environment and expelling high entropy (waste) back to the environment (Daly 1991). Daly and Cobb (1994) describes the process this way: "Since we neither create nor destroy matter-energy it is clear that what we live on is the qualitative difference between natural resources and waste, that is, the increase in entropy. We can do a better or worse job sifting this low entropy through our technological sieves so as to extract more or less want satisfaction from it, but without that entropic flow from nature there is no possibility of production" (196).

There are two *primary* sources of environmental low entropy: (1) the sun, and (2) the earth's terrestrial stocks of fossil fuels and minerals. As Georgescu-Roegen pointed out in his classic analysis *The Entropy Law and the Economic Process* (1971), the rise of industrialism has shifted humanity's dependence on energy currently coming from the sun to energy stored in terrestrial stocks. Most of the energy humans now use comes from nonrenewable resources, particularly metals and fossil fuels. This shift has enabled resource-intensive forms of urban-regional economic and demographic growth (Mumford 1972; Vitale 1983; Brown and Jacobson 1987).

Pre-Hispanic cities in the Valley of Mexico ran mostly on the organic energy accumulated in contemporary plant and animal matter. Unlike the cities of today, pre-Hispanic cities did not depend heavily on terrestrial stocks of energy stored in the form of fossil fuels. Mumford (1972, 141) captured this point when he wrote: "From the standpoint of their basic nutrition, one may speak of wheat cities, rye cities, rice cities, and maize cities, to characterize their chief source of energy; and it should be remembered that no other source was so important until the coal seams of Saxony and England were opened."[4]

The societies of Mesoamerica, including that of the Aztecs, did not have a stark urban-rural divide. Based on the close integration of the urban-rural dimensions of pre-Hispanic societies, Vitale (1983) labels their human settlements *agrarian cities*. There are lessons to be learned from what pre-Colombian cities accomplished in terms of city-hinterland integration.

Vitale (1983) argues that pre-Hispanic cities were like ecosystems in the sense that they neither consumed nor imported a great quantity of energy from outside their bioregion. Cities today are not at all like ecosystems because they typically import environmental low entropy from well beyond their immediate hinterlands. Today's cities are nodal points in a web that confounds any definition of an ecosystem. With regard to scale, Mumford (1972) observes: "Whereas the area of the biggest cities, before the nineteenth century, could be measured in hundreds of acres, the areas of our new conurbations must now be measured in thousands of square miles. This is a new fact in the history of human settlement" (150).

A number of scholars argue that in comparison to colonial cities or today's industrial cities, indigenous cities of the Valley of Mexico were more ecologically sustainable (Vitale 1983; Redclift 1987; Gudynas 1990). Such comparisons point to the pre-Hispanic city's reliance on the use and reuse of local natural resources including land, water, and biomass. A few examples are worth noting. In the pre-Hispanic cities of Teotihuacán and Tenochtitlán practically no waste was generated that could not be recycled or organically decomposed and returned to local nutrient cycles. Urban and agricultural practices tended to conserve resources while avoiding the discharge of pollutants into the air and waterways. Guillermoprieto (1990, 94) notes: "Aztec people could hardly conceive of waste: they used cornhusks to wrap food in and inedible seeds to manufacture percussion instruments. All organic waste went into the compost . . . with which the Aztecs compensated for their lack of agricultural land. Each street was swept clean every morning, and the day's cargo of excrement was deposited in a special raft tied at the street's end."

Human excrement in Aztec society was an important resource in a deliberate process of nutrient recycling. On this last point, Kandell (1988) notes that the waters in and around Tenochtitlán were kept relatively unpolluted by strict prohibitions against disposing human excrement in the city's canals and lake. Most of the excrement, which was composted in outhouses by the water's edge, was applied to the *chinampas* (floating gardens) and fields as fertilizer by workers in canoes (Kandell 1988, 46). This involved an indigenous application of knowledge that was passed down from generation to generation for centuries.

In stark contrast, human excrement is the most pervasive and health damaging "toxic waste" in most cities of the Third World today (Hardoy, Mitlin, and Satterthwaite 1992, 204). Brown and Jacobson (1987, 29) estimate that, worldwide, over two-thirds of the nutrients contained in human excrement are lost to the environment as unreclaimed sewage, often polluting aquifers, bays, rivers, and lakes. Presumed sewage contamination of the Ajusco aquifer has been a major point of contention in Mexico City for more than a decade. Leff (1990a, 53) notes that 30 percent of the Federal District's population lacks access to adequate sewage facilities. As a result, open air defecation ends up producing enormous quantities of pathogenic organisms that make people sick. Participants in Ajusco's colonia ecológica productiva movement argued that sewage composting as a means to prevent pollution and to close nutrient cycles is one of the fundamental tasks necessary to promote ecological sustainability. Brown and Jacobson (1987, 29) make a related point from a forward-looking perspective: "As the energy costs of manufacturing fertilizer rise, the viability of agriculture—and, by extension, cities—may hinge on how successfully urban areas can recycle this immense volume of nutrients."

Recycling helps replace *open loop systems* (one-way flows of resources *in* and wastes *out*) with *closed loop systems* (consume-process-reuse) (Smit and Nasr 1992). Along such lines there are some success stories. For instance, Smit and Nasr (1992, 142) point out that Mexico City pumps some of its sewage fifty miles to the northern part of the valley where it is used to irrigate over 100,000 hectares for livestock feed. As many as a hundred other cities in Mexico engage in similar practices (142). Brown and Jacobson (1987, 29) point out that the collection of human excrement (known as night soil) for use as fertilizer continues to be an important function in other cities, particularly in Asia: door-to-door handcarts and special vacuum trucks are used to collect night soil in many of the older neighborhoods of Seoul, South Korea, for recycling in the city's greenbelt. According to the World Bank, one-third of China's fertilizer requirements have been provided by night soil. This has helped maintain soil fertility for centuries (29).[5]

Shifting from open loop systems to closed loop systems is not simply a matter of selecting different technologies for intervening in the environment. As Norgaard (1994) points out "the mechanisms for perceiving, choosing, and using technologies are embedded in social structures which are themselves products of modern technologies" (29). Yes, technology may provide certain answers, but what's the question? Is the question ultimately one about reproducing the dominant socioeconomic and political order of things? Or is the question one that seeks alternatives that ultimately challenge social and political relations of power as well as mainstream conceptions of society-nature relations? Political ecology is concerned with the latter. Hence an awareness of coevolutionary dynamics is useful.

Norgaard (1994, 29) provides a coevolutionary perspective of urban concentration: "The pattern of people living in cities, the organization of people to serve multinational industrial enterprises, the centrality of the bureaucratic order, and the use of Western science for social decision-making have all coevolved around fossil hydrocarbon-fueled develop-ment." The use of petroleum has enabled cities to lengthen their supply lines and draw basic resources, such as food and raw materials, from dis-tant elsewheres (Brown and Jacobson 1987; Rees 1992). Relatively cheap oil and progrowth economic policies underpin the surge in industrializa-tion and urban growth that continues to radically transform developing countries.

In view of the built environments emerging, critics have long decried the results: "as congestion thickens and expansion widens, both the urban and rural landscape undergo defacement and degradation, while unprofitable investments in the remedies for congestion, such as more super-highways and more distant reservoirs of water increase the economic burden and serve only to promote more of the blight and disorder they seek to palliate" (Mumford 1972, 151).

Carley and Christie (1993, 31) refer to "a crisis of automobility." Expan-sive roadway networks (what Mumford refers to as manmade deserts) and megasprawl are most pronounced in places like the Los Angeles–San Diego metropolis, the Baltimore-Washington Metropolitan Area, the Southeast of England, and the Tokoyo-Osaka belt. But sprawling transportation grids are also being built out in the Third World around booming cities such as Bombay, Kuala Lumpur, Bangkok, and Mexico City (31). Lowe (1991) argues that the failure to see land use planning as a transport strat-egy is a tragedy; it has inadvertently given ascendance to the automobile as a factor shaping urban growth. Right now a pitched battle is occurring in Mexico City over plans to build another major highway. The plan is to complete a third ring around the city (given the growth of the city, the previous rings are no longer peripheral). The new highway will link up the completed outermost ring with the highways to Puebla, Querétaro, Pachuca, Cuautla, Toluca, and Cuernavaca (Legorreta 1995). Such a proj-ect would have a tremendous impact on the Ajusco region.

The coevolutionary perspective offers new insights into transportation problems and the "crisis of automobility." But it is not the principal task of this book to focus on transportation (see Davis 1994). Coevolution and political ecology offer insight into a wide range of issues concerning technology options, land use, infrastructure, and urban development. This book is primarily concerned with human settlements, grassroots activism, productive ecology, and empowerment. Each of these dimensions come together in the case of Ajusco where the colonia ecológica productiva movement took place. As I noted in chapter 1, the CEP movement called for adequate livelihood opportunities and income generation, the integration of urban-rural and society-nature relations, and collective strategies for community empowerment. At its zenith, the CEP movement had hundreds of families rallying under its "productive ecology" banner. At the same time, the movement had networked across the urban-rural divide and had successfully articulated multiclass alliances with independent researchers, university students, newspaper columnists, international agencies, and campesinos both locally and from the northern part of the country. Grassroots activism of this sort can and does generate experiences of social experimentation, social learning, and *empowerment* that are wellsprings of praxis and opportunity. To gauge the potential of such activism, the principle task of this book, it is necessary to take into account the fact that Mexico City's social and biogeophysical terrains have coevolved in the process of urbanization and history making. It is necessary to take a historical perspective of the constraints and opportunities inherent in the relations between human societies and the environment in the Valley of Mexico. Such a perspective calls attention to the inadequacy of the concepts of environmental determinism and cultural possibilism in history making (Garavaglia 1992).

The Valley of Mexico: Humans as Components of Ecosystems

To be precise, the Valley of Mexico is not a valley at all. The area in question is a closed hydrographic basin encircled by mountains. It has no natural drainage for its waters—one of the defining features of a valley. Nonetheless, convention has it to call the place a valley; I follow this convention.

Sanders, Parsons, and Santley (1979) document that the Basin of Mexico has a complex geological history. It was formed by volcanic activity during the late Tertiary and early Quaternary geologic periods. Prior to this time, the area now occupied by the basin was part of a shallow tropical sea. Beginning in the late Tertiary (65 million years ago), a series of seven phases of volcanism resulted in the elevation of the Basin floor to its present minimum elevation of 2,235 meters above sea level. This activity produced the surface topography visible today (Sanders et al. 1979, 88). There are no traces of the first phase. The second phase, which began about 30 million years ago, created the low ranges of hills in the northwestern corner of the valley. Phases 3 and 4 left a series of discontinuous topographical features, some of which Mexican geographers refer to as the Sierras Menores. Phase 5, dating from 1.8 to 5 million years ago, was one of major volcanic upheaval. During this epoch the lofty sierras that border the valley on the east and west were formed. Phase 6 was a time of widespread volcanic activity all over the valley. Phase 7 began approximately 1.8 million years ago; it formed the massive Sierra del Ajusco, which now delimits the valley to the south. The most recent eruption spewed from the Xitle Volcano around 2,200 years ago and blanketed the miniregion of Ajusco with a thick overlay of jagged lava rock (see chapter 1, figure 1.3).

Prior to the formation of the Sierra del Ajusco, the Central Mexican Basin was a true valley with a north to south downward slope that drained into the Balas River. The valley became enclosed as massive volcanic ridges built up around the western, southern, and eastern sides of a great central depression and as lava and ash blocked drainage outlets to the south. For thousands of years, alluvial detritus eroded from the surrounding hillsides and accumulated in the central depression. Volcanic activity, glaciation, and erosion have created a variety of soil types.

Today the basin is an elevated plateau ranging from 6,800 to 7,900 feet in altitude. Its boundaries are the Sierra de Nevada on the east, the Sierra de las Cruces on the west, and the Sierra del Ajusco in the south. To the basin's north lie a series of smaller, scattered ranges of hills. The plateau itself has an oval shape that extends sixty-eight miles north to south, and about fifty miles east to west. Its surface area is roughly 2,700 square miles, roughly 950 square miles of which are urbanized (Kandell 1988, 10) (see figure 4.1). About 45 percent of the total area of the basin (430,000 hectares out of 960,000 hectares) takes the form of a plain. Around 40 percent of this plain (270,000 hectares including lake surfaces) is level; the rest

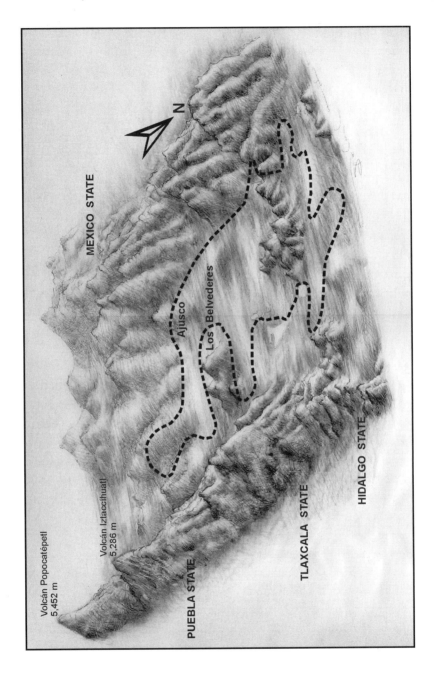

of the plain (160,000 hectares) is composed of foothills (Ibarra, Puente, Saavedra, and Schteingart 1987, 101)

Given its elevation and geographical location the basin has a varied climate, generally characterized as subtropical and tropical. Sanders, Parson, and Santley (1979) identify nine major environmental zones, each with its own inventory of faunal, floral, and other natural resources. Rainfall varies across the valley, a fact that is especially significant with regard to Ajusco. It rains almost two or three times as much in the Sierra del Ajusco as elsewhere in the valley. Annual precipitation increases from 400 to 600 millimeters annually in the center and northeast of the valley to 1,200 millimeters in the foothills of the mountainous countryside (Garvaglia 1992, 570). Most of the annual rainfall occurs during a few severe rainstorms from June through September. There is little or no precipitation the remainder of the year (National Research Council 1995, chap. 2). Given its permeable volcanic subsoil and the fact that it gets more annual precipitation than anywhere else in valley, the Ajusco watershed is considered essential for the replenishment of the basin's underground water supplies (SEDUE 1989).

The seasonally torrential rainfall on the slopes of Ajusco presently recharges aquifers at a rate estimated to be six cubic meters per second. This would account for 20 percent of the total amount of water pumped up from the valley's underground supplies destined for consumption in the Federal District (Schteingart 1987, 456). Based on calculations of the total annual rainfall in the area, the government estimates that the Ajusco watershed has the potential to absorb 1,000 liters of water per square meter of land per year (DDF 1989b, 5).[6] Ajusco's geomorphic characteristics and microclimate play a significant role in the valley's hydrological cycles. The formation of the Valley of Mexico is not complete, nor will it ever be. As the earth's crust cools, it will continue its tectonic shifting, quaking, and upheavals. The killer earthquake that rocked the valley in 1985 is a clear reminder of this simple fact. The quake measured 8.1 on the Richter scale

Figure 4.1
The Metropolitan Area of Mexico City within the Valley of Mexico. Source: This is an artistic version by Belinda di Leo, of a map that appeared in the *National Geographic*, 166, no. 2 (August 1984). The vertical scale is exaggerated two times.

and caused 20,000 casualties. It destroyed or seriously damaged 95,000 homes, 5 hotels, 704 schools, 859 businesses and factories, 345 office buildings, 41 hospitals, and 33 theaters (Cremoux 1991, 12).

At the end of 1994, the snowcapped Popocatepetl Volcano (roughly forty miles from the Ajusco volcano) spewed a massive cloud of black ash raising fears of an imminent eruption. Fifty-four thousand people had to be evacuated from the volcano's southern and southwestern banks (*San Diego Union Tribune,* December 26, 1994). People have since been allowed to return to their homes. But the intensifying seismic activity since March 1995 has raised fears once again. A report issued in June 1996 estimates that 400,000 people live under serious threat from volcanic activity (Petrich 1996).

The status of the valley's formation as "in process" is a view supported by the emergence of a nonequilibrium paradigm in the study of ecosystems. In his analysis of historic relations between human societies and the environment in the Valley of Mexico, Garavaglia (1992) argues that there is a "permanent tension between situations that could be described as homeostatic, i.e., generating a measure of stability, and periods of upheaval that undermine that stability. Of course, upheavals can also develop new adaptational capacities. The history of all ecosystems is one of constant tension between these two opposing forces. No ecosystems are really in a state of stable equilibrium, assuming, of course, that we apply a time-scale that exceeds the life span of a human being" (575). Over the past twenty years, the equilibrium paradigm has been seriously challenged by a growing body of scientific evidence suggesting it is flawed (Botkin 1990; Pickett, Parker, and Fiedler 1992). The emergent nonequilibrium paradigm allows for the study of humans as components of ecosystems (McDonald and Pickett 1993; Zimmer 1994). This inclusion enables us to avoid the impasse presented by the concepts of environmental determinism on the one hand and cultural possibilism on the other (Garavaglia 1992).

Marx observed a long time ago, "men make their own history, but they do not make it just as they please, they do not make it under circumstances chosen by themselves" (Marx, cited in Mills 1959, 182). The same can be said from an ecological perspective: people make ecosystems, but not just as they please, nor under circumstances of their own choosing. Without stating it in such terms, Garavaglia (1992) argues this point with reference to pre-Colombian societies that inhabited the Valley of Mexico. He suggests that the Aztec's complex urban-agricultural arrangement, which enabled very high levels of food production in a radically altered environment, "almost amounts to the specifically human development of a separate biome" (573).

To a certain extent, it can be argued that the location and form of the Aztec and other pre-Colombian societies in the Valley of Mexico were geographically or environmentally determined: access to water and fertile land were certainly defining elements. But such a conception can be misleading. Particular societies shape as well as adapt to their environment. Culture can be a biophysical force in its own right.[7] This was the case in the Teotihuacán and Aztec empires. Even more so, it is the case with the subsequent colonial and industrial societies that have populated and transformed the valley.

Comparing Pre-Hispanic, Colonial, and Industrial Urbanization in the Valley of Mexico: Lessons for the Present

The Valley of Mexico has been inhabited by human beings for at least 15,000 years (Kandell 1988). Beginning around A.D. 300 the valley saw the rise of societies that historians describe as belonging to the Classic period in Mesoamerica (Ponting 1991). The first was the Teotihuacán Empire, which at its height influenced all of Mesoamerica (Vitale 1983). It was replaced by the military empire of the Toltecs which, in turn, ended with the rise of the Aztecs. The city of Tenochtitlán, founded in 1345, was the political, commercial, and religious center of the Aztec Empire. It is upon the ruins of Tenochtitlán, which was destroyed in 1519 during the Spanish conquest, that present-day Mexico City is built.

Garza and Schteingart (1978, 52) suggest that it was for religious and political reasons that the leader of the Spanish conquest, Hernán Cortés, decided to build the center of the Spanish Empire on the ruins of Tenochtitlán. Around 1522 the reconstruction of the city was started. In 1548 the rebuilt city was named the "Renowned Imperial City of Mexico." Historians note that the biological and cultural consequences of the conquest, including the introduction of foreign diseases and harsh exploitation, practically decimated the indigenous population (Crosby 1972). In

1521 Tenochtitlán had as many as 300,000 people. Less than a generation later, at its inauguration as a Spanish city, it contained just 30,000 people including both the indigenous people and Spanish settlers—a reduction of 90 percent (Orozco y Berra 1973, 36).

When the Europeans first arrived in the Valley of Mexico during the early sixteenth century, it was heavily populated. Experts suggest that the population of the valley easily topped 1.5 million (Garavaglia 1992). A large portion of this population was urban. Sustaining such a large urban population, even today, would present an enormous challenge. It would require the mobilization of resources on a huge scale. In his excellent article on this subject, Garvaglia (1992) examines how successive societies mobilized the resources necessary for life in the Valley of Mexico. Historical-comparative analysis along these lines provides insight into the origins of the urban-ecological problematic manifest in the valley today.

Given the attributes of great biodiversity, the presence of large lakes within a naturally enclosed watershed, and the availability of a wide range of essential resources, the Valley of Mexico was especially well suited for human habitation during the pre-Hispanic period, although flooding and water quality were persistent problems (Berdan 1982). Natural springs, meltwater from snowcapped mountains in the Sierra del Ajusco, and run-off from the summer rains all flowed toward the center of the basin. Until the nineteenth century most of this water emptied into a series of shallow lakes that covered much of the basin floor (Sanders, Parson, and Santley 1979, 81). As well as providing an effective means of transportation, the valley's lakes contained a wealth of food sources, including five kinds of fish, frogs, polliwogs, freshwater crustaceans and mollusks, turtles, and various aquatic insects and their larvae (Orozco and Berra 1864, cited in Sanders, Parson, and Santley 1979). From the lake's surroundings, available food sources included reeds, wild rice, a blue-green alga called *tecuilatl,* and assorted waterfowl, including ducks, coots, and geese (Garavaglia 1992, 574). Endowed as such, the valley was a "center towards which various population groups gravitated and an area in which resources from a wide variety of ecological zones could be combined" (574). Although it is currently being diminished, biodiversity in Mexico City's ecological conservation areas is still impressive (Ramos 1995) (see table 4.1).

Table 4.1
Biodiversity in Mexico City's ecological conservation area

Insects	174 species, the majority of which are butterflies
Amphibians	15 species
Birds	220 species, more than 60% of which are migratory (e.g., ducks and herons)
Reptiles	32 species
Mammals	84 species, including a mountain rabbit *(teopringo)* in danger of extinction
Fish	8 species, 2 of which were introduced (trout and carp)

Source: Páramo, Bermeo, and López (1995).

The first agricultural settlements in the basin took place (between 1100 and 500 B.C.) where rainfall is highest: along the southern slopes of the valley including the foothills of Ajusco (Kandell 1988, 17). These early farm communities settled there to maximize their access to water for farming purposes. They gradually developed an irrigation system that combined the building of steplike terraces that prevented erosion, and they conserved soil moisture with innovative ways of storing and conveying water from mountain springs and rainfall (Garavaglia 1992, 570). Water has thus been a defining element in peoples' struggle for survival and patterns of settlement from the very beginning of the valley's history.

Teotihuacán: The Valley's First Planned City
The earliest permanent settlements were able to expand their production of maize and their population thanks to their practice of hillside terracing. But it was not until people figured out how to exploit the lake system at the bottom of the valley that the basin's first urban-demographic explosion took place. Such a breakthrough occurred around 100 B.C., when farmers learned how to make use of the valley's underground springs. The place was Teotihuacán, an area at the northeastern outskirts of what is presently the metropolitan area of Mexico City. Kandell (1988) describes the dynamic growth of Teotihuacán:

Single families or small groups acting together dug crude irrigation ditches to divert the spring waters into the fields. By this simple act, a permanent source of water for farming was secured for the first time in the Valley of Mexico. The momentous discovery lured other farmers to the valley floor. Steady harvests over the years encouraged a population boom. With more food available, more infants were allowed to live. A growth cycle was set in motion: more extensive irrigation led to greater food production; abundant crops could support greater numbers of people; the larger population demanded even bigger harvests; and to increase food output, more laborers were needed to build more irrigation projects. (18)

At its apogee (around A.D. 500), Teotihuacán may have been inhabited by as many as 150,000 to 200,000 people (Vitale 1983, 60). It was North America's first planned city. Stuart (1988) notes that "Teotihuacán was the most highly urbanized center of its time in the entire New World, and in size and splendor it surpassed Old World cities of later times" (136). The highly populated city was supplied with food (fruit and vegetables as well as the staple maize) from diverse local sources. Irrigation in the valley fields and hillsides plus a system of highly productive *chinampas* on the lakes and swamps proved to be exceptionally productive.

Tenochtitlán and the Chinampas

Reid (1985, 183) describes the *chinampa* form of agricultural production as an agriecosystem. She points out that this unique method of sustainable cultivation "optimized the exploitation of the Valley of Mexico's environmental resources, such as its hydrology, topography and especially the use of water and the soils" (183). There is evidence that *chinampas* existed as long ago as A.D. 100. However, the most vigorous development of this agriecosystem seems to have begun during the thirteenth century (Garavaglia 1992, 572). From the early fourteenth to the early sixteenth centuries, the Aztecs used *chinampas* on a massive and unprecedented scale in their capital city of Tenochtitlán (Vaillant 1972). They created an elaborate system of aqueducts to carry spring water from higher elevations in the southern part of the Central Mexican Basin to their city situated on land reclaimed from the saline Lake Texcoco (National Research Center 1995, 2). At its height, the island city of Tenochtitlán had 200,000 people. Tenochtitlán and nearby cities dramatically expanded after a dike was constructed across Lake Texcoco in the mid-fifteenth century (see figure 4.2).

Berdan (1982) points out: "Whenever conditions were favorable to chinampa cultivation, lakeside and island cities expanded their settlements by continually dredging new canals and building up the rectangular plots. When a cultivator built a house on the chinampa, the plot took on the

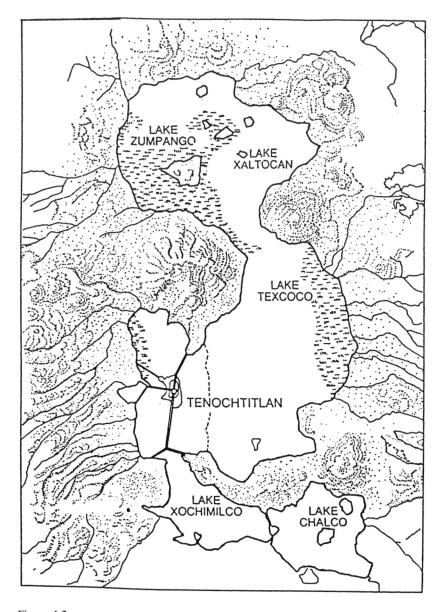

Figure 4.2
The Valley of Mexico showing the island of Tenochtitlán in Lake Texcoco, 1519.
Source: Jeremy Sabloff (1989).

added dimension of a residential site" (23). The advanced development of the "floating gardens" revolutionized agriculture in the Valley of Mexico. As Kandell (1988) states: "[t]he chinampas were an ingenious solution to a dilemma that had vexed the inhabitants of the Valley of Mexico for more than a millennium: since the low-lying lake waters could not be guided to the fertile land on slightly higher ground, then the land would be brought to the water" (28). Located on an urban grid pattern, the *chinampa* plots were usually adjacent to the dwelling units of the families that cultivated them. Berdan (1982, 21) notes that in Tenochtitlán the *chinampas* ranged in size from 100 to 850 square meters and were occupied by as few as two to three and as many as twenty-five to thirty persons, with an average of ten to fifteen persons per *chinampa*. These gardens were built up in the lakes from canal silt and marsh weeds.[8] They were regularly fertilized with human manure and irrigated by the canal.

Remarkably, *chinampas* continued to be a major source of vegetables for Mexico City until the 1930s (Garavaglia 1992, 573). On the basis of their agronomic potential alone, it makes sense to resurrect what remains of the *chinampas* in the valley. This much is even recognized by the current Mexican administration, which has launched a major initiative to restore what little remains of the floating gardens (Ramírez Saiz 1992; Rocha 1992).[9] More important though, as Redclift (1987) argues in his book about sustainable development, *chinampas* raise questions that go beyond measures of agronomic potential: "The accounts of pre-Columbian sustainable agriculture should also lead us to more fundamental questions about 'development' itself, and the role of the environment in the development process. Should we dignify with the term 'development' a process which leads millions of people to sacrifice their health and energies to survival? Perhaps an ecological alternative lies not so much in learning things we do not know, as in 'unlearning' things we do know" (110).

Redclift raises an excellent point. Here is where a historical-ecological perspective and the concept of entropy can be useful. First, a point of clarification: To suggest that the historical societies of the Aztecs and Teotihuicanos may offer lessons about ecological sustainability for the present does not mean that these societies were sustainable themselves. In fact, there is evidence that both societies gave rise to major environmental problems—some of the very same problems that threaten the valley today.

As Garavaglia (1992) points out, it is a mistake to identify the Spanish conquistadores as the only causal factor leading to the collapse of the Aztec Empire. Relations between Aztec society and the environment were in crisis before the Europeans ever arrived on the scene. The people were thus considerably weakened and susceptible to the crushing domination of the conquistadores. Similarly, the collapse of the Teotihuacán Empire was linked to problems arising from the overuse of irrigation and the consequent failure of the agricultural base (Ponting 1991, 78). The depletion of aquifers and the intense slashing and burning of hillside forests, either to expand acreage for farming or to extract wood for construction, contributed to the empire's collapse (Kandell 1988, 22). Deforestation of the hillsides around the city of Teotihuacán had diminished the watershed's capacity to absorb rain. Some of these very same processes are occurring in the zone of Ajusco today (Wilk 1991).

The Historic Shift in the City's Primary Source of Environmental Low Entropy: From Solar Flows to Terrestial Stocks

The recent eradication of *chinampas* in Ixtapalapa, as documented by Reid (1985), is a metaphor for the larger ongoing transformation of the valley. The last remnants of *chinampas* in Ixtapalapa (a southern ward in the Federal District) were covered over in 1980 to make way for the Central de Abastos (Mexico City's vast central supply depot) a facility teaming with fuel-guzzling, carbon-spewing, inbound and outbound delivery trucks. Some 50,000 trucks deliver 25,000 tons of food to Mexico City on a daily basis.

From the years of the Porfiriato—an important transitional period in Mexican history (1876–1911), during which Porfirio Díaz served eight times as the nation's president—up through Mexico's so-called miracle decades (1940s–1970s), Mexico City was especially favored in the national push for industrial modernization. During the Miguel Alemán Valdés presidency (1946–1952) the capital became the hub of a national transportation network. This development involved major investments in airport facilities and railway lines and a fourfold extension of asphalted highways (Kandell 1988, 494). By the 1940s, Mexico City had the largest consumer market in the country, the most numerous labor force, the greatest concentration of entrepreneurial and managerial talent, the banks and government agencies upon which business depended for credit and permits, and an advanced urban infrastructure (494).

The displacement of Mexico City's *chinampas* by the Central de Abastos in 1980 is simply another indicator of how the valley is undergoing a process of urbanization that depends less and less on flows of energy from the sun as its primary source of environmental low entropy. Instead, urban growth gets a rising portion of its energy from sources stored in the earth's terrestrial stocks. This broad shift has set in motion a series of dynamic trends, two of which are relevant to story told here. First, at the macrolevel, the diminution of solar-powered production systems and the expansion of fossil-powered production systems have become locked in a positive feedback cycle. As Bormann (1976) points out, "Increased consumption of fossil energy means increased stress on natural systems, which in turn means still more consumption of fossil energy to replace lost natural functions if the quality of life is to be maintained" (cited in Ehrlich and Roughgarden 1987, 606). The second dynamic trend concerns the progressive loss of bioregional self-sufficiency in the valley. For instance, during the time of the Aztecs, when harvests in the valley were bountiful, the *chinampas* alone provided between one-half and two-thirds of the food requirement of Tenochtitlán. Kandell (1988) notes that "Usually no more than 10 to 15 percent of Tenochtitlán's food came from outside the valley, because beyond the reach of the lakes goods could be moved only on the backs of men, who themselves had to be fed along the way" (59).

The past several decades of urban growth in the valley have taken a toll on agricultural productivity. In their study of growth and its consequences in the southern part of Mexico City, where Ajusco is located, Robinson and Wilk (1991, 1) point out that as urbanization spreads from the city center up the sides of the valley, it is displacing agricultural fields and the grasslands and forests that grow in the higher elevations. They estimate that 50 percent of the urban growth that took place in Tlalpan from 1959 to 1989 occurred on agricultural land. The changes contribute to an increasingly serious water shortage, excessive runoff and soil erosion, and scarcity of productive agricultural lands (2).

The effective demand for food in Mexico City—which constitutes 50 percent of the national total—has spurred the growth of agribusiness. Marketing to consumers in Mexico City and elsewhere, agribusiness has tended to produce and supply food products that consume an increasing amount of energy in the processing and packaging process (Zepeda and Pérez Cota 1988).[10] In the urban milieu, the by-products of this process add to the waste stream.

A rising portion of the waste stream is inorganic and in the form of glass, metal, and plastic (Castillo, Camarena, and Ziccardi 1987). One study reports that in 1950, Mexico City produced 370 grams of garbage per day, per capita (predominantly biodegradable). By 1992 this amount had risen to 1,000 grams. Whereas in the 1950s only 5 percent of the waste generated was nonbiodegradable, by 1992 the nonbiodegradable portion surpassed 40 percent (Mexico 1992). On the whole, The Federal District produces 11,000 tons of solid waste per day,[11] only 75 percent of which is collected (Barry 1992, 270). Another 7,000 tons are generated daily in the rest of the metropolitan area, for a total of 18,000 tons per day. It is projected that soon after the turn of the century 27,000 tons will be produced daily (48 percent in the Federal District; 52 percent in the contiguous municipalities of the states of Mexico and Morelos (Mexico 1992, 14).

The rising tide in Mexico City's waste stream is just one example that illustrates the pervasive and ecologically damaging shift that has occurred from closed loop systems to open loop systems (Smit and Nasr 1992). In terms of "throughput," González Salazar (1990, 68) cites evidence that each person requires, on a daily basis, roughly 300 liters of water, 2 kilograms of food, and 5 kilograms of fuel. The waste stream from this flow of resource use includes at least 250 liters of drainage, 1 kilogram of garbage, and 0.5 kilogram of atmospheric contamination per day, per person (68).[12] There is a growing awareness that this throughput of resources, as it currently takes place in the valley, is not sustainable (Castillo, Camarena, and Ziccardi 1987).

To convey the image of contemporary urban development as an open loop system that "wastes" resources, Rees (1992) employs the metaphor of an ecological black hole. He writes, "However brilliant its economic star, every city is an ecological black hole drawing on the material resources and productivity of a vast and scattered hinterland many times the size of the city itself" (125). Rees (1992) cites evidence that the primary consumption of food, wood, fuel, waste-processing capacity, and so on, co-opts on a continuous basis several hectares of productive ecosystem for each inhabitant. This observation puts the question of urban-ecological sustainability in a new light. Rees argues that rather than ask what population a particular region can support sustainably, we should pose the critical question: "How much land in various catagories is required to support the region's population indefinitely at a given material standard?" (125).

The rise of agribusiness and the availability of cheap imported food has contributed to the demise of the *chinampa* form of production; it has also undermined the practice of traditional ejidal and communal subsistence farming (Zepeda and Pérez Cota 1988). This observation, as I will discuss in some detail further on, is crucial with respect to understanding the urban-agrarian conflict that underlies the Ajusco case.

City-Hinterland Linkages and Nutrient Cycles

Given that the continuous availability of food surplus and other resources from hinterlands is so crucial to the rise of cities, the ongoing reliability of said surplus and rural resource flows must figure significantly into any discussion of sustainability in the urban context. The observations that follow suggest that there should be support for urban agriculture, one of the principal arguments in the colonia ecológica productiva proposal.

The growth of agribusiness is part of a historic shift wherein agriculture has been promoted as an "adjunct to industry" (Nuccio 1991). Nuccio relates this shift to the globalization of production. By the 1970s, a new mode of agricultural production in Mexico began to emphasize commercial contracting and technological packaging of whole crops industries such as strawberries, asparagus, cucumbers, and tomatoes. Nuccio (1991, 110) argues that "the reduction of agriculture to the adjunct of industry and the internationalization of agribusiness to the production of exportable fruits and vegetables and of feedgrains contributed to Mexico's crisis of the 1980s." More specifically, it led to policies that held down the price of tortillas to subsidize the food imports into the fast-growing cities. This measure depressed the price of corn and undermined the viability of rural livelihoods thereby adding to the flow of urban-job seekers.

In addition to spurring rural-to-urban demographic flows, agriculture as an adjunct to industry has put Mexico on a "food security tightrope" (Rhodes 1991). The cultivation of old land races (traditional varieties of grains, fruits, and vegetables that have their origins in wild plants first discovered thousands of years ago) is being abandoned in favor of cultivating modern "improved" varieties. This large-scale shift to improved (genetically altered) varieties, often cultivated in single variety monocultures, has greatly increased the potential for widespread crop failure. The trend toward planting one strain instead of many varieties creates a situation in which modern plant breeders have less and less margin for error (Rhodes 1991, 83). The green revolution of the 1960s gave the world "miracle" seeds that were resistant to insects and disease and that yielded additional tons of grains per year. However, as Rhodes (1991) points out, the miracle seeds were not perfect and the risks associated with raising them have increased: "Opportunistic insects and viruses mutated and unlocked the genetic resistance of the new seeds. The pests sent scientists scurrying, searching for genes to withstand the threats. They have been successful so far. Meanwhile, the old varieties and wild plants are disappearing from many places, replaced by improved crops that are genetically uniform" (88).

The loss of naturally diverse germ plasm, much of it not yet studied, is called genetic erosion. Of course, there are more factors causing genetic erosion than agricultural practices. Rhodes (1991) estimates that by the middle of the next century, one-quarter of the world's 250,000 plant species may vanish as a result of deforestation, the shift to monocultures, overgrazing, water control projects, and urbanization (79).

As this is not the place to elaborate on all the problems associated with modern agriculture and genetic erosion, the point to stress here concerns city-hinterland linkages. In terms of its capacity to continue producing surplus, how secure is the rural production process that has enabled the phenomenal growth of Mexico City? In short, there is mounting concern that it is not secure for both ecological and social reasons (Barry 1995).

In her article about global warming and climate change in Mexico, Liverman (1992) argues that declining conditions in rural Mexico are already associated with increasing unemployment, social unrest, and migration to cities and the United States (75). In addition to the problems stemming from genetic erosion and the transformation of agriculture into an adjunct of industry, Mexico's food security is at risk because of the large-scale shift out of basic grain production into production of export forage crops. Furthermore, there is scientific evidence that recent years in Mexico have seen delayed rainy seasons and hotter summers.[13] Liverman (1992) notes that the perceived climate changes have been linked to deforestation and to the prediction of global warming associated with the greenhouse effect. In view of these changes, Liverman (1992) draws attention to the resource-fulness of Mexican farmers who have developed a variety of technical and social means to cope with the added stress. However, Liverman adds, "population growth, environmental change and most importantly eco-nomic and political transformations are now reducing the effectiveness of these traditional mechanisms and increasing the vulnerability to drought. More recent adaptations to drought include shifts into wage labor and out-migration" (1992, 75).

Also of growing concern are the activities of timber companies and ranchers, within as well as outside the Valley of Mexico, that have degraded biologically diverse, soil-protecting forest into eroded and desic-cated landscaped (Liverman 1992; Wilk 1991c). The state of Mexico, which includes seventeen municipalities that make up the MAMC, is being deforested at an average annual rate of 2,650 hectares (Programa Estatal de Protección al Ambiente 1996–1999, Gobierno del Estado de Mexico). From a rural perspective, such concerns have prompted researchers to advocate reductions in poverty, more efficient use of water, redistribution of land and resources, and support for undercapitalized but efficient meth-ods of traditional agriculture. The deforestation crisis has led some authors to argue that there can be no urban-ecological sustainability without first securing rural sustainability. Along these lines, Goldrich and Carruthers (1992) argue that "the revitalization of peasant communities and the social and ecological reconstruction of the countryside offer the most promising strategy for reversing the cycle of poverty and ecological destruction, thus transforming crisis into opportunity" (98). More specifically, Goldrich and Carruthers (1992) argue: "Underused lands and rural unemployment provide a point of departure for an alternative, ecologically sustainable development strategy built on two mutually enhancing bases. First, small-scale production to meet basic needs can restore viability to village and small-town life, promote food self-sufficiency, and contribute to an agricultural, demand-led strengthening of internal markets. Second, the idling of the Mexican workforce and the migration it creates can be reduced through widespread labor-intensive environmental reconstruction" (110).

The position articulated by Goldrich and Carruthers (1992) makes a lot of sense; it even has growing support in theory and practice (see Fox and Gordillo 1989). However, the degree to which rural revitalization can help ameliorate contemporary urban and regional development problems is often overstated. The creation of alternative rural livelihoods is crucial, but on its own inadequate to meeting the tidal swell in Mexico's economically active population. It is important to take into account city-hinterland linkages and to strive for greater integration between the two regions.

One area that shows potential for mixing otherwise separate urban and rural activities is food production. As I have already noted, there is growing interest in the potential of urban agriculture (for food as well as fuel production) and in methods of resource reuse and recycling. These themes were advanced by the professionals and ecological squatters in Ajusco's colonia ecológica productiva movement. Along similar lines, Smit and Nasr (1992) argue that "urban agriculture is the largest and most efficient tool available to transform urban wastes into food and jobs, with by-products of an improved living environment, better public health, energy savings, natural resource savings, land and water savings and urban management cost reductions" (152). Furthermore, Smit and Nasr state that

[C]ities can be transformed from being only consumers of food and other agricultural products into important resource-conserving, health-improving, sustainable generators of these products. In particular, agriculture in towns, cities and metropolitan areas can convert urban wastes into resources, put vacant and underutilized areas into productive use, and conserve natural resources outside cities while improving the environments for urban living. . . . [S]ustainable cities require an economic process to close the open loop system where consumables are imported into the urban areas and their remainders and packaging dumped as waste into the bioregion and biosphere. Thus, the "through-put" of resources by towns and cities needs to be reduced. (141)

Interest in urban agriculture and the notion of closed loop development stems from an awareness of problems of ecology and development rooted in the contemporary division between urban and rural functions. There is a strong case for the argument that contemporary urban development must begin to close the loop on resource use. Along these lines, Douglass (1987, 37) contends that trends in the use of the world's resources are undermining the very basis for sustaining cities; thus, it is imperative to promote urban resource ecologies (systems involving solar and biomass energy, water reclamation, composting, urban gardens, and organic and inorganic resource recycling). This call for urban resource ecologies recognizes that "cities will be increasingly pressed to renew themselves, provide food in their hinterland, regenerate their resource base, and to recycle their own wastes. This may see an urban life that oddly revives the past—composting, vegetable gardens—while using the best of new technologies" (Douglass 1987, 37).

As Mexico City enters the twenty-first century, it will be obligated to develop energy efficiency and to regenerate or sustain its resource base, including local ecosystems. Certain aspects of the closed loop, urban-ecological systems of past societies examined in this chapter suggest some possible approaches. For instance, closed loop production processes can be usefully incorporated into housing environments; that is, benefit can be derived from an approach to housing as a place for production and income/resource (re)generation as well as a place for consumption and reproduction (Mena Abraham 1990). This argument was advanced by the proponents of the colonia ecológica productiva initiative in Ajusco. This book will examine the merits of that argument but not in a way divorced from the valley's biogeophysical and historical context, nor from the power relations that run through its socioeconomic and political structures. The next two sections of this chapter thus draw together environmental, historical, and sociopolitical dimensions of two elements fundamental to ecology and development in the Valley of Mexico: water and land.

The War on Water

Water has become such a critical planning challenge in Mexico City, particularly with respect to irregular settlement and the ecological conservation of Ajusco, that it is necessary to give it careful consideration. Dynamics leading to today's water crisis in the Valley of Mexico can be traced back over 400 years. Comparing Spanish colonial society to the pre-Colombian society of the Aztecs, Garavaglia (1992) notes that a radical transformation took place with respect to water. He describes how the Spaniards, who had seized control of the Aztec island city of Tenochtitlán in the early sixteenth century, chose to battle the water rather than live with it.

The Aztecs had worked out a productive urban-hydraulic system in Tenochtitlán that utilized *chinampas* and that enabled them to work and live with the water. Garavaglia (1992) asserts that the people of pre-Columbian societies had learned to accept the water's many benefits together with its threat of floods. Water was considered to be the source of almost all the good things in life—reflected in the important role played by the water divinities in the Aztecs religious pantheon. Garavaglia (1992) notes that the attitude of the Spaniards (most of whom arrived from parts of Spain where water was in very short supply) was the exact opposite of the Aztecs. The Spaniards waged war against the lake system from the first moment they confronted it: "Amid the bloodshed of the first invasion of Tenochtitlán, the [Spaniards] destroyed causeways, canals, locks and strips of cultivated land. Next it was the turn of the beautiful gardens with fountains kept by the chieftains and nobility in the towns of the valley" (Garavaglia 1992, 574). Although this initial destruction took a heavy toll, the subsequent and gradual drying out of the lake system proved to be even more disruptive. Garavaglia describes the impact:

[T]he slower process (which was, in the long run, more destructive of the environment that the inhabitants of the valley had slowly remodelled over a period of several millennia) was the gradual but uninterrupted task of drying out the valley's lake system. The invaders had no desire to live with the water and decided to find an outlet for it from the endorheic basin in order to have done with the flood and . . . all the rest. This was the background to the Drainage of Huehuetoca project, which dragged on for several centuries before it began to speed up the drying out of the basin by giving it an outlet into the River Tula. (574)

The 1846 discovery of potable groundwater under artisian pressure set off a well-drilling furor (Orozco y Berra [1854] 1973). From the time of that discovery to the present, "a combination of increasing ground water extraction and artificial diversions to drain the valley resulted in the drying of many natural springs, shrinking of lakes, and a loss of pressure and subsequent consolidation of the lacustrine clay formation on which the city is built" (National Research Council 1995, chap. 3). All of these changes have had a profound impact on ecology and development in the valley. Besides undermining the *chinampa* area as a viable source of livelihoods, water extraction and diversion processes have caused exposure of land at the bottom of the lakes and swamps to forces of erosion. In the dry seasons, winds blow the dust and detritus lying on the surface into the air, causing dust storms. Health-threatening dust clouds *(tolvaneras)* have been a recurrent problem in the valley, especially in places like Netzahua-coyotl, where over 2 million people built their settlements on desiccated land that used to be covered by nitrogenous swampland and Lake Texcoco. The desiccation of the lake system over the past four centuries is shown in figure 4.3.

The valley's high altitude and subsoil conditions, together with ongoing deforestation,[14] rapid population growth, and the valley's ever expanding built-up area including pavement and buildings are all major factors con-tributing to the water crisis. Urban and agricultural demand for water now exceeds the supply made available by the valley's hydrological cycle. As the draw on water from the valley's aquifers has increased (to forty cubic meters per second), the rate at which it is replenished has decreased (to twenty-three cubic meters per second). Urban sprawl has diminished the land's capacity to absorb runoff generated from a rainstorm. Over the past decade, the decline in the recharge capacity of the subwatershed in Tlalpan, particularly in the foothills of Ajusco, has received considerable attention. In one study for the Lincoln Institute of Land Policy, David Wilk (1991c) analyzed land conversion and total runoff generated by the different land uses in Tlalpan's northern subwatershed (which includes Los Belvederes in Ajusco).

Taking into account land use changes from 1959 to 1989, Wilk found that there has been a geometric increase in runoff. From 1959 to 1985, the runoff increased from 0.291 inches (14.5 percent of the rainstorm) to 0.410 inches (20.5 percent of the rainstorm). By 1989, Wilk estimates, the runoff reached 0.460 inches (23 percent of the rainstorm). Underscoring the sig-nificance of these figures, Wilk notes:

These figures are quite significant if we consider that (a) less than forty percent of the watershed is urbanized, and yet, the runoff generated as a percent of total rainstorm is considerably higher than the average for the Valley of Mexico, and (b) the annual average increase in runoff during the last four years is almost three times higher than the annual average increase in the previous twenty-six years. The net loss of surface water due to urban coverage has, therefore, the potential to affect climatic conditions and restrict the availability of groundwater resources in the future. (1991c, 8)

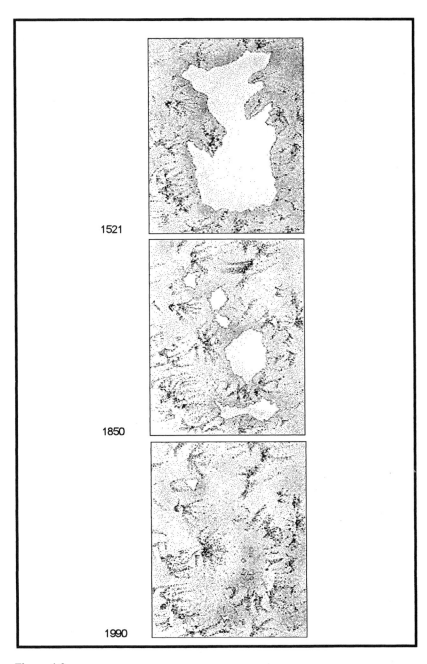

1521

1850

1990

Figure 4.3
The desiccation of the Valley of Mexico, 1521–1990. Source: Comisión Metropolitana (1992).

The dramatic increase in the total runoff from 1985 to 1989 corresponds to the consolidation of the built environment, including the pavement of streets, throughout Los Belvederes. This trend has continued with the spread of irregular settlements into new parts of the ecological reserve in Ajusco Medio between 1989 to 1996 (notably Tlalmille, Actocpa, Primavera, Verano, Los Zorros, Chimilli, and Héroes de 1810 among others) (Quadri de la Torre 1993b, 37).

This section of the chapter has drawn attention to the historical roots and dimensions of Mexico City's water crisis. The next section provides a historical-comparative perspective with respect to another vital aspect of ecology and development in the valley: land.

Land and Livelihood

The Spaniards' "war on water" was indicative of a more pervasive set of European attitudes and practices with respect to society-nature relations. Garavaglia (1992, 575) makes this point well, noting that the people of Mesoamerican societies had a completely different kind of relationship with the environment. Garavaglia is careful not to romanticize the society of Mesoamericans or to suggest that their societies were models of urban-ecological sustainability. Indeed, he cites evidence that the valley was undergoing relatively serious environmental degradation prior to the arrival of the Spanish conquistadores.

Nonetheless, when compared to those of the Europeans, Mesoamerican accomplishments are noteworthy; they offer lessons for the present. Garavaglia (1992, 575) notes that pre-Colombian societies "did not seem to be out to dominate nature but rather to form part of it and merge with it. This approach is in stark contrast to the Western aim of controlling and subjugating nature." This contrast is evident in the war on water as well as in patterns of resource use that continue to this day. Along these lines, Redclift (1987) notes the growth of Mexico City dictated a quite different pattern of resource use from that prevalent in Mesoamerican societies. He writes:

In the space of four centuries an ecological system built on sustainable agriculture (chinampas and terrace cultivation) had been replaced by one in which labor and land become first separated and then recombined (the hacienda community) under a new technology (plough and animal traction). The objective, to ensure control over economic surplus, was achieved at the cost of destroying the indigenous ecological and cultural systems. . . . Mexico City became the material representation of a new kind of development, in which in the short term economic growth and the carrying capacity of the ecological system increased, but which ultimately undermined the long-term stability of the resource base on which the rural population depended. (1987, 111–112)

By the eighteenth century, the establishment of haciendas (large estates) in the Valley of Mexico created a situation in which indigenous people were deprived of land and pushed into increasingly marginal areas where subsistence production was practiced (Redclift 1987, 111). This historical dynamic underpins the land use conflicts and urban-ecological problematic in Ajusco today.

Understanding the process of ejido urbanization is fundamental for grasping the means by which land is allocated in Mexico City (Díaz Barriga 1993). From 1916 to 1961, seventy-five ejidos were created within the Federal District, totaling 19,273 hectares. By 1976, 80 percent of the ejido lands (15,358 hectares) had already been absorbed by urban growth (1993, 2). Between 1940 and 1975 approximately 50 percent of the city's expansion occurred on ejido, or communal, land. In 1983, of all the land in Mexico City that was still possible to urbanize, approximately 80 percent had communal, or ejidal, tenure, 16 percent had private tenure, and 4 percent had public tenure (Calderón 1986, 11).

One of the large ejidos in Ajusco is called San Nicolás Totolalpán (henceforth abbreviated SNT). Prior to its becoming an ejido, SNT was a pueblo dating back to 1535. The historical context out of which this ejido emerged illustrates the larger changes in land use and livelihood structure that have taken place in the valley over the past several hundred years. In 1563, Viceroy Luis de Velasco of "New Spain" granted land use rights to the indigenous population of SNT (Rivera 1987, 24). Land grants *(mercedes de tierra)* of this sort were meant to protect indigenous forms of communal property and to thereby stabilize the indigenous labor forces that had been greatly weakened by the exploits of Spanish conquistadores and by the church (Mendieta 1985, chap. 7; Cockcroft 1983, chap. 1). Rodríguez and Rodríguez (1984, 32) in their historical investigation of land redistribution in Tlalpan, found that these titles gave the indigenous population "a means to protect themselves from the constant encroachment of the Spanish, who always managed to appropriate for themselves the best land." The property titles gave the indigenous population the opportunity to legally protest. To the present day these titles play a key role in the politics of land allocation in Ajusco.

From the mid-1800s on, inhabitants of SNT began to lose land and other common resources, including water supplies and timber, to the owner of the Hacienda Eslava (namely, Fernando de Teresa). They also lost land and resources to owners of the Haciendas El Arenal and La Canada, as well as to the owners of La Magdalena, a thread factory (Schteingart 1987, 459). Faced with these losses, campesinos of SNT supported Zapata during the Mexican Revolution (1910–1917). After the revolution, conflicts between SNT and nearby haciendas and factories continued. The campesinos protested and on July 23, 1923, the pueblo officially became an ejido by presidential decree.

The ejido of SNT was expanded in 1935. In 1939, it covered 1,354 hectares. By 1987 this area had been reduced to 758 hectares as a result of expropriations (Rivera 1987, 60). Presently, the ejido is composed of a large forested area, some agricultural fields, and a steep rocky area, where the settlements of Los Belvederes are located. Figure 4.4 shows ejidatarios

Figure 4.4
Ejidatarios of San Nicolás de Totolalpán, 1993. Note: The ejidatarios are digging a hole within which they will place a post to mark the boundary of their ejido. The boundary separates their ejido from the urban area of Los Belvederes.

of SNT digging a hole within which they will place a post to mark the boundary of their ejido. The boundary separates their ejido from the urban area of Los Belvederes.

In sum, the struggle for land in the zone of Ajusco dates back more than four centuries. Although the protagonists in the contest for control have changed over time—from indigenous groups versus the Spanish crown and church, to ejidatarios versus hacienda and factory owners, to more complicated arrangements involving ejidatarios with new social agents (including colonos, clandestine subdividers, and a host of new state agencies in charge of urban development)—the contest for control has commonly involved recourse to legal mechanisms and strategies. Increasingly, ecological arguments have figured into the politics of land allocation. Along these lines, Mexico City's shifting legal and institutional terrain represents new state-society relations. This point is pursued in chapter 5.

5

State-Society Relations and the Production of Human Settlements

Mexico's government has been the most stable regime in modern Latin American history (Cornelius and Craig 1988, 2). The same political party—the PRI—has governed the country since 1929. Mexico's political system has been viewed as an oasis of democracy. However, as Fox and Hernández (1992, 165) point out, this oasis has been a frustrating mirage for many Mexicans. In practice, the Mexican government lacks adequate checks and balances between the legislative and judicial authorities, both of which are dominated by the executive branch—notably the president, who operates with relatively few restraints on his authority. The electoral process has a long history, but it is rife with fraud and manipulation. Representatives affiliated with the PRI have managed to dominate continuously both houses of the federal legislature (Cornelius and Craig 1988, 15). There is a long history associated with the rise and fall of social movements for political rights and "effective suffrage," but the results have been mixed.

Overall, the Mexican state has successfully sustained its domination over the channels that have linked it to civil society at the national level. Yet, the state represents an ensemble of interests from within as well as from outside the regime. The state is not under the control of any one segment of society in particular, although the middle class and entrepreneurs do have greater influence. As Cornelius and Craig (1988) point out, Mexico's state is indeed strong, but it does not so dominate civil society that it is completely autonomous:

The Mexican state cannot rule in open defiance of the rest of society (as, for example, the Chilean military regime has done most of the time since it came to power in 1973) . . . the state is not able to manipulate all opposition groups all the time, and the strongest of them at least can try to bargain with the government. The regime also strives to be inclusionary, incorporating the broadest possible range of social, economic, and political interests within the official party, its "mass" organizations, and opposition groups whose activities are sanctioned by the regimes (15–16)

Along similar lines, Fox and Hernández (1992, 167) point out that the government has adroitly managed to control most social and economic organizations, such as trade unions, peasant organizations, and business associations. Joining such organizations is often obligatory—that is, if the prospective member wishes to keep his or her job or position. And leadership positions in such organizations are typically reserved for party loyalists.

With respect to maintaining such tight control, the Mexican state's "success" can be attributed to its skillful combination of "carrots and sticks": government responses to popular movements for social reform and democracy have typically combined partial concessions with repression. When access to material gains is provided, it is often at the price of political subordination (Fox and Hernández 1992, 167). At the same time, the state has demonstrated a remarkable capacity to preempt manifest conflict through political mediation and bargaining (Ward 1986). Cornelius and Craig (1988, 16) observe that "As potentially dissident groups have appeared, their leaders usually have been co-opted into government-controlled organizations, or the state has established new organizations as vehicles for emerging interests." Nowhere is this more evident than in the processes of political and social control exercized in Mexico City.

The Federal District

The Federal District is the seat of federal powers in Mexico. It is divided into sixteen politicoadministrative units called delegaciones or wards. Tlalpan, which contains the zone of Ajusco, is one of the Federal District's southern wards. The people living in the Federal District do not elect their representatives. In place of an elected mayor, the Mexican president appoints an official called the *regente*. The official justification for this arrangement is that the Federal District is a territory accountable to the entire republic. Because Mexican citizens do directly elect the nation's president, it is assumed that the president can rightfully appoint a mayor (the *regente*) to govern the Federal District. The president also appoints each of the ward's local administrators (comparable to city hall authorities), who are called *delegados*.

The regente heads the Federal District Department (comparable to a municipality's city hall). The regente has considerable influence (Ward 1989). Once inside the Federal District Department, this official fills key positions with select loyalists. Many delegados are nominated by the regente for confirmation by the executive. The regente also holds a high-ranking cabinet office. In view of this structure of political representation, or, more accurately, the lack of representation, Ward (1989, 308) argues that "Few places in the democratic world have less local democracy than Mexico City." This comment applies to the Federal District and to the contigious municipalities of the state of Mexico. Candidates for political office in the state of Mexico are usually selected by the president, governor, organized (compliant) labor groups, and the PRI itself. As Ward (1989, 310) notes, "The PRI orchestrates the election on behalf of candidates who emerge from this nondemocratic process and from behind closed doors."

Mexico's City's official agenda may embrace key tenets that have emerged out of the Earth and City summits, yet one of the most important ideals—that of participatory democracy—continues to be illusive. But that may be changing. The convention that grants Mexico's president the right to serve (in effect) as governor of the Federal District suffered a serious blow as a result of the 1988 elections. According to the final vote tallies, Salinas received very little support from the people in the metropolitan area. He gained only 27 percent of the vote in the Federal District whereas the opposition candidate, Cuauhtémoc Cárdenas, garnered 49 percent. The system of political representation has become the subject of heated debate in Mexico's upper and lower houses of Congress.[1] In his 1995 State of the Union address, President Zedillo announced that the people of the Federal District will eventually obtain the right to formally elect their own local government. This may be done by transforming the Federal District into the nation's thirty-second state (Hernández 1996).

In the meantime, Mexico City residents may participate in a hierarchically ordered neighborhood *consultative* structure that has existed in the Federal District since 1977. It includes one consultative council *(consejo consultivo)*, 16 neighborhood councils *(juntas de vecinos)*, 2,000 resident associations *(asociaciones de residentes)*, and 44,000 block chiefs *(jefes de manzanas)* (Ward 1990; Aguilar 1988). A more recent addition to the consultative structure is the Federal District Assembly of Representatives (Asamblea de Representantes del Distrito Federal ARDF), which comprises sixty-six elected officials. Established as part of the recent constitutional reform for the democratization of the Federal District (Carmona Lara 1991, 39), the ARDF acts as a "watchdog" assembly that demands greater accountability from city hall. It also proposes to Congress and to the city administration new laws, policies, and regulations. In terms of dealing with environmental concerns, the ARDF has been responsible for a number of significant initiatives concerning transportation planning and water use (Carmona Lara 1991, 39–40).[2] The prospects for the ARDF to speed any kind of democratic opening has to be understood in the context of Mexico's ongoing political restructuring.

Mexico's Political Restructuring

The stability of the Mexican political system, which has been characterized by its enormous capacity to control opposition and popular discontent through corporatist and populist mechanisms, can no longer be taken for granted. The crisis of the past decade has reduced the government's resources and legitimacy, thereby eroding its capacity to maintain traditional populist policies and to continue disciplining corporatist organizations. This is not to say that the populist legacy of the Mexican state can be written off. Rather, the political foundation of the "populist pact" of the past—which had been based on a unity of organized labor, peasant organizations, and popular sectors loyal to the dominant political party (PRI)—is undergoing significant changes. One of the most important changes is the transition from an emphasis on protest toward efforts to build actual social and political alternatives. Foweraker and Craig (1988a, 2) argue that "the proliferation of popular movements outside of (but also within) the traditional postrevolutionary, corporatist structures is one of the most significant developments in Mexican politics in the past twenty years." There are, however, no guarantees that the changes under way will necessarily conclude in favor of grassroots democratic forces or, for that matter, sustainable development.

The struggle for social transformation is constrained by the general level of organization and consciousness in the broader process of political conflict. The success of struggle depends not only on changes that may occur within the arena of electoral politics but also on the ongoing politicization of popular groups coalescing within civil society. Power relations are historically rooted deep in social structures and institutions; they are not reconstituted "above" society as a structure that could be radically transformed simply because some force or other wills it (Jessop 1985). In this respect, it is necessary to eschew the model of Leviathan in the analysis of power (Foucault 1983).

Opposition movements have indeed shaken the PRI apparatus to its core. However, as Peter Marris (1982) points out, the power to disrupt social and political relationships is not symmetrical with the power to establish new relationships and new social meaning. People have certain expectations forged over time about politics, politicians, the institutions of society, and how these all function. Popular groups with democratic agendas must oppose not only the state's coercive apparatus and judicial system, but also the fragmenting and depoliticizing ideologies of consumerism and possessive individualism, as well as more traditional codes of behavior including *clientelism*, *paternalism*, and *caciquismo*, which are deeply embedded in civil society. There is no necessary or automatic linkage between the democratization of social relations at the grassroots and the democratization of the state (Mainwaring 1987).

It is not the aim of this chapter to fully account for the shifts in constituencies that make up Mexico's political system,[3] but instead to stress that the populist arrangements of Mexico's past are under pressure, in part because popular urban movements reject such measures. As Foweraker (1989, 109) points out, "After 1968, the struggles of civil society, directed to a broad and implicitly democratic set of demands, discovered organizational forms and strategic capacities which the state has found difficult to counter and contain." The defining element of popular mobilization's most successful strategic initiatives has been its formation of political alliances. These alliances have become especially important given that civil society in Mexico suffers multiple sectorial, regional, political, and cultural cleavages and that "political control has traditionally been assured by the clientelistic relations which reinforce the divisions between its many and various constituencies" (Foweraker 1989, 111). The result has been an intensification in the contest to form the legal and institutional framework through which the organized actors of society assert their political agendas.

One of the areas where the impact of these changes can be seen is in the arena of state-society relations governing the urban poor's access to land and processes of irregular settlement. In the three decades following 1940, the informal (self-help) production of low-income settlements—based largely on illegal and semilegal means of access to land—alleviated an otherwise critical housing shortage (see table 5.1).

Two sets of forces now appear to be constraining informal modes of access to land. First, after a generation of hyperurban growth, the available supply of unbuilt land in Mexico City has become more scarce and as a consequence, more actively contested among potential users. Second, property development has become both more subject to state control and more commercialized. Compounding these constraints is the impact of the economic crisis and the politics of austerity.

These factors have made it increasingly difficult for the state to sustain what Castells (1983, 175) calls urban trade unionism or *urban populism*: "a process of establishing political legitimacy on the basis of popular mobilization supported by and aimed at the delivery of land housing and public services."[4] Consequently, an increasing number of popular groups are operating outside clientelistic channels of mediation. The devastating earthquakes in Mexico City in 1985 acted as a catalyst that significantly heightened the level of political mobilization of civil society (Ramírez Saiz 1986a; D. Rodríguez 1986).

Table 5.1
Housing production by sector, 1951–1980 (in thousands)

	1951–1960		1961–1970		1971–1980	
	Dwellings	%	Dwellings	%	Dwellings	%
All sectors	1,150.0	100.0	1,877.0	100.0	3,929.7	100.0
Public	62.0	5.4	175.0	9.3	735.8	18.7
Private	331.0	28.8	503.0	26.8	639.5	16.3
Informal	757.0	65.8	1,199.0	63.9	2,554.4	65.0

Source: Schteingart (1993, 293, table 13.2).

Some of the most powerful popular movements in Mexico City, including the National Coordinator of Popular Urban Movements (*Coordinadora Nacional del Movimiento Urbano; Popular CONAMUP*) and the Assembly of Barrios (*Asamblea de Barrios*) have focused their energy on the production of healthy housing environments for the urban poor. One of the most dramatic successes along these lines is the (re)construction of low-cost housing in the downtown section of Mexico City following the 1985 earthquakes (Connolly 1990; Eckstein 1990; Ward 1990). Grassroots groups with a focus on housing for the poor have had a significant impact on the political process in Mexico (Cornelius, Gentleman, and Smith, 1989). The personification of this heightening activism and its promise is colorfully represented in the image of a hero: Superbarrio (see figure 5.1). A large and growing literature documents the activities of popular movements in response to urban crisis (Portes 1989; Díaz Barriga 1990; Coulomb and Duhau 1989).

One of the most significant characteristics of recent social movements in Mexico City and around the world is the depth of their critique, along with the extent to which an alternative approach to development is sought (Walker 1990; Slater 1985). The critique stems from a view that the crisis in Mexico and throughout Latin America is not only economic and even less so only financial—it is a crisis in the mode of development with far-reaching ecological, as well as social, ideological, and political dimensions (Sunkel 1985; Aguilar 1985; Friedmann 1987, 1992; Sunkel and Giglo 1981).

In Mexico City, popular groups such as those working to build cooperative housing in El Molino (Suárez Paréyon 1987), or those engaged in promoting organic waste recycling systems and productive ecology in Ajusco (K. Pezzoli 1991; Schteingart 1987; Mena Abraham 1987), recognize that the urban housing crisis goes beyond issues of land allocation and management into broader issues of livelihood opportunities, developmental sustainability, and ecological balance.

Of course, popular groups do not have a monopoly on the discourse that draws attention to the profound depth of the political as well as ecological crisis in the capital. PRI officials also sound the alarm. For instance, the delegado of Tlalpan, Francisco Ríos Zertuche, made the following statement in a rather dramatic speech before the ARDF: "In our city we

Figure 5.1
Superbarrio.

have arrived at the limits: space, healthful air, water, environmental well-being, the capacity to protect and provide security for the population in the face of natural catastrophes, and the resources necessary to bring to completion any projects of revitalization or reorientation are all exhausted. Thus crisis is produced in the model of the city, in the apparatus of management designed for it, and in the paradigms through which it is sustained."[5]

During his speech Ríos Zertuche noted that Tlalpan has been severely impacted by the rapid expansion of irregular settlements. In fact, Ríos Zertuche claimed that out of all of the wards in the Federal District, Tlalpan, especially the miniregion of Ajusco, has the most severe urban-ecological problems. Moreover, the problems and the demand making by popular organizations in the region have placed the ward in an especially vulnerable situation. Tlalpan, Ríos Zertuche argued, is besieged by "constant conflicts." However, he notes that through "arduous work" his delegacion has contained much of this conflict by channeling it into "institutional outlets."

Popular groups that are mobilized to claim and defend land in Tlalpan and throughout Mexico are not new actors; they have been a significant element in the politics of urban land allocation since the 1950s (Perló 1981). However, the channels and institutional forms through which popular groups have historically acted as demand-making associations involving clientelistic relations with the state are undergoing significant changes (Ramírez Saiz 1986a). From the 1950s to the early 1970s, state-society relations in Mexico City could be characterized in terms of clientelism and co-optation (Cornelius 1975). During the course of the 1970s, state intervention in the irregular settlements processes spawned a multitude of agencies, often with overlapping functions. By the end of the 1970s these agencies were becoming more technocratic in their approach and in their handling of demand making (Ward 1986).

As links between the state and community groups have become more structured, the clientelistic and particularistic criteria for the allocation of housing resources have tended to be displaced by more formal, technocratic criteria (Ward 1986). In situations of conflict, popular groups have responded to this shift in criteria by making technical arguments a part of their countervailing strategy for gaining access to resources. Increasingly, such popular responses are not simply reactive. As the case of the colonia ecólogica productiva movement illustrates, they sometimes entail proactive initiatives. Indeed, researchers point out that an increasing number of popular groups have learned to use the government's legal and technical language to propose viable alternatives corresponding to their demands.[6]

Constructive initiatives promoted by popular groups have thus far been possible only at the level of neighborhoods or locally targeted plans, not for entire cities or regions, nor for the country as a whole (Ramírez Saiz 1990b, 234). Nonetheless, they represent a significant advance in extant forms of urban social struggle (234). The shifts in Mexico's legal and institutional terrain involve changes that present both new constraints and new opportunities.

A principal argument of this book is that state intervention in land use in Ajusco is more a reactive strategy to control popular struggle than a "rational" attempt to improve the urban environment. The creation of new institutions, programs, and laws aimed at protecting the environment in the periphery of the city has more to do with social control than it does with growth control or development, much less "sustainable development." However, it would be mistaken to argue that the state's ecological arguments and environmental policies are completely devoid of legitimate concern for the environment and public well-being, or that the politics of containment count solely as technologies of power employed by the state to control popular groups. First, legitimate concerns do have at least a part in the scenario. Second, politics of containment apply to more than popular groups. The state also acts to control developers, as well as itself—mediating contradictory and competing factions within its own shifting internal organization (Jessop 1982). It is a complicated subject for inquiry. My approach to the subject is to progressively build up a middle-range explanatory framework. I turn here to provide background on the legal and institutional context of housing, particularly with respect to "irregular settlements."

Types of Housing Production and Human Settlements

The Research Center for Development (Centro de Investigación para el Desarrollo, A.C.; CIDAC) notes that Mexico's urban housing market is divided into an upper, middle, and lower segment. The upper segment is accessible to those who earn ten times or greater the minimum salary. Private developers targeting this income level have been able to realize profits from investments in housing.[7] The middle segment of the housing market is accessible to those who earn between 2.5 times and 10 times the minimum salary. By far, people in this income range have accrued the most benefits from publicly subsidized housing programs and lines of credit. Since the early 1980s, the most important programs targeted at this middle segment have been the Fund for Banking Operations and Discounts to Housing (Fondo de Operación y Financiamiento Bancario a la Vivienda; FOVI) and the National Institute for the Fund for Workers' Housing (Instituto del Fondo Nacional para la Vivienda de los Trabajadores; INFONAVIT). FOVI targets those who have stable incomes of between three and ten times the minimum salary. INFONAVIT is a fund available to members of Mexico's labor unions (Schteingart 1989b).

The lowest segment of the housing market, which is the focus here, involves people whose earnings range from 2.5 times the minimum salary to less than the minimum salary. Most families fall into this category and thus are not able to purchase housing in the formal sector. In Mexico City, 47 percent of the families lack the income necessary to purchase or rent a dwelling unit available in the formal real estate sector (CIDAC 1991, 29). This segment of the housing market has the most severe deficiencies in urban services. At a national level, the 1990 census reports that 25.2 percent of all housing units do not have an indoor toilet, 20.6 percent lack potable water, 36.4 percent lack drainage, and 12.5 percent do not have electricity.[8]

There are several types of low income housing. The first two types, which I mention only briefly in order to provide historical context, are (1) *vecindades* (inner-city apartment buildings), and (2) the *ciudades perdidas* (translated as "lost cities," meaning inner-city slums). The great majority of *vecindades* have one or two levels with a central patio; they were built before 1940 and are located in central parts of Mexico's larger cities. Most have between twenty to fifty units composed of either one or two small rooms; bath and washroom facilities are shared among units. Not all units in *vecindades* are low income. In some cases, rents are relatively low because of rent control, but in others rents are higher (COPLAMAR 1982).

During the first few decades of this century, private sector *vecindades* housed most of the urban population. However, since 1940 the *vecindad* share of the total housing stock has greatly declined.[9] Investment in such rental units became unprofitable. In many cases, disinvestment and over-crowding have led to deplorable conditions (COPLAMAR 1982). As Gilbert and Varley (1991, 34) explain, "growing numbers of people, a changing investment climate, a modified response by the state to housing and land issues, changing technological restrictions on urban growth, and changing tastes in accommodation and urban living were all producing a situation which encouraged the widespread shift from renting to legal or *de facto* home-ownership."

First emerging as a significant feature of Mexico's urban landscape in the 1940s, around the same time that *vecindades* stopped keeping pace with demand for low-income, inner-city rental units, *ciudades perdidas* are densely crowded slum sections, lacking potable water and drainage, located in various parts of the inner city. "The majority of the dwellers are persons of very minimal resources, who at times rent waste materials in order to construct their housing" (COPLAMAR 1982, 28).

In addition to the two types just discussed, there are three other types of low-income housing environments in Mexico City, each of which involves "self-help." The nomenclature describing self-help settlements varies (Burgess 1985). Use of the term here includes (1) government-sponsored *sites-and-services*, and progressive housing projects, (2) colonias populares—by far the most common type of housing environment, and (3) colonias de paracaidistas—settlements in formation by way of land "invasion." Such invasions have become less common. In each of these three environments, houses are typically built by the initial owner-occupants. Over an extended period, the houses and neighborhood are upgraded. Of course, upgrading does not always occur; on occasion the government eradicates the settlement in question. It is estimated that 65 percent of the 2 million housing units standing in the Federal District in 1986 were created by means of self-help (DDF 1986, 40).

The only significant government program that has targeted those in the lowest segment of the housing market is the Trust Fund for Popular Housing (Fideicomiso Fondo de Habitaciones Populares, FONHAPO). FONHAPO received only 4 percent of the allocation of federal resources for housing programs from 1983 to 1988, but it was responsible for approximately 15 percent of the housing starts over the same period. By comparison, FOVI had 50 percent of the resources and realized only 30 percent of the housing starts (CIDAC 1991, 99). The difference can be attributed to the type of assistance. Most of the FOVI resources went into the production of more costly finished housing units, whereas most FON-HAPO funds went toward building *vivienda progresiva* (progressive hous-ing,—a simple core structure that can be gradually expanded by the owner-occupant), and *sites-and-services* (projects in which beneficiaries get lots—equipped with access to water, drainage, and electricity—upon which they may build their own dwelling units from scratch).[10]

Although sites-and-services and other progressive housing schemes are applauded by many researchers and by international development agen-cies, the government has sponsored few such projects in Mexico City.[11] Part of the reason for this lack stems from dynamics rooted in agrarian politics and institutional arrangements for dealing with land expropriation and land tenure regularization. In order to establish something like a sites-and-service project, the government would require land reserves. But there are contradictions within the state that throw up obstacles to the creation of such reserves. Although the call to establish land reserves is clearly emphasized in National Development Plans (Planes Nacionales de Desar-rollo, 1988–1994; 1995–2000), it is in the interest of certain state agen-cies—namely, the Secretariat of Agrarian Reform (Secretaría de Reforma Agraria; SRA) and the Commission for the Regularization of Land Tenure (CORETT)—to see that the irregular settlements process continues instead (Azuela de la Cueva and Duhau 1987, 59–60). The land transactions that follow the legalization of lots in irregular settlements provide these agen-cies with "grist for their mills." There is no incentive for them to undercut their "business" by assisting in the creation of land reserves.

The success of land tenure regularization schemes has been the undoing of any prospect for creating significant land reserves (Azuela de la Cueva and Duhau 1987, 59–60).[12] Additional reasons the government has failed to create land reserves for housing in Mexico City have to do with (1) the politics of containment, which place emphasis on establishing ecological reserves (Duhau 1991b), and (2) the neoliberal push for land privatazation and investment in big capital projects as a stimulus for urban development

(Pradilla Cobos 1995a). The lack of adequate land reserves helps to explain the dramatic growth of the city's colonias populares.

In the Federal District, Tlalpan has one of the greatest concentrations of low-income housing and deficiencies in urban services. In 1993, Ríos Zertuche (the delegado of Tlalpan) reported that his delegación had serious deficits in drainage and running water.[13] These deficits are mostly concentrated in the sprawling colonias populares of Ajusco Medio and Ajusco Bajo.[14] Ríos Zertuche notes that more than 200,000 people (roughly one-half of Tlalpan's total population) currently live in inadequately serviced, yet rapidly expanding colonias populares. Within the entire Federal District, 1,040 colonias populares lack drainage, creating conditions that negatively impact approximately 5 million people (Leff 1990a, 53).

The bulk of Mexico City's housing was produced through such irregular (self-help) settlement during the 1950s, 1960s, and 1970s. Gilbert and Varley (1991, 25) point out that during this period "Mexico was an archetypal Latin American society, its cities carefully divided into discrete social areas, affluent and attractive suburbs coexisting, albeit at a distance, with vast areas of poorly serviced unattractive low-income settlement." The expansion of self-help settlements has been dramatic. It is estimated that in 1947 only 2.3 percent of Mexico City's total housing stock was self-built. By 1952 this had risen to 22 percent and by 1975 more than 50 percent of the metropolitan area's population lived in one variety or another of colonia popular (Connolly 1982, 150). Over the past decade, an estimated 50,000 housing units per year have been produced by the informal sector in Mexico City (Hidalgo 1995). The local organization of colonias populares and their particular relationship to the state have been firmly integrated into the political system at large. Underlying the system of resource allocation by the government's dominant political party, the PRI, are strict but unwritten laws of exchange that demand votes and support through subordinated political participation.

The rapid expansion of irregular settlements has happened in practically every large Mexican city. Consequently, there has been a radical shift in the tenure of housing. During the 1940s most of Mexico's people still lived in rural areas, and most city dwellers rented their habitation. Presently, the great majority of people (71.3 percent) live in urban areas, and most are homeowners.[15] This shift in tenure resulted from a number of economic and technological factors, "but the single most important reason is a public policy that simultaneously depopulates the countryside (in support of commercial agriculture) and fosters urban home ownership via squatting or low-cost purchase of small plots and self-help construction" (Walton 1991, 629).

In sum, there are seven types of housing environments in Mexico's largest cities. The first two—(1) middle-class public housing, and (2) middle- and upper-class private developments—are not accessible to the great majority of city dwellers. Most urbanites have taken up residence in self-help settlements, including (3) sites-and-services and progressive housing projects, (4) colonias populares, and (5) invasions. Many others live in (6) vecindades, or (7) ciudades perdidas. Table 5.2 gives rough estimates of the percentage that these categories contain of Mexico City's total 1990 urban housing stock.

Economic Crisis and Land Management

The self-help mechanisms by which the urban poor have historically gained access to land, and gradually built and upgraded their dwellings and communities, have come under stress. In order for irregular settlement to con-

Table 5.2
Urban housing environments and their share of total urban housing stock in Mexico City, 1990

Type of housing environment	% share of total urban housing stock
Higher-income housing	
Middle-class public housing	10
Upper- and middle-class private developments	15
Lower-income housing	
Vecindades and *ciudades perdidas*	20
Colonias populares, sites-and-services, progressive housing, and invasions	55
Total	100

Sources: Estimates based on data from COPLAMAR (1982, 28–30), INEGI (1991), CIDAC (1991), Gilbert and Varley (1991), and Selby et al. (1990).

tinue taking the edge off an otherwise critical housing shortage, the urban poor must continue to have access to land, be it through legal or illegal means. Yet, access to land for the urban poor in Mexico City is increasingly constrained. The principal reason can be summed up as follows: Developers and speculators are consolidating control over unbuilt land supplies at urban fringes (Schteingart 1983; Dunkerley 1983). Consequently, the concept of land as a commodity is becoming more established. At the same time, the state is enforcing increasingly sophisticated administrative controls, including zoning laws and land use regulations. The era of extensive availability of land for low-income settlements has come to an end (Angel, Archer, and Wegelin 1983; Oberlander 1985; Habitat 1983, 1996a; Doebele 1987).

It may be that the threat of land commodification as a force undermining the capacity of informal, small-scale development has been exaggerated (Ward and Macoloo 1992). What is certain, however, is that after a generation of hyperurban growth in Mexico City, the available supply of unbuilt land has become increasingly scarce and as a consequence, more actively contested among potential users. Land is essential to all human activities. And, as Doebele (1987, 7) argues, "Like water in dry regions, it is a commodity that cries for public management and control." This sentiment is reflected in *Agenda 21,* one of the key documents to come out of the 1992 Earth Summit. *Agenda 21* calls for an integrated approach to the planning and management of land resources.[16] Text in the document reads:

Expanding human requirements and economic activities are placing ever increasing pressures on land resources, creating competition and conflicts and resulting in suboptimal use of both land and land resources. If, in the future, human requirements are to be met in a sustainable manner, it is now essential to resolve these conflicts and move towards more effective and efficient use of land and its natural resources. Integrated physical and land-use planning and management is an eminently practical way to achieve this.[17]

Although the promotion of integrated physical and land-use planning and management certainly makes good sense, the performance record of governments as effective managers of land has been poor. This is especially true in the case of Mexico City, considering that a large proportion of urban expansion occurred through incorporation of collectively held land. This situation, which provided favorable conditions for the government to control the land market and urban growth patterns, has served instead to obstruct orderly urban development and to encourage land speculation (Garza and Schteingart 1978; Legorreta 1992; Schteingart 1993).

The problem lies partly in the fact that every piece of land is unique and that therefore, land does not lend itself to the uniform procedures of bureaucracies. More significant, the special value of land, which is heavily dependent on socially created demand and publicly provided services, has created situations ripe for clientelism and corruption. In this respect, certain interests are served by tolerating, and sometimes by facilitating, the expansion of irregular settlements. For instance, the illegal alienation of land—by opening supplies otherwise not accessible to the majority of low-income people—has helped the government maintain political legitimacy and social stability. Also, it can be argued that the illegal nature of many land transactions keeps housing costs down, thereby helping to suppress the wage levels necessary for labor to reproduce itself (Connolly 1982).

In this context, it is not hard to understand why there has been political resistance on the part of dominant groups to changes that would fundamentally alter the nature of land markets and their accessibility. As Wayne Cornelius (1976, 265) remarked over two decades ago, "the political influence enjoyed by real estate interests, professional speculators, the construction industry, and their allies within the official party-government apparatus is such that most initiatives in this area can be blocked, or their implementation subverted in such a way as to greatly weaken their impact."

Land problems are inherently complex both in theory and in practice. Simple solutions are suspect. Dunkerley (1983) makes this point with respect to the widely advocated solution of public ownership. He notes that even public ownership of land or detailed control of its allocation cannot necessarily ensure efficient, equitable, or harmonious patterns of urban land development (4). An emergent constraint on access to land has been the state's *politics of containment,* defined earlier as the legal and institutional means aimed at controlling the horizontal expansion of irregular settlements on the basis of ecological arguments. In addition to mounting concern about protecting watersheds and space for agroforestry activities, a number of ecological disasters have also prompted the state to implement the politics of containment. For instance, in 1984 a PEMEX gas storage plant exploded, causing great loss of life and property damage in the communities surrounding the plant. The explosion took place in Tlalnepantla, an industrial area in the northwestern part of the city. Tlalnepantla was practically uninhabited in the 1950s, when PEMEX selected it as the site for one of its major gas storage facilities. But, as Kandell (1988, 565) observes, the benefits of the urban services put in place to service the plant (e.g., transportation, electricity, water grids) attracted thousands of squatters to the surrounding area. By the early eighties, several hundred thousand people had settled within a two-mile radius of the PEMEX plant. The same thing has happened around a score of other gas storage depots throughout the metropolitan area, thereby exposing many more hundreds of thousands of people to the threat of disaster. Kandell describes the disaster that struck Tlalnepantla:

At 5:42 A.M. on November 19, 1984, a truck parked in the Tlalnepantla plant exploded. Within seconds, four huge tanks holding more than 1.5 million gallons of liquefied gas erupted. Balls of flame shot up into the air, raining fiery debris on the shantytown. More than five hundred inhabitants perished, most of them so badly burned that they could not be identified. About five thousand people were injured, and a hundred thousand were evacuated. Sixty-six acres had been razed. (1988, 565)

The problems arising from such haphazard urban growth have prompted the government to stiffen its policy prohibiting land invasions and the formation of new colonias populares. Consequently, evictions have increased (Gilbert and Varley 1991, 49). As the horizontal expansion of Mexico City becomes increasingly constrained, there is evidence that parts of the city are experiencing a sort of *urban implosion*—defined by the Center for Housing and Urban Studies (Centro de Estudios Urbanos y de Vivienda; CENVI), as "demographic densification combined with a restricted territorial expansion."[18] CENVI reports that Mexico City's urban implosion has led to the increased subdivision of lots, the proliferation of rooms for rent, and overcrowding. As a response to the increasing demands for rental space, CENVI estimates that 35,000 rental units are added to Mexico City's informal housing stock each year. Most of these units are in the consolidated colonias populares of the city's periphery. This popular sector supply of rentals falls outside the auspice of government planning and the tax base; it an unregulated supply without any codes that specify construction norms or health and safety standards.[19]

In sum, conventional mechanisms of informal urban land development are increasingly constrained by socioeconomic, political, and environmental factors. The urban poor find it harder to gain access to land; the alternatives to irregular settlement are inadequate. As these trends continue, they are likely to exacerbate the segregation of the city into a number of protected, well-serviced upper-income sections on one side, and many more poorly serviced, increasingly overcrowded and unhealthy low-income housing environments on the other side. Such a dynamic is occurring in Tlalpan, especially in the miniregion of Ajusco, where ecological arguments and the politics of containment have found their most conflictive testing ground.

6

Ecological Arguments and the Politics of Containment

The legal regime that targets the environment in Mexico is an ensemble of laws, regulations, and standards implemented by a combination of federal, state, and local government authorities. Environmental planning in Mexico takes place within a civil law system that relies largely on administrative institutions and measures for interpreting and enforcing its laws (EPA 1992). In contrast, the U.S. legal system relies on juridical institutions and litigation as well as administrative regulation and review (9). Dispute settlement and law enforcement within Mexico's civil law tradition depends largely on administrative mechanisms and negotiation between parties. Consequently, there is greater relative power vested in the executive governmental bodies to take unilateral actions, and there is greater use of administrative rather than judicial authority to achieve enforcement (9).

The antecedents of environmental law and planning can be traced to Article 27 of Mexico's founding Constitution, which incorporated the concept of natural resource conservation (Carmona Lara 1991, 26). The current basis of Mexico's environmental protection program is the 1988 General Law of Ecological Balance and Environmental Protection (Ley General del Equilibrio Ecológico y la Protección al Ambiente known as the General Ecology Law Ley General de Ecología). It is beyond the scope of this chapter to provide a comprehensive review of this complicated legal and institutional terrain.[1] Rather, the aim here is to document the emergence of environmental planning in Mexico City with a focus on the origins and content of the *politics of containment*. The review is chronological with reference to the major turning points in the urbanization of the Ajusco greenbelt (see table 6.1). The emphasis here is on the period spanning the 1930s to the early 1990s. The situation through the mid 1990s is covered in part 4.

Table 6.1
Key legal and Institutional elements pertaining to the politics of containment

1936	President Cárdenas decrees that there be a Parque Nacional Cumbres del Ajusco.
1976	The Law of Urban Development in the Federal District is published.
1978	The Coordinating Commission for Land and Cattle Farming in the Federal District (COCODA) is created.
1980	The Federal District Master Plan composed of the General Plan for Urban Development and sixteen partial plans, one for each Ward, is approved. Ecoplan of the Federal District is created.
1982	A mounted police station at a site along the urban-ecological perimeter in Ajusco Medio is constructed.
1983	President de la Madrid presents the Program for the Development of the Metropolitan Area of Mexico City and the Central Region, and to execute it he establishes the planning committee for the Development of the Federal District (COPLADE) and Ecological Plan of the Federal District. The General Office of Urban Restructuring and Ecological Protection (DGRUPE) is created.
1984	The Program for the Conservation of Ajusco is announced. The Program for the Urban Restructuring and Ecological Protection of the Federal District (PRUPE) is announced. The Subdirectorate of Ajusco is created and its Program to Contain Urban Expansion initiated.
1985	The Program delimiting the Ecological Conservation Line is established. Measures to regularize land tenure in Los Belvederes are initiated.
1986	The General Program for the Urban Development of the Federal District is published. The Coordinating Commission for Rural Development in the Federal District (COCODER, formerly COCODA) announces its Land Use Management Program.
1987	A revised version of the General Program for the Urban Development, including the partial plans for each of the sixteen Wards is published.
1988	The General Law of Ecological Equilibrium and Environmental Protection is enacted. National Development Plan (1988–1994) is established. The National Solidarity Program (Pronasol) is established. The National Program for Ecological Protection is announced.
1989	Decree to establish the Ecological Park of Mexico City in Ajusco is made. The Program for the Management of the Ecological Conservation Zone is announced.
1992	The new Secretariat of Social Development (SEDESOL) is created.

1993   The Federal Human Settlements Law is revised.
1996   National Development Plan (1995–2000) is adopted.
        The Environmental Law of the Federal District is enacted.

Source: Compiled by the author from various government documents.

The Origins of Urban and Environmental Planning

When Mexico City lost its municipal status in 1928 it became the responsibility of the Federal District to establish the norms and regulations for urban planning in the capital. Over the period of 1928 to 1953, several important pieces of planning legislation were passed. In 1933, the Zoning and Planning Law of the Federal District and of the Territories of Baja California (Ley de Planeación y Zonificación del Distrito Federal) was passed. With modifications made in 1936, 1940, and 1953, this law provided the basis for planning until 1970. A planning commission was established, as were Federal District consultative councils and their local subdistrict equivalents *(delegaciones)* (Garza 1987, chap. 9).

From 1953 to 1970, there was practically no new planning legislation or initiatives (Ward 1986, 223). Around the early 1970s, Mexican environmental policy began to emerge as a distinct body of law. In their excellent overview of this subject, Mumme, Bath, and Assetto (1988) point out that under the leadership of President Luis Echeverría Alvarez (1970–1976), several significant initiatives were undertaken. Most notably, there was an enactment of an ordinary statute *(ley ordinaria)* entitled the Federal Law for the Prevention and Control of Environmental Contamination (Ley Federal para Prevenir y Controlar la Contaminación Ambiental), and, by presidential decree, the Subsecretariat of Environmental Improvement (Subsecretaría de Mejoramiento del Medio Ambiente) was created. Around the same time, the Constitution was reformed (art. 73, sec. 16) to incorporate the principle of preventing and controlling pollution (Carmona Lara 1991, 30).

These initiatives were in good measure prompted by Mexico City's rapid expansion, which had given rise to profound problems of urbanization and pollution. Up until the early 1970s, however, it was uncommon to hear of organized public activity protesting environmental conditions (Mumme, Bath, and Assetto 1988). Before the 1970s, Mexico City newspapers cast the issue in terms of the pains of urbanization, seldom reporting about these conditions in ecological terms (NEXOS 1983). Mumme, Bath, and Assetto (1988, 11) explain, "The principal impetus for governmental involvement in environmental issues came from a select group of academic and official research institutions attuned to international intellectual currents."[2] It was not until later that popular groups became a significant force pushing an explicitly environmental agenda.

Urban physical planning emerged in Mexico during the 1970s and was formally established in 1978. Prior to this implementation, planning was concerned with economic planning that comprised a process of mutual adjustment led by a central guidance cluster of four major institutions: the presidency, Ministry of Finance, Central Bank, and National Finance Institution. A single, centralized planning agency to fulfill this purpose was rejected by President Cárdenas (1934–1940) and was never again taken up.

Due to the increasing complexity of urban problems, physical planning gained importance on the political agenda during the 1970s. Politicians discovered that planning could serve as an ideological tool to enhance state control (Ward, 1986, 219). However, the status and authority of Mexico City's Planning Department (Departamento de Planeación del Desarrollo Urbano) has been constrained in two ways: first by its institutional structure and second by its low budget. As Ward (1986, 225) points out, at the institutional level there has been a proliferation of agencies and bureaucracy, often with overlapping or competing responsibility for planning. Competition between bureaucratic factions is standard fare in Mexico's governmental system. In terms of funding, planning has not been a major priority (227).

The 1936 Presidential Decree

Official measures aimed at protecting the ecology of the Valley of Mexico go back as far as the 1930s. The zone of Ajusco has stood out from the beginning. In 1936, President Lázaro Cárdenas issued an executive decree declaring the zone of Ajusco (and much more) a national park, Parque Nacional Cumbres del Ajusco. It is worth quoting a section of this decree because its content—aimed at securing ecological equilibrium—is so similar to the kind of arguments underpinning more contemporary environmental policies (see box 6.1). President Cárdenas called for a vast expropriation of land for the park, but the decree was never fully implemented. Around the same time, two other national parks were created within the Federal District: (1) the Cerro de la Estrella (initially 1,100 hectares, subsequently reduced to 143 hectares), and (2) the Parque Tepeyac (initially 312 hectares, subsequently reduced to 294 hectares). These parks are 3 of the nation's 197 officially designated ecological reserves (Páramo, Bermeo, and López 1995).

In 1947 President Miguel Alemán greatly reduced the extension of the Parque Nacional Cumbres del Ajusco, when he issued a decree that permitted the paper factory Loreta y Peña Pobre to exploit the zone's forest resources. Cárdenas had intended to set aside 920 hectares for the Ajusco park; today it has 181 hectares. It wasn't until the 1970s, with the emergence of physical planning, that any significant steps were actually undertaken in the name of ecological conservation.

The Emergence of Physical Planning during the 1970s

The First Law of Urban Development of the Federal District 1975
In December 1975, the first Law of Urban Development of the Federal District (Ley de Desarrollo Urbano del Distrito Federal LDUDF) was passed. The LDUDF established the legal basis to make mandatory the existence of a comprehensive citywide plan and subplans. Specifically, the law spelled out the requirement to set forth a master plan *(plan director)* that must include a general plan as well as sixteen partial plans, one for each of the *delegaciones*. The urban area was to have a proper zoning system—for which the ultimate responsibility resides in the Federal District Department (DDF).

The LDUDF created what has been one of the most important juridical instruments for urban planning and land use regulation in the Federal District (Gil Elizondo 1987). It laid down the juridical foundation for numerous plans, policies, programs, and institutions that have aimed, in part, to contain urban growth and environmental degradation. With legal support from the DDF Organic Law (Ley Organica; art. 13), the Constitution (art. 27), and the DDF Civil Code (Ley Organica; art. 830), the

Box 6.1
Excerpt from the 1936 Presidential Decree that Designated a Large Part of Ajusco as a National Park

---

Considering that mountain ranges of the national territory form the principal divisions of our valleys, occupied by populous cities, and that at the same time such ranges . . . contribute, in a considerable manner, to river flows, underground springs, and small lakes . . . thereby sustaining the valley's hydrological regime (that is as long as there continues to be forest cover, as there should be in order to prevent erosion of hillsides and to maintain climatic equilibrium), then it is therefore necessary that these mountain ranges, their forests, pastures, and cultivated fields, be efficiently protected, with the objective to form a sufficient protective ground cover guaranteeing the stability of climatic and biological conditions—a form of forestry conservation that cannot be achieved . . . if private interests linked to communal, ejidal or private property prevails.

. . . Ajusco, given its height and extension, as well as the diversity of its landscape, constitutes, within the national majestic panorama, a monument of exceptional and grandiose beauty. . . . Its lofty peaks are covered with trees adapted to a cool climate . . . and in the lower hillsides there is a fauna inhabited by wildlife giving Ajusco the virtues of a veritable nature museum. Considering all the mountain ranges of the national territory, . . . the Sierra de Ajusco is without doubt one of the most wondrous and one of the most significant for its great geological contrasts and for its proximity to the most populated center in the Republic. . . . At all costs this land must be protected against degradation, maintaining its forests in good shape . . . guaranteeing the provision of potable water . . . for the population of the Federal District.

Source: México (1936).

LDUDF constituted the legal foundation for government officials and state agencies to prohibit on the basis of ecological arguments the expansion of irregular settlements.

General Law of Human Settlements, 1976

Constitutional amendments to articles 27, 73, and 115 laid the basis for the Mexico's Congress to pass the General Law of Human Settlements (Ley General de Asentamientas Humanos LGAH) in 1976. The LGAH established the basis for consistent and integrated state intervention in the planning of human settlements. Laws prior to this were at best regional and sectoral in their concern with spatial aspects of planning (Ramírez

Saiz 1983). The LGAH was born out of a diffuse parentage (different groups from within the federal administration, as well as local government officials) with the general intent "to give planning a greater political status; in other words, to convert planning into means to guide government" (Antonio Azuela de la Cueva, cited in *El Día,* July 21, 1986).

The LGAH was the first law of its type to incorporate into planning activity the principle of coordinating different levels of government, and it identified the levels of responsibility for policymaking—federal, state, conurbation, and municipality (Ward 1986, 222). It also specified the need to bring into concert public, private, and social initiatives, while emphasizing the necessity to include community participation in the elaboration of plans (Antonio Azuela de la Cueva, cited in *El Día,* July 21, 1986).

The Ministry of Human Settlements and Public Works (Secretaría de Asentamientos Humanos y Obras Públicas; SAHOP) was established under President López Portillo (1976–1982) in order to consolidate planning activity. In response to worsening domestic and environmental conditions and to increased domestic and international pressures for change, environmental considerations were written into the López Portillo administration's National Plan of Urban Development (Plan Nacional de Desarrollo Urbano; 1978), National Plan of Industrial Development (Plan Nacional de Desarrollo Industrial; 1979), and Global Plan of Development (Plan Global de Desarrollo; 1980). As part of the National Plan of Urban Development, a new General Directorate of Urban Ecology (Dirección General de Ecológia Urbano) was created within SAHOP and was charged with programing environmental considerations into future urban planning (Mumme, Bath, and Assetto 1988, 16).

First Reckonings with the Rural Dimension, 1978

In 1978, creation of the Coordinating Commission for Land and Cattle Farming in the Federal District (Comisión Coordinadora para el Desarrollo Agropecuario del Distrito Federal COCODA) constituted an institutional innovation that aimed to rectify the urban bias in planning (DDF 1982, México 1982). COCODA's institutional charter spells out that (1) in order to prevent urbanization of land in the rural zone of the Federal District, then (2) there must be greater vigilance and a stricter enforcement of the law to prevent such development, and (3) such vigilance must occur at the same time as rural-based activities and be provided with incentives and support. The premise is that by fostering the economic viability of rural settlements, the penetration of urban expansion can be staved off. The target area for this strategy was the agrarian communities in the south of the city, especially those of Tlalpan, including the zone of Ajusco. This prorural strategy was not significantly funded until the late 1980s (Wilk 1991c).

Approval of the Federal District's First Master Plan, 1980

In 1980, approval of the General Plan for the Urban Development of the Federal District (henceforth referred to as the Master Plan), and of the partial plans for each of its wards, actualized the mandate decreed by the Urban Development Law of the Federal District (1976). The Master Plan constituted what continues to be the basic instrument of planning in the Federal District; it is complemented by diverse financial, administrative, and legal instruments aimed to control land use and foster development (México 1982). Its juridical support stems from the Constitution (art. 27, 73, 115), the General Law of Human Settlements, the presidential decree that approved the National Plan for Urban Development, and the LDUDF. The 1980 Master Plan became the normative instrument of land use that established the primary zoning of the Federal District, as well as the official policies, objectives, and strategies aimed at fortifying urban development and guaranteeing ecological protection.

In terms of growth control, the 1980 Master Plan stated the objectives were to (1) control urban expansion in the Federal District, particularly in the Southwest, South, and Southeast; (2) densify the urban area in order to optimize the occupation of areas presently underutilized in terms of infrastructural capacity; and (3) orient demographic growth to zones inside the area acceptable for urban development (México 1982).

In terms of establishing a "politics of conservation," the 1980 Master Plan called for the establishment and maintenance of a buffer zone. It divided the Federal District into four zones: (1) a zone for urban use, (2) a zone reserved for future urban growth, (3) a buffer zone at the edge of the city, where land use transition from rural to urban would be subject to restrictions, and (4) a zone for ecological conservation. Shown in figure

6.1, the buffer zone was a fringe varying from between 1 and 3 kilometers in width, covering an area of 194 square kilometers, including 5 square kilometers of rural settlements within the area. The Master Plan stated that "[t]he function of the zone, at the outer edge of the urban area, is to establish a gradual union between the zone for urban use and the area for ecological conservation where urban land use is prohibited" (109). Land uses permitted in the buffer zone included habitation for rural settlements, agriculture and cattle ranching, parks, and limited services (e.g., facilities for recreation and tourism). The buffer zone ran through the middle part of Ajusco. But as I noted earlier it did not prevent the growth of irregular settlements, notably Los Belvederes, from taking place there.

According to the 1980 Master Plan, Ajusco Medio was classified as within the buffer zone. On the basis of this zoning, the area's rapidly expanding irregular settlements were outlawed. This merely overdetermined their irregular status. The settlements location already put them in violation of a number of laws and regulations (including those governing agrarian property, the subdivision of private property, and planning norms that set an altitude limit for mountainside development).

On April 20, 1982, President López Portillo, with authority drawn from Article 89 of the Constitution, expedited the DDF Regulatory Rule for Zoning (Reglamento de Zonificación). This rule legally established the norms by which the DDF would exercise its powers concerning zoning and land use allocation in accordance with the Urban Development Law of the Federal District (México 1982, 89–132). The language describing the objective of the buffer zone shifted from the weaker formulation, *to separate the city*, to the stronger wording: *to contain urban growth* (91).

On June 28, 1982, the delegado of Tlalpan, Gonzalo Aragón, emphatically declared, as he and many other political functionaries had repeatedly done over the previous several years, that "absolutely no new settlements will be permitted in Ajusco" (*Excelsior*, June 29, 1982). To put teeth in these declarations the construction of a mounted police station was completed in Ajusco Medio right along the boundary between the legally designated urban area and the buffer zone. The construction, including stalls for sixteen horses, was ordered by then chief of police, Arturo Durazo Moreno, who owned a palatial estate just four miles from the site. The police station was supposed to act as a disincentive for irregular settlers.

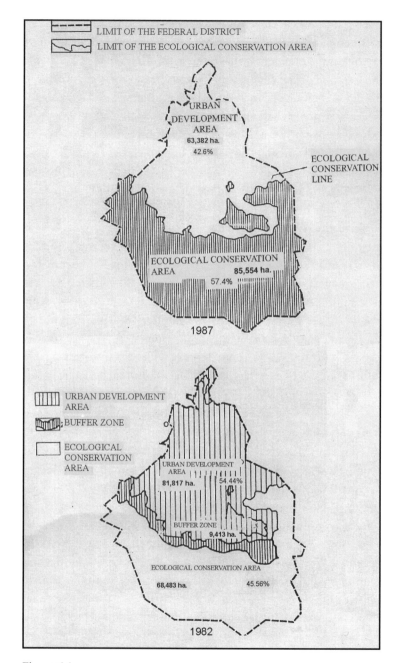

Figure 6.1
The primary zoning of the Federal District, 1982 and 1987. Source: DDF (1987, 72).

It was manned around the clock by a contingent of policemen, who, with automatic rifles slung over their shoulders, would periodically patrol the zone on horseback. Evidently, the effect was minimal, as irregular settlements continued to expand within the immediate vicinity of the station. From the beginning the buffer zone was a failed planning concept (Aguilar and Olvera 1991). It was eventually removed when the Master Plan was revised in 1987 (see figure 6.1).

De la Madrid and Institutionalization of the Politics of Containment, 1982–1988

Since the start of the de la Madrid administration, the traditional dominance of economic planning has been reasserted at the expense of a downgrading of physical planning. Yet, Miguel de la Madrid Hurtado's tenure as president also marked the beginning of a drive to strengthen the status of environmental policy in Mexico. In 1982, de la Madrid combined the portfolios for housing and environment into a single cabinet-level ministry, the Secretariat of Urban Development and Ecology (Secretaría de Desarrollo Urbano y Ecología; SEDUE). This restructuring of the cabinet indicated the heightened political authority attached to environmental concerns (Mumme, Bath, and Assetto 1988, 18). Also, the administration undertook an intensive set of actions aimed at mobilizing and sensitizing the public to environmental concerns (SEDUE 1982).

During 1983 a series of state and regional conferences brought together local political leaders from PRI's sectors, government officials, scholars from the state universities, and citizens groups to discuss the new environmental program and identify environmental problems.[3] Paralleling these conferences, the ecology and environmental committee of the House of Representatives convened a series of four regional conferences focusing on legislative issues in the areas of air quality, water quality, energy development and soil contamination (Mumme, Bath, and Asetto 1988, 19). These conferences received wide press coverage that dramatized environmental themes. The platform of PRI also incorporated a commitment to environmental improvement. In his 1983 State of the Union address, President de la Madrid specifically mentioned the seriousness of ecological problems.

Ecological arguments became firmly established in the creation of new institutions, laws, programs, and plans. At the level of the Federal District, an ecoplan *(plan ecológico)* and an ecological program *(programa ecológico)* were put into effect during July 1983. These steps represented an attempt to incorporate technical measures into development planning that would take into account ecological factors. In 1983, the Commission for the Limits of the Federal District (Comisión de Limites del Distrito Federal) was established in order to deal more coherently with the city's limits (i.e., setting and enforcing them). It should be noted that much of this effort remained on paper only. For instance, in a careful study of the impact of zoning, Aguilar (1987, 32) found that "in reality the concept of the buffer zone is not related to anything, it just exists as a fantasy on the paper of planners." Aguilar examined urban expansion into the buffer zone from 1980 to 1984. He did a survey that asked people residing in the irregular settlements if they knew it was a "buffer zone." Ninety-two percent of those surveyed said no. In conclusion, Aguilar's found that at the local level the planning norms were weak and irrelevant for controlling physical expansion: "The regulations were totally unconnected with the manner in which urban land is appropriated, especially in the case of the illegal occupation of land. . . . Planning of this sort is an ideological facade" (37).

Program for the Development of the Metropolitan Area of Mexico City and the Central Region

On October 27, 1983, President de la Madrid presented the Program for the Development of the Metropolitan Area of Mexico City and the Central Region (Programa de Desarrollo de la Zona Metropolitana de la Ciudad de México y de la Región Centro) and to execute it he established the Planning Committee for the Development of the Federal District (Comité de Planeación para el Desarrollo del Distrito Federal COPLADE-DF). COPLADE was composed of 31 commissions, including the Special Commission for the Sierra del Ajusco (Comisión Especial de la Sierra del Ajusco). During the ceremony in which de la Madrid presented the new program and planning commissions, he stated that Mexico City "suffers severe environmental contamination and considerable deterioration in its ecological equilibrium" and that "the pressure for land and the absence of a long range politics to orient human settlements has made difficult the adequate provision of public services, and has made housing very expensive" (*El Mercado de Valores,* no. 45, November 7, 1983, p. 1138).

At the same ceremony, Carlos Salinas de Gortari (who was then the secretary of budget and programing and who became president of Mexico in 1988) announced that the program defined "a new politics of land use" that would permit "a greater control of urban expansion by "more intensely exploiting investments and by establishing ecological and agricultural reserves vital for the city" (1139).

Also created in 1983 was the General Directorate of Urban Restructuring and Ecological Protection (Dirección General de Reordenación Urbana y Protección Ecológica), a new top-level office within the Federal District Department mandated to deal with the capital's urban planning and environmental problems. The official policy line issued from the DDF at this time was that the irregular settlements of Ajusco Medio were polluting the environment and would have to be eradicated. The director of the Directorate of Areas and Territorial Resources (Dirección de Areas y Recursos Territoriales; DART), Hidalgo Cortés, argued that "the industry of invasion" must be stopped and that in the extreme case more than 10,000 families would have to be relocated out of the zone of Ajusco Medio (*El Día,* October 2, 1983).

Official opinion regarding Ajusco was demonstrated in a legal document (dated August 1983) called a *dictamen pericial* (expert judgment). This judgment was part of a complex legal proceeding known as a *juicio de amparo* (court injunction trial). It was elaborated by a branch office within SEDUE in response to complaints formally registered by a grassroots leader from one of the threatened settlements in Los Belvederes. The judgment found that the inhabitants of Los Belvederes disturbed the ecology of the zone and that they should not be permitted to remain. The judgment served as a rallying point around which grassroots groups articulated countervailing arguments.

The Program for the Conservation of Ajusco
On March 19, 1984, the press secretary of Tlalpan announced that a large portion of the population of Los Belvederes would have to be relocated to other parts of the city. Later in the year, on May 28, 1984, the regente of the Federal District reversed this position when he announced the Program

for the Conservation of Ajusco (Programa de Conservación del Ajusco). This program stated, for the first time, that there would not be a massive eradication of the irregular settlements already established in Ajusco Medio. Lines between the legally designated urban area and the area for ecological conservation were redrawn so that those colonias already rooted in (Los Belvederes) could eventually be legalized. This exception was supposedly going to be the absolute last, and again there were declarations that no more urban settlements would ever be permitted in the zone. The delegado of Tlalpan, Gilberto Nieves Jenkin, asserted that "the government will be much more aggressive against those who speculate with the irregularity of land tenure" (*El Día*, June 28, 1984).

In his presentation of the Ajusco conservation program, Rodrigo Moreno Rodríguez, then general secretary of the government of the Federal District (presenting on behalf of the regente) stated that

As we can all see, the anarchic and unruly expansion of the city has reached truly dangerous limits. This places us in the dilemma of having to choose between two paths: either we self-destruct by allowing things to continue as they are, or we attempt, the community and the authorities together, to arrest this degenerative tendency and to regularize the city's future growth—thereby guaranteeing, at least, that tomorrow will not be worse than today. Without exaggeration we can assert that to continue increasing this overcrowding and the progressive dehumanization of our coexistence, we are heading down the path towards collective annihilation. . . . The city administration has decided to take the second path, strictly controlling the future growth of the city. (México 1984, 2–3)

An outcome of the conservation program was the construction of the Subdelegación of Tlalpan in Ajusco. According to its founding charter, the mandate of the subdelegación (subward) was to "respond institutionally to the complex and intensifying social, political, juridical and ecological problems of the zone" (Subdelegación del Ajusco 1985c). More specifically, this frontline state agency was mandated to achieve "politically stability" and "to maintain the physical presence of the Capital's authorities in the zone of Ajusco."

Program for the Urban Restructuring and Ecological Protection of the Federal District

On September 24, 1984, the first version of the Program for the Urban Restructuring and Ecological Protection of the Federal District (Programa de Reordenación Urbana y Protección Ecológica PRUPE) was announced. PRUPE's authors intended it to provide broad guidelines indicating the de la Madrid administration's position. In PRUPE's first document for public circulation, the fact that the buffer zone had failed to serve its purpose was acknowledged: "Urban expansion has nullified the buffer zone, a situation which demands that decisive and forceful action be undertaken. If such action is not undertaken, it is foreseeable that life in the city [will fall into] generalized chaos, inside a framework where the natural environment's deterioration increases" (DDF 1985, 24).

PRUPE contradicted the Program for the Conservation of Ajusco by asserting that the entire Sierra del Ajusco (including Los Belvederes) should be expropriated and kept clear of urban use for the sake of ecological equilibrium.[4] Much of the content of PRUPE was ambiguous, and it stirred up a great deal of discontent and controversy. In 1985 it was revised, and the call for expropriation was dropped. The elaboration of PRUPE was intended to enhance the state's capacity "to assert control over land use, to assure ecological equilibrium, and to restructure areas of irregular growth" (DDF 1985, 5). PRUPE outlined the need to reform the Urban Development Law of the Federal District (i.e., simplify its numerous regulations). The intent was to empower the state to intervene (with recourse to sanctions and fines) in those areas of the city best suited for ecological conservation. In this respect, one of the most important targets was said to be the zone of Ajusco. PRUPE called for the creation of a special environmental police force called *ecoguardias* (ecology guards). The base of operations for the *ecoguardias* is now in Ajusco Medio.

The PRUPE initiative is a good example of a flawed planning process. It was severely criticized, and as irregular settlements continued to expand into the ecological reserve during the 1980s, there was no coherent policy about how to maintain the so-called reserve. The PRUPE debacle illustrates a problem with planning that continues to this day. Preoccupied with drawing lines and enforcing special zones, such programs lack grounding in the real politics of land allocation and urban growth (Aguilar and Olvera 1991). Wilk (1991c, 18) suggests that this tendency may be waning. With reference to the city's southern fringes, he argues that since the late 1980s, one can see an increasingly conscious effort to link plan making and implementation through a more programatic response in the region. For instance, physical planners of the DDF are tending to spend more of their time and resources in executing and coordinating programs. It may be that project design and execution will gain an edge over blueprint, indicative planning. But the question still remains, Planning for whom and in whose interest? The process is best understood in terms of political mediation and preemptive reform.

The first version of PRUPE placed a great deal of emphasis on its call to establish natural reserves for the ecological protection of Mexico City. The PRUPE document argued that "to recuperate ecological equilibrium is an undeniable condition to achieve Mexico City's survival" (25). From this perspective, it called for the expropriation of 77,000 hectares for ecological conservation and the creation of the special force of ecology police mentioned above. The idea was to make the land the patrimony of the Federal District. According to the authors of the initiative, "[o]nce the areas in question became the property of the Federal District, an integral ecologically oriented administrative system will be established for them, supported by a body of strategically distributed ecology guards, whose objective it will be to insure that the use and destiny of the unpopulated land will be for environmental improvement and ecological equilibrium" (DDF 1985, 14). Support for this position was said to come from experts on the matter: "urban technicians declare that the urbanized portion of the Federal District's total land area should not exceed 40 percent. If urban development expands beyond this proportion, there will be a severe and increasing risk of not only rupturing ecological equilibrium, but of doing so with irreversible consequences" (DDF 1985, 12).

PRUPE's approach to the urban-ecological problematic was widely critiqued by academics and environmental organizations. Serving as a summary of the kind of criticism that actors in the city launched against PRUPE, box 6.2 lists the concerns directed against the first version of PRUPE by the National Ecological Alliance (Alianza Nacional de Ecología) based in Mexico City. Other environmental actors on the scene were also critical. The League of Agrarian Communities of the Federal District opposed the expropriation proposal (*El Día*, August 23, 1985). On the other hand, a middle-class environmental organization, the Movimiento Ecologista Mexicano, supported the expropriation (*UnoMásUno*, February 22, 1985).

Box 6.2
The National Ecological Alliance's Critique of the Program for the Urban Restructuring and Ecological Protection of the Federal District, (PRUPE), 1984

---

1. It dealt only with the Federal District when in fact the concerns are multijurisdictional.
2. It focused on symptoms, not causes, and failed to direct actions at decentralization. Instead, it promoted continued investment in the capital.
3. The scope of the program was twenty-five years, but there was no guarantee of avoiding sexennial disruptions.
4. It failed to link itself to the National Development Plan (1983–1988) or to the National Plan for Urban Development. Thus it was institutionally disarticulated.
5. The program did not include explicit measures against industrial pollution.
6. It did not include measures to actually educate the public.
7. There was no planned input based on popular participation. The COPLADE system was not mentioned.
8. No mention of a budget was made. Paying for the infrastructure policy would create problems—aggravate the vicious cycle of expelling the urban poor toward the periphery.
9. Expropriation limits were not clearly specified. The only clear figure was the call to create positions for 6,000 ecology guards.
10. It lacked a clear delineation of administrative responsibility for implementing the program and was unclear in its relation to federal entities such as SEDUE, SARH, SRA, or DDF.
11. The program aimed to promote ecological politics on the basis of declarations and wish lists and on lofty plans presented in sensational terms.
12. The zoning approach emphasizes the need for ecological protection in the rural area, but urban ecology is just as crucial.

Source: This summary is drawn from a mimeograph without a date (see *Uno-MásUno*, September 20, 1984).

Capturing the sentiment among the academics that were critical of PRUPE, Jorge Legoretta noted that PRUPE did not get at root causes; it was essentially a class-based piece of political propaganda that promoted residential segregation by displacing the poor from the best locations in the city (*UnoMásUno*, February 8, 1985). In light of this scathing criticism, the government revised PRUPE. The consultation process was carried out from September 1984 to May 1985. It involved five stages: presentation and dissemination, public hearings, analysis and discussion, conferences with social groups, and the consolidation and incorporation of suggestions (DDF 1985, 12).[5] On August 24, 1985, the second version of PRUPE was announced. This document formed the basis for the General Program for Urban Development of the Federal District (Programa General de Desarrollo Urbano del Federal District), 1987 to 1988.[6]

PRUPE's revisions reflected earlier criticism. On the basis of planning instruments specified in the Law of Urban Development in the Federal District (Ley de Desarrollo Urbano del Distrito Federal), PRUPE spelled out a framework: (1) to regularize and control the physical expansion of the capital, reorganizing human settlements and extending the coverage of public services, and (2) to recover ecological equilibrium—improving the technology of garbage treatment, establishing strict controls regarding toxic emissions, and creating zones for natural reserves. Also, in line with the National Development Plan and the Program for the Development of the Metropolitan Area of Mexico City and the Central Region, PRUPE provided strategies: (1) to transform the current pattern of urban development, (2) to control urban expansion and prohibit the development of new irregular settlements, (3) to establish new forms of administration, with community participation, for the protection of green spaces, (4) to decentralize industry—especially those forms most harmful to quality of life and natural resources (DDF 1985, 8).

In an important change of position as far as the fate of irregular settlements in Ajusco was concerned, PRUPE declared that in its efforts to protect the environment, city government should favor mechanisms that foster development in rural areas of the Federal District and that strengthen the administration's control over land use—only as a last alternative should the option of expropriation and/or relocation be used (DDF 1985, 14). Following this line, PRUPE dropped the notion of expropriating 77,000 hectares. But by 1988, renewed emphasis would again be placed on expropriation.

By 1985 the decision was made to regularize the settlements of Los Belvederes and to include this part of the city in the legally designated urban area. Initial steps were undertaken by DART and by the Tlalpan ward beginning in 1985. *Constancias* (certificates of possession, not of ownership) were issued to the holders of 4,803 lots. On a metaphorical level, regularization was justified by the authorities as a measure to stop, or at least to slow down, the spread of an infection or disease: like a cancer, irregular settlements must be cut out, stopped from spreading. If they have already become an inextricable part of the urban corpus, then they must be treated (regularized). Initiating land tenure regularization along the outer limit of the legally designated urban area was viewed as analogous to cutting a fire line, thereby consolidating a perimeter of control, in an effort to ward off or at least slow down the ecological firestorm of urban sprawl.[7] But this explanation does not go very deep. The politics of regularization have more to do with social control than it does with a well-implemented policy of ecological conservation.

The "Green-Line" Program

During the second half of the 1980s, the politics of containment acquired new momentum. The comprehensive strategies of the early 1980s were succeeded by more specific growth management policies. As Wilk (1991c, 9) points out, physical limits to growth were proposed (including a "green-line" program for the southern fringes referred to as the Programa de la Línea de Conservación Ecológica) that involved considerable investment in regulation and enforcement.[8] The Physical Planning Division (Subdirección de Planificación Física) of the Department of the Federal District has concentrated a large portion of its budget on the green-line program.[9] This program was part of the Federal District's new urban and regional development plan that was approved for the period of 1987 to 1988.

General Program for the Urban Development of the Federal District, 1987–1988

On July 31, 1986, the General Program for the Urban Development of the Federal District (1987–1988) was presented to the public.[10] This Program has been the city's master plan for nearly a decade (Páramo, Bermeo, and López 1995). The program reclassified the entire Federal District into two parts: one for urban development (43 percent of the total land area) and one for ecological conservation and rural development (57 percent of the total land area). The program eliminated the buffer zone that was in the master plans of 1980 and 1982. Although the thirteen settlements of Los Belvederes in Ajusco were incorporated into the legally designated urban area, the entire area to the south of these settlements was zoned for ecological conservation and rural development.

The newly delimited ecological conservation area contained 36 rural settlements and 475 irregular settlements (Páramo, Bermeo, and López 1995). Since 1987, 27 new irregular settlements have taken hold, increasing the total to 502. Select parts of the ecological conservation area—where irregular settlements had been in place for at least ten years—were designated as special zones of controlled development (Zonas Especiales de Desarrollo Controlado ZEDEC). Thirty-eight ZEDECS were established in twelve of the Federal District's wards. ZEDECs have turned out to be highly problematic.[11] Even so, the government has renewed its emphasis on this land management approach in its 1995–2000 Program for the Urban Development of the Federal District (Zedillo 1995).

At about the same time that the 1987–1988 General Program for Urban Development was announced, the Coordinating Commission for Rural Development (Comissión Coordinadora para el Desarrollo Rural, COCODER; formerly COCODA) announced its Program for Rural Development and the Protection of Natural Resources, as well as its Program for Land Use (Programa para el Desarrollo Rural y la Protección de los Recursos Naturales), to be applied in the seven predominantly rural delegaciones in the south of the Federal District, including Tlalpan. These programs proposed mechanisms to limit the advance of irregular settlements, to conserve and recuperate natural resources, to promote interinstitutional coordination, and to safeguard areas of the Federal District identified as best suited for agriculture, ranching, forestry activities, and the protection of wildlife (*UnoMasUno,* March 21, 1986). In terms of juridical support, the programs drew, inter alia, on the Integral Rural Development Law of the Federal District (Ley Integral de Desarrollo Rural). Despite the seemingly pro-rural stance of these initiatives, representatives from the thirty-two pueblos of the rural area in the Federal District (including presidents of ejidos and members of *juntas de vecinos*) formed a common front to protest the subplans of rural development. They argued that the plans were made without their input and that the plans allow subdivision that would encourage urban sprawl and put an end to their livelihood.[12]

Faced in March 1986 with the ongoing illicit subdivision of land zoned for ecological conservation in the part of Ajusco Medio referred to as El Seminario, the delegado of Tlalpan, David Ramos Galindo, solicited the assistance and intervention of the Federal District's attorney general of justice (*procuraduría general de justicia del DF*). The stated objective was to root out and punish the fraudulent real estate developers that have continued to operate in the zone. Galindo emphatically expressed that "not even one single subdivision more will be permitted in this perimeter" (*El Día*, March 24, 1986). Also, in response to the propaganda tactics of the fraudulent subdividers, Galindo said that his delegation would begin a campaign, including the placement of huge signs throughout Ajusco Medio, to educate the public about the illegality of land transactions in the zone and about the risks involved.

At the front line, the director of the Subdirectorate of Ajusco (the subdirectorate, a satellite office of the Tlalpan Ward, will be discussed in greater detail in chapter 10), Raúl López Garnica, declared that "they will not permit any more invasions in Ajusco Medio" (Garnica 1987, 1). Massive desalojos were executed in the zone of El Seminario during September and October of 1987. El Seminario is on the outskirts of Los Belvederes, an area of approximately 600 acres lying within the ecological reserve. Over the fourteen-month period (September 1987 to November 1988) the state carried out eight massive desalojos in the area.

In late 1988, another series of desalojos eradicated more than 1,000 provisional housing units from El Seminario. Nevertheless, the area continued to be settled and resettled (see figure 6.2). There was extensive coverage of these desalojos in all the major newspapers in Mexico City. Authorities argued that besides endangering the so-called ecological equilibrium of the city, El Seminario had become "a refuge of central Americans dedicated to assaulting commercial centers, residences, and passersby" (*El Día*, November 5, 1988). Official policy maintained that any and all irregular settlement in the ecological zone will be eradicated, even upper-class residential estates (*El Día*, November 7, 1988). These are precisely the same kinds of observations that were made about the settlements of Los Belvederes in the late 1970s. In effect, little has changed.

New Limits to Old Problems, Late 1980s to Early 1990s

Past failures to strictly enforce limits on horizontal urban growth encouraged the government to develop new programs, new plans, and, yes, new limits. Since the late 1980s a number of new regional and national initia-

Figure 6.2
Women evicted from Lomas del Seminario, one of the irregular settlements in El Seminario, Ajusco Medio. Note: Shown next to their roadside dwelling unit, the women await an opportunity to resettle in the foothills from which they were forcibly removed.

tives have drawn closer attention to the importance of environment-development linkages. In June 1988, the National Program for Ecological Protection (Programa Nacional para la Protección Ecológica) was announced. It was designed to address the economic tendencies that give rise to ecological problems and to link society's demands for environmental improvement with the action of public institutions.

Priorities listed in the 1988 ecological protection program included the establishment of parks and land reserves, the establishment of water conservation policies, the elimination of highly polluting activities, and the application of corrective measures for the rehabilitation of areas in which there has been a marked deterioration of the ecology (DDF 1989b). The National System of Protected Natural Zones (Sistema Nacional de Zonas Naturales Protegidas) set aside sixty-eight protected areas throughout Mexico.[13]

In 1988, the government passed the General Law of Ecological Equilibrium and Environmental Protection (Ley General del Equilibrio Ecológico y la Protección del Medio Ambiente). This detailed law contains 137 pages of text and includes the key chapter 4, "Ecological Policy," chapter 5, section 1, "Ecological Planning," as well as sections concerning "the ecological regulation of human settlements." The role of this law in the Federal District has been more symbolic than real (Mumme 1992, 133; Greenpeace México 1995). Mexico's economic crisis has undermined the enforcement of environmental laws, and implementation continues to be ad hoc.[14]

To deflect criticism of its poor track record (with respect to real accomplishment as opposed to intentions), the Salinas administration concentrated its efforts "on a few demonstration projects that appear to project a bona fide concern for environmental improvement" (Mumme 1992, 133). The same approach was taken in dealing with the urban-ecological problematic in the city's southern fringes during the late 1980s (Azuela de la Cueva 1989c).

The National Solidarity Program (Programa Nacional de Solidaridad: Pronasol) was started by President Salinas as soon as he took office in 1988. As the centerpiece of the government's strategy "to eradicate poverty," its stated objective has been to create and improve basic services and facilities, including housing, for citizens living in low-income urban neighborhoods as well as in the nation's poorest rural and indigenous communities. The solidarity program has been funded from (1) sources that were made available after the foreign debt was renegotiated; (2) the divestiture of public enterprises;[15] (3) federal, state, and municipal contributions; and (4) resources provided by beneficiary communities (Dirección General de Comunicación Social 1991). More than 50 percent of the public sector's capital expenditures were channeled through Pronasol in 1990 (Barry 1992, 99)

As the government described it, "*Solidarity* has adopted an innovative approach in that it encourages communities to take part in solving their most pressing problems. The priorities of the program in each community are thus determined by the beneficiaries themselves, who organize their own communities and elect their own representatives" (Dirección General de Comunicación Social 1991, 101). Activities of the program involve support for housing construction, infrastructure projects, and public services, as well as job creation, health care, education, nutrition, and food distribution. In Tlalpan—the ward of the Federal District where Ajusco is located—the Pronasol program financed during the period of 1990 to 1992 public works projects (water and drainage) at a rate three times greater than that accomplished from 1983 to 1989. Much of this spending was concentrated in the conflictive zone of Ajusco Medio.

In 1989 the Mexican government (Federal District Department) announced a highly touted decree to expropriate 727 hectares of land in the El Seminario part of Ajusco Medio for the purpose of establishing a new ecological park of Mexico City.[16] The formal justification for the decree (reported in the *Diario Oficial,* June 28, 1989) is no different than that used by President Cárdenas to support his decree targeting the same area nearly a half-century earlier. In my interview with Jorge Gamboa de Buen, head of urban planning for the Federal District, I was told that this was a first step. The 727-hectare expropriation would eventually be supplemented by expropriations of more than 700 additional hectares. Gamboa noted that the strategy calls for expropriating both private property and agrarian property. He told me that the government decided to first expropriate 727 hectares of private property because it was easier to do so; private property involved less entanglement with the politics surrounding the agrarian property regime. The "trick," Gamboa said, is to pay a fair price. The value of the land in this part of Ajusco Medio according to the public registry is only 600 pesos/per square meter (approx. $US.20 per square meter); the real market value is 35,000 pesos/per square meter (approx. $US12 per square meter) (1993). The official minimum wage at this time was 13,300 pesos per day (approx. $US4.30). Gamboa said they plan to pay $11,000 pesos per square meter (approx. $US3.60 per square meter). The area for the proposed park of 1,500 hectares is shown in figure 6.3. Its alignment is along the outer edge of Los Belvederes.[17] To support this initiative the government announced the Program for the Management of the Zone Subject to Ecological Conservation.[18]

Committing to Sustainable Development?

Since the late 1980s, official discourse in Mexico has steadily picked up on the notion of promoting sustainable development. In his fourth State of the Union address (November 1992), President Salinas declared that "Environmental protection is an essential part of social development policy. My administration has therefore made a commitment to *sustainable*

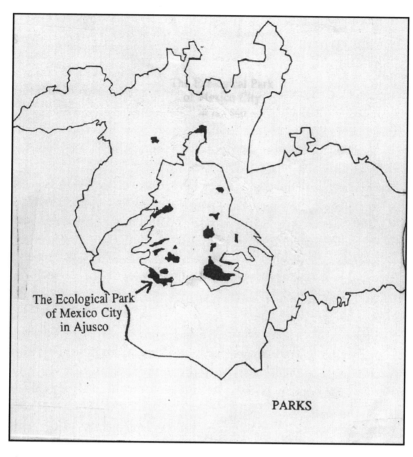

The Ecological Park
of Mexico City
in Ajusco

PARKS

Figure 6.3
Parks within the Metropolitan Area of Mexico City. Source: Cremoux (1991, 15).

development which requires that we make changes in modes of production and that as a society we live in harmony with the environment" (México 1992, 38). This position was reflected in the policy speak of Mexico's Secretariat of Social Development (Secretaría de Desarrollo Social; SEDE-SOL). Created in May 1992, SEDESOL took the place of SEDUE. The establishment of SEDESOL combined numerous different agencies, but it also created two new ones: the Attorney General's Office for Environmental Protection (Procuraduría Federal de Protección al Ambiente; PRO-FEPA) and the National Ecology Institute (Instituto Nacional de Ecología;

INE). The intent was to divide up environmental responsibilities between enforcement and legislation.[19] With regard to balancing urban and regional development, "the idea that guides the actions of SEDESOL is that of *sustainable development*, which demands a rational use of natural resources and which is conscious of not limiting the quality of life of future generations" (SEDESOL 1993, 4). SEDESOL's budget for environmental programs in 1992 was $US68.2 million, a significant jump up from only $US3.9 million for environmental programs under SEDUE in 1989 (Colosio 1992).[20]

Recent changes to the Nation's Human Settlements Law (Ley General de Asentamientos Humanos) unanimously approved by the full House on July 2, 1993 also include measures to incorporate the notion of sustainable development: for example, a strengthening of the municipality that may now contract with the private sector in the allocation of public services, and an empowerment of the capacity of planning vis-à-vis integrated land policy and territorial reserves aimed at avoiding real estate speculation and the illegal occupation of land.[21]

The theme of sustainability has also gotten attention in Mexico's National Development Plan and National Program for Urban Development (Plan Nacional de Desarrollo, and Programa Nacional de Desarrollo Urbano). These documents have placed emphasis on the need to establish an urban-rural balance in the country. One of the strategies listed in the National Development Plan (1988–1994) was to channel economic activity toward places optimally suited in terms of resource availability, especially water, and to discourage the growth of overpopulated zones (including Mexico City) and of zones that have severe resource deficiencies (SPP 1989, 110). In the National Program for Urban Development (1988–1994), housing and human settlements were portrayed as integral components of the larger urban milieus that encompass watersheds, air basins, sensitive lands, and ecosystems. The housing crisis was characterized as rooted in economic conditions, but also as part of a larger set of environmental problems that in some cases are reaching critical thresholds.

During the past presidential administration, Salinas claimed that much was done to reduce pollution through citizen participation. He announced that his administration made headway in the nation's capital by (1) improving automotive fuel quality, (2) initiating the construction of two new Metro (subway) lines, (3) implementing a reforestation program,[22] (4) taking firm measures to protect groundwater recharge zones, and (5) increasing the city's ecological reserves by nearly 3,000 hectares. These last two points had a direct impact on the politics of land allocation in the Ajusco greenbelt. Despite the government's stepped-up efforts, the newly designated ecological conservation area and park continue to be penetrated. And it is not just irregular settlements that cross the line. Besides the urban poor, wealthy families, the state itself, and rural actors have also penetrated the so-called ecological zones.

For a long time, upper-class residential development has taken place in areas designated for ecological conservation. As time goes on there will probably be fewer cases in which upper-class elites build single isolated estates and mansions. The number of robberies from these secluded estates has gone up dramatically in the past several years. As a result there has been a trend toward developing exclusive upper-class residential complexes encircled by high stone walls. These complexes, such as Jardines en la Montaña (Gardens in the Mountain), have their own security guards. Access is highly restricted. Consequently, an even more dramatic polarization of residential space by income is under way. Along such lines, Figuereo Osnaya reports that upper-class estates are illegally developed within the Cuajimalpa part of the ecological reserve (close to Ajusco) (*Uno-MásUno*, July 18, 1992). Real estate agents selling luxurious estates in Cuajimalpa advertise in these terms:

We're inside . . . we offer mansions as little as 800 meters from the beginning of one of the ultimate ecological reserves of the Federal District: Desierto de los Leones; for all practical purposes the development is enclosed in a part of this reserve, surrounded by woods and still virgin hillsides where one can still breath pure and fresh air. . . . The project is located on a 3,800 square meter site, and because it is *in the ecological reserve zone,* there are only six exclusive residencias." (*UnoMásUno,* July 18, 1992)[23]

There is also evidence that certain factions within the government itself violate the ecological reserve. Porras Robles reports a case in which one government agency battles another. Specifically, the delegado of Iztapalapa curses COCODER officials and others for a chain of corruption in the illicit subdivision of the ecological reserve (*UnoMásUno,* August 21, 1992).

In terms of conditions within the Federal District's rural areas, there are a number of serious problems. According to the spokesperson of COCODER, fire recently burned 2,847 hectares in the ecological conservation area, one-quarter of which was forest. The COCODER representative said that the fires were deliberately set so that people can settle there (Sosa 1991). In other parts of the ecological reserve, Durán (1991) reports that ejidatarios chopped down more than 5,000 trees in a radius of two kilometers and made holes 1.5 to 3 meters deep to extract topsoil. The topsoil is sold to people to use in their residential gardens. This underground commercial activity is suspected to be supported by the agrarian and delegational authorities. It is estimated to bring in 15 million pesos per day. A flotilla of trucks make multiple return trips—up to sixty per night. Each trip carries a load worth an estimated 250,000 pesos (approx. $US75.00). A grand total of 480 tons of black earth is extracted every day.

Finally, it should be noted that irregular settlement in the ecological reserves is also fueled by the growth of rural populations in the metropolitan area. The ward of Magdalena Contreras is a case in point. The *ejidatarios* there say they are congested, living four or five families on a single plot of land. They want space for housing, yet their land is designated as part of the ecological reserve.[24] In sum, there are multifaceted demands on the resouces that make up the ecological reserve. However, the government has defined the task principally in terms of curtailing the growth of low-income irregular settlements through its politics of containment. The ecological arguments are fundamentally driven by political expediency. Nowhere can this be seen more clearly than in the case of Ajusco.

III

Mexico City's Contested Ecological Reserve

7

The Expansion of Irregular Settlements into the Greenbelt Zone of Ajusco

A Highly Contested Terrain

Irregular settlements have formed in Ajusco in much the same way that they have formed throughout the metropolitan area. The urban poor have used one of the following means: (1) purchased their lots from ejiditarios who, technically speaking, did not have the legal right to sell, (2) entered into private purchase agreements with landlords or real estate developers who then defaulted on the adequate provision of services and land registration, or (3) squatted on the land. Such arrangements in Ajusco and throughout Mexico City have been accompanied by a high degree of collective self-provisioning and popular mobilization aimed at the delivery of public services, infrastructure, and security of tenure. The distinctiveness of Ajusco derives from the particular way in which historical land development patterns, popular mobilization, and ecological arguments have combined under new conditions of urban expansion. The miniregion of Ajusco has become *a focal point and testing ground for the state's politics of containment.*

Much of Ajusco is forested and dedicated to rural uses such as farming, cattle raising, and forestry. However, conditions within Ajusco's rural area, as well as dynamics engendered by the proximity of a rapidly expanding metropolis, have fostered the rapid transition of land use from rural to urban. As occurs in the outskirts of fast-growing cities throughout Mexico, the legacy of agrarian reform and agrarian politics have complicated urban policy and planning in Ajusco by creating a situation ripe for illicit land transactions. As I noted in the last chapter, it is not just low-income people who have staked out claims for land in Ajusco. The zone's greenery, clean air, and panoramic vistas have attracted real estate developers and higher-income groups interested in upper-class development. Historically, economically better-off groups have concentrated in the southwestern part of the city and in the case of Ajusco, dynamics of spatial segregation are intense. Competition for land in the zone is exaggerated, and contesting forces are polarized.

The rapid transition from rural to urban land use in Ajusco has involved fraudulent schemes by real estate developers, illegal sales of communal and ejidal property, land invasions and violence, mass eradication of incipient low-income settlements, popular resistance and opposition movements, widespread corruption, and deepening ecological disruption. The preceding factors have prompted the government to gain greater control over Ajusco's development and land use conflicts. Official arguments concerning the fate of Ajusco have been stated in the extreme—as if one more inch of urban sprawl will trigger an ecological disaster. But it is important to distinguish political rhetoric from actual beliefs, attitudes, and expectations. From what I was told by both higher- and lower-level planners and government officials, it is clear that few really believe that the function of the greenbelt is to absolutely prevent urban expansion.[1] Rather, it is to slow it down. There is an acceptance—at least until recently—that the urban sprawl *(mancha urbana)* will grind forward no matter what; it is the pace and content that have to be controlled.

Until the early 1970s, Ajusco was still a sparsely populated rural area. But by 1979, approximately one-quarter of its land area was urbanized with a population of more than 150,000 people. The bulk of the inhabitants were low-income families, living in self-built housing that lacked an adequate public infrastructure and services. It was during the period of extremely rapid growth in the early 1980s that I visited Ajusco for the first time. I was directed to Ajusco by community activists in another part of Mexico (Ajuacatlán, Querétaro) where I had been doing field research with the Cooperative without Borders (Cooperativa Sin Fronteras). Activists in the cooperative had described to me some of the community organizations active in the most recently urbanized (illegally settled) part of Ajusco. Specifically, they told me about a well-organized settlement called Bosques del Pedregal in which the people had formed a self-help cooperative called the Cooperative Society of Bosques del Pedregal in Struggle (Sociedad Cooperativa, Bosques del Pedregal en Lucha). At the time, Bosques del Pedregal was one of the thirteen contiguous irregular urban settlements in the part of Ajusco Medio called Los Belvederes. Los Belvederes covered 431 acres of land and was divided into 219 blocks containing 5,000 lots of approximately 200 square meters each. The population of Los Belvederes expanded rapidly from a few hundred in 1976 to roughly 50,000 by 1988 (see figure 7.1).

The people I was working with in the Cooperative without Borders suggested that I look up Ernesto Bravo, one of the principal organizers of the Cooperative Society of Bosques del Pedregal in Struggle. Getting a reference to Ajusco from hundreds of miles away was my first indication that the zone had a reputation well beyond its immediate borders. In August 1983, I left the pueblo of Ahuacatlán and traveled by bus to Mexico City to locate Bosques del Pedregal in Ajusco and to look up Ernesto Bravo. I had no exact address because the settlement in question was not officially sanctioned nor recorded on any map. I figured I would travel to the southernmost part of the city and from that point seek more specific directions. From the capital's central square (the Zócalo), it took me about one hour traveling by subway (the metro) to get to Universidad—the last southernmost stop on line 3. From that end point numerous taxis, mini vans, and buses ferried out to the farthest reaches of the city. I did not know it at the time, but a newly won bus route conveyed people from the Universidad metro terminal all the way to Los Belvederes, where Bosques del Pedregal is the last stop (a bus trip that lasts forty-five minutes to an hour).

When I arrived at the Universidad metro terminal, I hired a cab and asked the driver to take me to the new place called Bosques del Pedregal. The driver delivered me to an upper-class residential subdivision recently constructed in what was supposed to be a part of the National Park of Bosques del Pedregal. I said, "no, this can't be!" The spacious residential estates in that area were obviously very expensive. I told the driver I was looking for an illegal settlement of low-income people. Ultimately, via a deeply potholed, bumpy stretch of dirt road at the outermost periphery of the urban area, I was delivered to the right place. Remarkably, both settlements were in the same general vicinity. In this way, I witnessed for the first time the dramatic spatial segregation by class that characterizes urban development in the southern part of the city. Enclaves of fully ser-

Urban expansion
Ajusco area

1970

1979

1985

1988

Open space

Railroad

Ecological
conservation
line

viced commercial real estate and luxurious housing are developed in close proximity to a sprawl of lower-income irregular settlements that lack most basic urban services.

Life at the Limits

When I arrived for the first time to the colonia Bosques del Pedregal, I set about searching for Señor Ernesto Bravo. It was a bright Sunday morning and many people were out working on projects of all sorts. Throughout the settlement I could see groups of fifty to sixty people. They were collectively laboring to level the terrain for housing and to clear paths for rudimentary roads. One group was leveling an area for children to play on. The steep ruggedness of the volcanic terrain made this work very difficult; it required picks, sledgehammers, and a lot of sweat. The land juts up from the lower lying part of Ajusco at inclines ranging from fifteen to more than forty-five degrees (see figure 7.2). More than one-half of the settlements in the area are situated on inclines thirty degrees or steeper. This makes access arduous, even more so given the sharpness and instability of the rocky surface. Despite such difficulties, most everyone I saw that bright Sunday seemed to take part in the collective projects—the children and the elderly as well as young men and women.

Remarkably, I had no trouble finding Señor Ernesto Bravo. I simply asked people where he lived. They were able to point me in the right direction, and I soon arrived at his house. When I first met Bravo, I told him that I would like to study urban growth and community dynamics in the so-called ecological reserve. He and several others responded by questioning my motives while they took me for a long walk around the area. The sights I took in for the first time that day, and soon there after, were impressive. They serve as a dramatic reference point for the changes I have witnessed in this zone over the past decade.

The dirt road that served as the main access to Los Belvederes was cut into the upper part of the hillside where the settlements were located. This

Figure 7.1
Urban expansion into the zone of Ajusco, 1970–1988. Source: Detenal Aerial Photography, Flight number R313, scale 1:40,000, Photo number L-8, June 1979 and Flight number R1036-31, June 1970.

Figure 7.2
Topographical contours and road grid of Ajusco Medio, 1985. Source: Covitur,
Aerofoto R-2934 F79, Flight Nov. 1985.

road formed the outermost periphery of the city; it divided the urbanizing from the as yet unurbanized area. Armed police on horseback regularly patrolled the road and the surrounding area. The police had a base of operations on the main access road. It was a large red brick structure surrounded by a chain-link fence topped with barbed wire. In addition to living quarters, the structure had ten stalls for horses. Adjacent to this police base was a forty-foot-long cylindrical tank supported in a horizontal position on cement blocks. The tank was supposed to be used for water storage and distribution to people of Los Belvederes, but that never happened. Upon the tank, in huge white letters, the Cooperative Society of Bosques del Pedregal in Struggle had painted a message for all the community to see:

THE MILK STORE IS THE BEGINNING OF OUR STRUGGLE, ORGANIZED WE WORK AND STRUGGLE, JOIN THE COOPERATIVE. "TODAY WE DO THE DIFFICULT, TOMORROW WE WILL DO THE IMPOSSIBLE."

The police fortress juxtaposed with this insignia of popular struggle, painted by people pushing up against the urban-ecological limit, was a striking contrast. It clearly illustrated that Ajusco was a tension laden frontier with respect to state-society power relations. Upon my introduction to the zone of Ajusco, I learned that a significant chunk of land inside the boundaries of Los Belvederes was still used for agricultural production (milpas). Most of it, though, had already been subdivided for urban use.

In terms of housing conditions, the average lot measured approximately 200 square meters and was occupied by 5.6 residents. The settlement's overall density was about 150 inhabitants per hectare. The most common type of dwelling unit was a small, single-room structure constructed with wood or volcanic rock for walls and with branches, wood slats, and/or bituminized cardboard for a roof (see figures 7.3 and 7.4). In terms of public services and facilities, my first tour through Los Belvederes did not have many places to stop. There were deficiencies in all areas, including a critical need for schools, medical clinics, and commercial and recreational facilities. These deficiencies, in addition to the insecurity stemming from illegal land tenure, motivated the settlers to mobilize and collectively provide for themselves.

Water was delivered by special water trucks called *pipas*. Each family was allocated 250 gallons per week. This worked out to less than thirty

Figure 7.3
Dwelling units in Bosques del Pedregal, constructed of discarded lumber and wooden pallets, 1987.

liters per person per day. The World Bank estimates that at least fifty liters per person per day are necessary to ward off health-related problems (Falkenmark and Suprato 1992). Water deliveries were made to stations at select high points throughout each of the settlements. Water stations are level platforms, accessible to the *pipas,* upon which sit dozens of fifty-gallon open-top metal drums—each one of which is linked to a household downhill, by a garden hose through which the water can be transported by gravity (see figure 7.5). For the most part these hoses are supported to hang overhead on their descent downhill. On average the hoses extend 300 meters in length, representing an enormous cost to the consumer. Miles of hose and hundreds of brightly painted metallic drums are visible throughout the zone. Although such a system may be unsightly, in 1983 the institutional commitment to deliver water marked a major step toward the consolidation of Los Belvederes.

Another visually striking feature of Los Belvederes in the early 1980s was the infrastructure for the distribution of electric power. Wires could

Figure 7.4
Dwelling unit in Bosques del Pedregal, constructed of volcanic rocks stacked without cement, 1987.

be seen running overhead in all directions forming a dense and precarious webbing (see figure 7.6). Each household tapped into power sources to the north of the settlement. Power failures were constant.

My first walk around Ajusco in the summer of 1983, led by Ernesto Bravo and other community activists, took me through but also beyond the urbanized area further up into the as yet unsettled foothills. In this way, I was introduced to the striking beauty of Ajusco's undisturbed ecology; its pristine brooks, wildflowers, greenery, and wealth of biodiversity. As one ascends the foothills, the air becomes cool and crisp with a fresh scent of pine.

On clear days, the panorama from the heights of Ajusco is breathtaking. The sprawl of development fills the great Valley of Mexico like an ocean. The views are especially impressive on clear evenings when the gray city sprawl of daytime fades into a sea of colorful night lights. It is no wonder then that Ajusco is also a coveted site for those able to build expensive estates and upper-class housing projects. Indeed, the contest for the land

Figure 7.5
Girl helping lower water to her home, Bosques del Pedregal, 1985.

between the rich and the poor had a central place in the story line that community organizers first gave me as I began my fieldwork.

After that eventful first visit, I made repeated trips to Ajusco. For more than a decade I have compiled a record of the zone's development and have witnessed the process firsthand. Living in the settlement of Bosques del Pedregal with a family of four for six months (1983–1984) gave me insight into how difficult it is to establish a household and settlement under such rugged circumstances. I also learned that the urbanization of Ajusco in many ways is a typical story: irregular settlement made possible by high levels of popular mobilization and collective self-provisioning under conditions of great sacrifice and hardship.

The Politics of Containment and Grassroots Environmental Action

The case of Ajusco has captured attention from around the world. At certain junctures, external agents from foreign countries (e.g., the Interna-

Figure 7.6
Man hooking up wire to supply his dwelling unit near Bosuqes del Pedregal with electricity.

tional Development and Research Center, Canada; a German religious organization; the Inter-American Foundation, United States) have played a significant role in the zone's development. With technical and financial support from external agents outside of Mexico, as well as within Mexico, the zone of Ajusco has proved to be a fertile field for grassroots mobilization and social experimentation.

During the mid-1980s, when threatened with the mass eradication of their settlements, several grassroots organizations in Ajusco mobilized themselves to resist. Besides availing themselves of well-worn strategies of marches and demonstrations, these groups generated bottom-up proactive environmental action that aimed at promoting appropriate technology and a "sustainable" form of urban development. The intent of such action was essentially to stay the state's hand, which was ready to dislodge them. But it was also more than this: these groups skillfully captured a great deal of attention and support in local and international arenas with their proposal to transform their communities into colonias ecológicas productivas. In terms of an alternative development, very little has actually materialized out of Ajusco's environmental movements. However, extensive upgrading in Los Belvederes has occurred. On each of my return visits to the zone every year between 1983 and 1988, and then again in 1993, I witnessed remarkable legal as well a physical transformations. For instance, in 1986, the government began distributing thousands of land certificates that constituted a first step toward the legalization of land tenure throughout Los Belvederes. In 1987, the installation of central power lines improved immensely the previously chaotic electrical system.

Although I had witnessed dramatic changes over the years, on my return visit to Ajusco in January 1993 I was nonetheless amazed. Nearly five years had elapsed since the last time I had been there to conduct extensive fieldwork. Again, the first thing I did was navigate myself to the house of Ernesto Bravo. This time I found not a shack, but a two-story concrete dwelling. In itself, this is not surprising. It is commonplace for self-help housing and services to be upgraded over a five-to-ten-year period (see figures 7.7 to 7.10). What did surprise me is the dramatically increased overall density of the settlement. By 1993, it had lost all appearances of a frontier outpost on the periphery.

The bumpy dirt road I first traveled along in 1983, the road that divided the urban from the rural, was no more. Actually, it was there—but it was now paved and it had become a busy thoroughfare lined with shops and people buying and selling goods of all sorts. The one-room provisional bungalow that had been the kindergarten in Bosques del Pedregal is now a massive two-story building with dozens of classrooms and a large level play area. The secondary school has undergone the same transformation. The streets throughout Los Belvederes are now all paved, the infrastructure for water and sewage is nearly complete, phone lines are in, and the area is about to get postal service.

I was particularly struck by the obviously large investment that the state is placing in the water and sewage lines. During the early 1980s, while the government was threatening to eradicate the settlements based on ecological arguments, officials consistently asserted that even if Los Belvederes were to be incorporated into the legally designated urban area, the terrain

Figure 7.7
Housing unit prior to upgrade, 1983.

Figure 7.8
Same housing unit, upgraded ten years later, 1993.

Figure 7.9
School playground before leveling, Bosques del Pedregal, 1986.

in Ajusco was too steep and too rocky for the installation of traditional water and sewage lines and development was declared to be cost prohibitive. The same officials announced that the streets could never be paved either, at least not in the traditional manner. To do so would cause runoff and erosion that would interfere with the recharging of the underground aquifers. Yet in 1993, there it was—paved streets, and the beginning of water and sewage lines.

Pronasol made these public investments in Ajusco possible. Even as these investments quell social unrest in what used to be the most conflictive parts of Ajusco, however, new tensions build. With the status of Los Belvederes secure, newly formed illegal settlements proliferate. The urban frontier pushes further and further into the ecological reserve. The pristine brooks seen during my first visit in 1983 now foam with suds from people washing their clothing. Land erosion, deforestation, loss of biodiversity, and aquifer contamination continue largely unabated.

Figure 7.10
School playground after leveling, Bosques del Pedregal, 1987.

A Problematic Dichotomy

Since the late 1980s, the Mexican government has shown greater resolve to definitively eradicate all illegal settlements occurring within the ecological reserve. The Federal District Department has monitored the 157 kilometers of the line of ecological conservation from twenty-three watchposts manned by eighty *ecoguardias* (Sosa 1991). Yet, even with all the measures implemented—miles of barbed wire and chain-link fencing that mark the urban-ecological perimeter; the growing force of *ecoguardias;* billboards warning against trespassing and illegal land transactions; tighter government controls over land use and construction permits for all projects in rural areas; and repeated eradications of incipient settlements—the urban sprawl continues.[2]

The only part of the ecological reserve in this area that is totally protected from urban encroachment is a small seventy-seven-hectare section of land surrounded by barbed wire and chain-link fencing. Within this highly restricted section, declared to be an ecological park, ecologists and biologists are busily documenting the number and type of species of flora and fauna. One of the principal ecologists in charge of the park, Jorge Soberón, says, "We don't intend to merely reforest the place, we intend to bring about its ecological restoration to a natural state. This will take 5 to 10 years" (Soberón 1990). This work is important, especially with respect to its well-financed public education aspects.[3] But in the scheme of things, the park is only a tiny piece of land, a token ecological reserve. It represents an approach to ecological conservation that avoids any conception of humans as components of ecosystems. Most of the area surrounding the park continues to be developed irregularly with no serious regard for ecological parameters. Meanwhile, a relatively tiny chunk of land gets enormous press for its return to nature. This highly problematic situation illustrates a misleading dualism that has roots going back to the colonial period.

During one of my visits to Mexico City (January 1993), I asked Jorge Gamboa de Buen, the director of urban development in the Federal District *(coordinador general de reordinación urbana y proteccion ecólogica)* how it could be that settlement expansion continues to press beyond the urban limits, further and further into the ecological reserve. Gamboa de Buen responded that the ecology on the other side has no constituency. He also said he is going to propose that the Delegación of Tlalpan be divided into two jurisdictions, one for the urban area and one for the ecological area. He argued that the conservation of Ajusco cannot be achieved if private interests linked to communal, ejidal or private property prevails. For this reason, the government expropriated 729 hectares to establish—once and for all—an effective greenbelt to contain urban growth. But even the most cursory glance at history leads to the recognition that such strategies are not new and that they have failed to work.

The case of Ajusco illustrates the manifold problems with the government's top-down approach to planning for ecological sustainability in the urban milieu. But my concern is not to simply critique such initiatives; I would also like to examine the interplay of top-down and bottom-up initiatives. In this way, Ajusco's grassroots environmental movement of the 1980s figures prominently in the story that follows. I agree with Castells

(1983, xvi) that if urban research is to respond to the questions of our times, then we must analyze the relationship between people and urbanization and "such a relationship is most evident when people mobilize to change the city in order to change society."

Whatever changes may be necessary to attenuate urban environmental problems, they are not likely to result simply from polite appeals to state agencies, dominant classes, or technical bureaucracies. Such changes will ultimately also depend on the effective mobilization and strategies of popular groups aimed at asserting and defending their collective interests. Of course, such groups don't necessarily have all the answers; nor do they always have a united front. Care should be taken not to romanticize the capacity or intentions of popular groups. Deep social change is not always part of the agenda. Nevertheless, these entities can be a locus of creative energy. As Ramírez Saiz (1990b, 234) points out, although popular urban movements are often reactive, they also have an ability to undertake constructive initiatives.

As it turns out, the grassroots environmental action examined in this part of the book met with some success insofar as the popular groups succeeded in maintaining their presence in the zone. They won the right to negotiate for legal titles for the land they occupied, and they managed to secure the most basic urban services. But they were not successful in implementing some of their ambitious ecological projects. From the beginning, relations with the state were adversarial. The significant support the community activists received for their ecological projects (including construction of a community-operated sewage treatment plant and solid waste recycling facility for 120 families), all came from outside Mexico.

Popular mobilization reached a peak in Los Belvederes during the early 1980s, when the threat of eviction loomed over everyone there. By the early 1990s the grassroots environmental movement in the zone has dissipated; in effect it had been demobilized through the process of political mediation and incorporation of its goals into the dominant institutions of society. Figures 7.11 and 7.12 provide a visual metaphor of this shift. The cooperative's faded sign calling for struggle is in the process of being whitewashed by the state. To set the stage for an analysis of these dynamics, I turn here to provide historical background.

Figure 7.11
"The milk store is the beginning of our struggle," 1983.

Historical Context

The number of irregular settlements and the urban population in Ajusco
Medio rapidly increased from the late 1970s to the late 1980s (see figures
7.13 and 7.14). Over the thirty-year period of 1950 to 1980, Tlalpan's
population increased elevenfold, from 35,000 to 385,000. During the
1950s, the average annual increase in the number of people living in Tlalpan was 6.5 percent. During the 1960s, this figure rose to 9.2 percent and
between 1970 and 1980, the rate increased to an explosive 9.4 percent
(compared to a rate of 3.8 percent for the entire metropolitan area) (Ibarra,
Puente, Saavedra, and Schteingart 1987, 310). Robinsons and Wilk's
(1991) careful study of land conversion in Tlalpan (see table 7.1) found
that between 1959 and 1985, more than 4,600 hectares were converted
for urban uses, at an annual rate of 177 hectares (roughly the surface area
of all thirteen settlements of Los Belvederes combined). From 1985 to 1989
more than 1,300 hectares were added to the urban area, at an annual

Figure 7.12
The cooperative's message whitewashed by the government, 1993.

growth rate of 344 hectares. Wilk's (1991c) results, presented in table 7.2, show the effects of urbanization from 1959 to 1985. Almost half of this growth took place on agricultural lands. Over the same period, 40 percent of forested lands were converted for agricultural use. In a survey completed in 1985, Robinson and Wilk (1991, 2) found that 47 percent of the farming in Tlalpan was in areas of relatively poor agricultural soils. They observed that this shift to the use of land less suitable for farming has created severe ecological problems. In terms of urban growth, Robinson and Wilk found that more urban uses were introduced on forest-land over the four-year period 1985 to 1989 than in the previous twenty-six years, when only 3.9 percent of forested land was urbanized (2).

Actions (and just as important, inactions) of the state have fostered the expansion of irregular settlements into Ajusco Medio. For example, the construction of the scenic highway Circuito del Ajusco in 1974 opened up the zone. According to several researchers, the then secretary of the highway department, Armando Bracomontes, ordered the road to be constructed so that he could get to and from his residential estate. In any case, construction

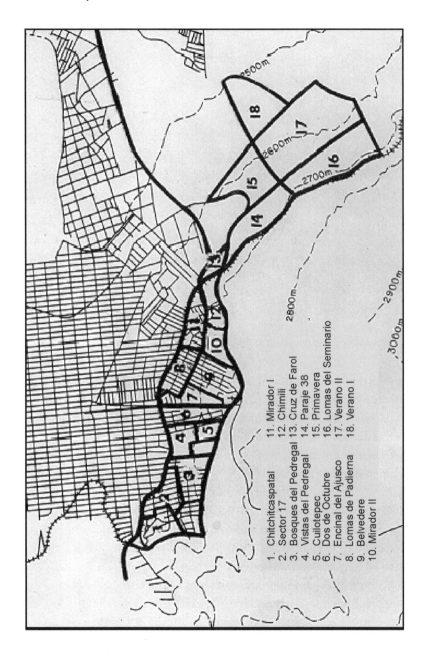

1. Chitchitcaspatal
2. Sector 17
3. Bosques del Pedregal
4. Vistas del Pedregal
5. Cuilotepec
6. Dos de Octubre
7. Encinal del Ajusco
8. Lomas de Padierna
9. Belvedere
10. Mirador II
11. Mirador I
12. Chimili
13. Cruz de Farol
14. Paraje 38
15. Primavera
16. Lomas del Seminario
17. Verano II
18. Verano I

of the highway led to a wave of land speculation that brought in its wake both upper-class and low-income development. The state has also constructed offices for its own use in the zone (e.g., the Colegio Militar, Canal 22, Televisa) and directly promoted the construction of institutions of higher education including the Colegio de México and the Universidad Pedagógica.

Other actions of the state, such as the blatantly contradictory and discretionary enforcement of the zoning laws favoring higher-income groups, have given momentum to irregular settlement by undermining the legitimacy of official policy and by giving popular activists an edge in their claim for land.[4] For instance, in the vicinity of where the government took steps to eradicate irregular settlements during the early 1980s (supposedly because the area is vital to sustain the ecological equilibrium of the city), the government also authorized development for higher-income groups (e.g., the amusement park Reino Aventura and the residential development Jardines en la Montaña).[5] One of the factors that has complicated the situation is the historical patterns of land use and property ownership in the region.

Contested Property Lines and Rural Conditions

Rural communities have been present in the zone of Ajusco since preColombian times.[6] Much of the area is still forested and dedicated to rural uses such as farming, cattle raising, and forestry. However, conditions within Ajusco's rural area, as well as dynamics engendered by the proximity of a rapidly expanding metropolis, have fostered the transition of land use from rural to urban—despite the politics of containment. Rural conditions that have made Ajusco prone to urbanization include historical land tenure arrangements, the characteristics of the land itself, and the ways in which production on that land has been carried out.

In 1940 most of Ajusco pertained to the ejidos San Nicolás Totolalpán de Eslava, Tlalpan, Padierna, and San Andrés Totoltepec (see figure 7.15).

Figure 7.13
Los Belvederes and El Seminario inside Ajusco Medio, 1988. Source: Delegación del Tlalpan (1992), DDF (1989b), and field research by author.

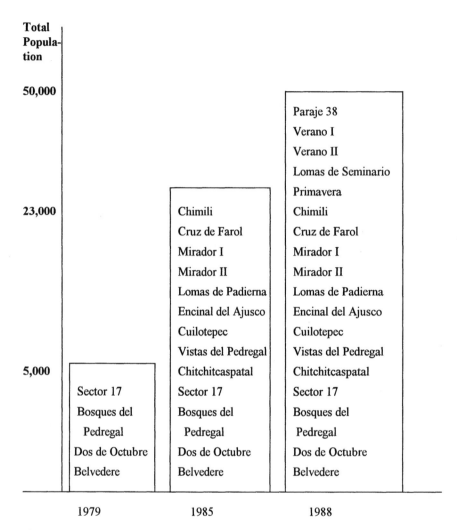

Figure 7.14
The Expansion of Irregular Settlements in the Zone of Ajusco Medio, 1979–1988.
Note: Given the lack of good demographic data, these population figures are rough
estimates. The increase over time reflects the demographic densification, or infilling, of existing settlements as well as the additional inhabitants taking up residence
in the newer settlements. Sources: The 1980 estimate was drawn from newspaper
accounts, oral testimony of residents, and from aerial photographs. The 1985
figure was reported in a census undertaken by DART (DDF 1985). I obtained the
estimate for 1988 from officials in the Subdirectorate of Ajusco.

Table 7.1

Hectares of land converted from agroforestry use to urban use, 1959–1985, 1985–1989

	Agriculture	Forest	Grassland	Shrubs	Total
1959–1985					
Hectares	2,243.26	181.45	783.63	1,406.77	4,615.11
Percentage	48.61	3.93	16.98	30.48	100.00
Annual average	86.28	6.98	30.14	54.11	177.51
1985–1989					
Hectares	681.37	298.75	120.36	276.81	1,377.29
Percentage	49.47	21.69	8.74	20.10	100.00
Annual average	170.34	74.69	30.09	69.20	344.32

Source: Robinson and Wilk (1991c, 2).

Table 7.2

Land conversion in Tlalpan, DF, 1959–1985

1985 land use (acres)	Percent of 1959 land use				
	Agriculture	Forest	Grasslands	Shrubs	Total
Urban (12,090)	48.6	3.9	17.0	30.5	100
Agriculture (21,050)		40.3	50.0	9.7	100
Grassland (7,510)	29.9	60.8		9.3	100
Forest (32,780)	31.0		32.2	36.8	100
Shrubs (2,190)	65.0	28.2	6.8		100

Source: Wilk (1991c, 45, table 11).

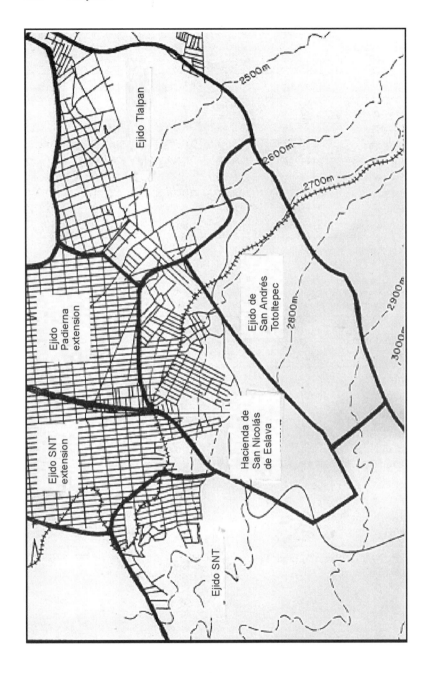

There were also significant extensions of private property referred to as the Hacienda de San Nicolás de Eslava and *pequeñas propiedades* (small properties). By 1988, the land tenure pattern had significantly changed in the northern part of the zone (see figure 7.16). Although figure 7.16 illustrates boundaries that clearly separate one property regime from the other, in practice these boundaries are hotly contested. Ejidatarios and developers have made overlapping claims, creating an incentive for each to sell the contested land before the other (*El Día,* June 24, 1987). There have also been heated disputes over property ownership within the areas generally acknowledged as private property (see figure 7.17).

Disputes to determine the macroboundaries of ownership that supposedly existed prior to subdivision for irregular settlement in Ajusco Medio have been mediated by the secretary of argarian reform and the Tlalpan Ward in a way that has created opportunities for graft and fraudulent deals and made the place ripe for illicit land transactions. Although such transactions are illegal, legal mechanisms (such as the *amparo,* an important juridical device brought into play in the power relations between contesting actors) have figured prominently into the politics of land allocation.

The condition of obscure boundaries is important to understand, but it contributes only a small part to an explanation of irregular settlement. To dig deeper one must also consider conditions in ejidos prior to their urbanization. The ejidos were first created during the 1920s as a result of the agrarian reform that expropriated land from haciendas in the area. However, the land allocated to the ejidos was for the most part rocky, steep, and of poor agricultural quality. Thus the land did not provide a sufficient base for the rural communities to sustain their livelihood. For this reason, and due to population growth, ejidatarios solicited expansions (*ampliaciones*) of their ejidos. The procedure involved soliciting the president of Mexico to expropriate additional land in their favor.

Most expansions of ejidos in the zone of Ajusco were granted during the administration of President Cárdenas (1934–1940). However, even

Figure 7.15
Land tenure pattern in the zone of Ajusco, 1940. Source: Tesorería del Distrito Federal, Subsecretaría de Ingresos Locales, Tlalpan (map E 14A39-53).

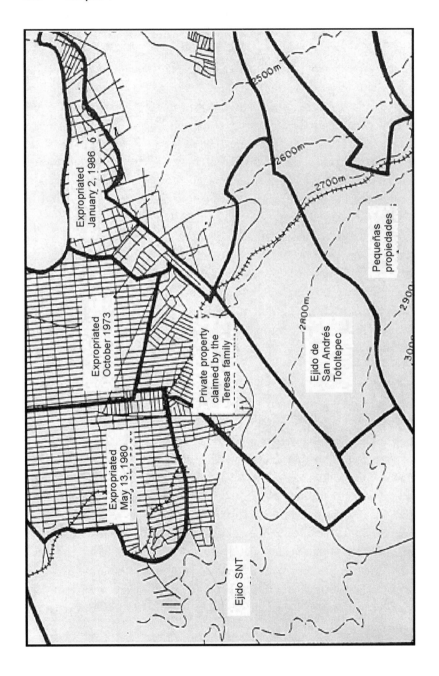

with subsequent expansions the ejido fell short of requirements necessary to make it an economically viable institution (Cockburn 1986).[7] This is largely because landlords of haciendas and private ranches of the area maintained control of the best agricultural land (Lugo and Bejarano 1981). Beginning in the 1950s the ejidatarios' conflicts over land were no longer oriented toward expanding their ejidos. Rather, they struggled for the highest indemnification for expropriation and for permission to use the land for residential purposes (Cockburn 1986, 740). This led to a reform in agrarian legislation and to the institutionalization of *zonas urbanas ejidales*. According to the Agrarian Reform Law, these zones were principally intended to meet the needs of ejidatarios working to reproduce their rural livelihood. However, the law permitted ejidatarios to sell extra lots in the urban ejido zones to neighbors (*avecindades*)—people who did not belong to the ejido but were supposedly necessary or beneficial to its operation (art. 91, 93).

The establishment of urban ejido zones had two important effects. First, it created a juridical loophole through which ejidatarios could sell their land. As Varley (1985, 72) explains, the law concerning such zones, with its vague language defining *avecindados,* provided an ideal opportunity for ejidatarios to sell their land to great numbers of low-income people completely alien to the ejido. Ejidatarios have been able to carry out such illicit land transactions with a *facade of legality* sufficient to reassure purchasers. Second, the institutionalization of urban ejido zones created a mechanism whereby the state could regularize or legalize unauthorized settlements on ejido land.[8]

Additional factors that have led to the illegal sale of ejido land stem from the ejido's declining productivity due to the spiraling cost of inputs, limited access to credit, lack of an agricultural infrastructure, impoverishment and erosion of the soil, and problems related to intermediation in the commercialization of their products (DDF 1985, 58). These conditions have made the illicit sale of land a much more remunerative prospect than alternative rural activities such as planting oats or maize. Meanwhile, the resources

Figure 7.16
Land tenure pattern in the zone of Ajusco, 1988. Source: Carta topográfica 1:250,000, Dirección General del Territorio Nacional, 1982.

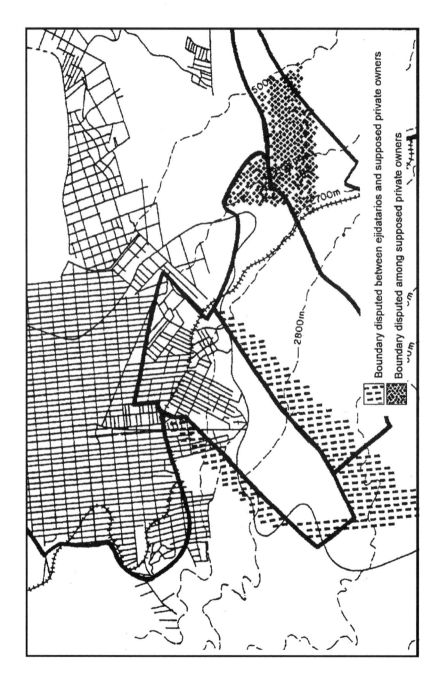

aimed at bolstering rural productivity, as a key strategy in the politics of containment, have been marginal. A little fertilizer and some technical training does not go very far against the kinds of structural imperatives that have led ejidatarios to continue selling land.

The level of enforcement employed to restrict illegal sales of ejido land has been minimal. Initial efforts to rectify the situation were embodied in the inception of the Commission for the Regularization of Land Tenure *(Comisión para el regularación de la tenencia de la Tierra, CORETT)*. According to CORETT's original charter, its main purpose was to regularize illegal settlements on ejido land and to actually act as a developer of low-income subdivisions by acquiring suitably located ejido land (*Diario Oficial,* August 7, 1973). However, as Ward (1989, 142) points out, strong opposition from the ejidal sector led the state to cancel CORETT's powers as a developer.

During de la Madrid's term as president (1982–1988) he emphatically denounced the rampant fraudulent subdivisions of ejido land, and he called for policies that would stimulate the creation of land reserves. Legal reforms and new *convenios* (agreements) in 1983 and 1984 among SEDUE, the SRA, and CORETT established the legal basis for expediting the identification and expropriation of ejidos whose land would likely be absorbed by future urban growth. Nevertheless, as Ward (1989, 142) argues, "even though these may be major steps toward the acquisition of rural land for long term urban development, in the absence of additional preventative measures they are unlikely to succeed in controlling land alienation by ejidatarios in the short or medium term." Indeed, as Ward suggests, quite the opposite may occur: "faced with the danger of government intervention to prevent them from illegally selling land, [policies along these lines] are likely to stimulate ejidatarios to sell or to seek ways to slow the process of handover to government" (142).

At the root of this problem is a fundamental disjuncture between the legacy of agrarian reform and agrarian politics on the one hand and urban development policy on the other (Azuela de la Cueva 1987, 533). Basically,

Figure 7.17
Macroboundary disputes in the zone of Ajusco, 1988. Source: Author's field research.

the process of irregular settlement on ejido land is dominated by agrarian politics, a control that urban policy has not been able to supersede. On another level, rural land in Ajusco has been susceptible to urbanization due to the overexploitation of its timber by private paper companies. In 1947, a concession was given to the paper factory Loreta y Peña Pobre to exploit the zone's timber.[9] At the same time ejidatarios and *comuneros* exploited the wood for their own use and sale. Practices of reforestation were initially practiced, but intensified clear cutting by the paper companies eventually began to devastate the forest. The ecological equilibrium was disturbed as fires and disease further eroded the productivity of timber as a viable and sustainable rural activity (Durand 1983, 55; COCODA 1982).[10] These consequences have intensified the problems of ecology and development in the area.

Only recently has an awareness of environment-development linkages begun to influence policy in rural parts of Ajusco (evidenced, for instance, in the design and implementation of agroforestry conservation programs). The relationship that rural livelihoods, technology, and resource use have to the sustainability of the environment is more readily recognized. Such recognition has to be brought into the urban milieu. Clearly, there is a contradiction between (1) the increasingly sophisticated politics of containment—embodied in new laws, plans, programs, and institutions—that have directly aimed at preventing the expansion of irregular settlements in Ajusco since the late 1970s, and (2) the fact that such expansion has occurred anyway. Have the politics of containment reflected a genuine concern for the ecological virtues of Ajusco? Or is it the legal and institutional maneuvering of political mediation aimed at enhancing the state's social control in the zone? Have the ecological arguments been employed as a ruse in an effort to eradicate the low-income population and allow upper-class development? What is the context of state-community relations, and how have these relations changed over time?

In response to these questions, this book argues that the state's activities in Ajusco have been conservative and reactionary; they have been more concerned with political mediation and quelling social unrest than with efforts to promote proactive ecologically integrated planning. However, this does not mean that the state's ecological arguments and environmental policies are devoid of legitimate concern for the environment and public well-being, nor does it mean that the politics of containment solely count as technologies of power employed by the state to control popular groups. As I noted earlier, legitimate concerns do have a part in the scenario. And the politics of containment apply to more than popular groups. The state also acts to control developers, as well as itself—mediating contradictory and competing factions within its own shifting internal organization. The next chapter takes a closer look at all of these actors and the balance of forces among them.

8

Key Actors and the Overall Balance of Forces

The actors that have had a hand in the urbanization of Ajusco can be grouped into four categories: (1) ejidatarios, (2) popular organizations and external agents, (3) developers, and (4) agencies of the state. None of these categories can be defined as homogeneous or as operating with uniform interests. Rather, each is a heterogeneous collectivity often experiencing internal contradictions and rival factions. In the presentation that follows each of the four categories of actors is first introduced from a broad perspective. Then, specific details about each of the main actors within the categories are provided.

Ejidatarios

The rural identity of ejidatarios distinguishes them from the other actors in the zone. Although most ejidatarios in Ajusco are actually employed in the urban economy, they are primarily identified as engaged in farming, ranching, and other rural activities. The ejidatario's most fundamental claim to land in Ajusco—based on usufruct rights granted to them through the Agrarian Reform Laws—has been that they use the land both as a productive asset and as a place for settlements. A single perspective cannot be attributed to all ejidatarios, but one common view is that urban encroachment (notwithstanding the fact that ejidatarios often foster such encroachment by illegally selling land) threatens their ejidos and their rural livelihood.

Although there are two groups of ejidatarios in Ajusco Medio, namely those from San Andrés Totoltepec and San Nicolás Totolalpán de Eslava, I focus only on the latter, for two reasons. First, the entire irregular settlement named Bosques del Pedregal (Bosques, for short) and part of Dos de Octubre are situated on the second group's land. Bosques and Dos de Octubre were among the first settlements to become established among the thirteen colonias that now make up Los Belvederes, and it is these settlements, along with the colonia Belvedere, that receive the bulk of attention in this book. Second, these ejidatarios have run the gamut from being the "principal enemies" of colonos in the late 1970s and early 1980s, to participating, more recently, as an ally in an alliance with a multisettlement coalition of colonos. The alliance was united in opposition to certain state agencies and developers.

Ejidatarios of San Nicolás Totolalpán

Before San Nicolás Totolalpán (SNT) became an ejido, it was a pueblo dating back to 1535.[1] During the early 1980s the ejido was inhabited by 450 ejidatarios, some of whom cultivated corn, potatoes, beans, and oats on approximately 500 hectares of nonirrigated land (Schteingart 1987, 464). Due to climatic conditions, erosion, plagues, lack of resources, and financial support, among other factors, the productive yield from the land was very low. Nearly all ejidatarios had to supplement their income by working in the urban economy. In his investigation of the economically active population of SNT, Rivera (1987, 99) found that in 1986 only two laborers fully earned their livelihood within the ejido.

The main authority inside the ejido is the General Assembly of all ejidatarios. Beneath this is the Ejido Commission (Comisariado Ejidal)—composed of a president, secretary, and treasurer—and the Vigilance Council (Consejo de Vigilancia), which is supposed to oversee the activities of the Ejido Commission. Ejido leaders often fail to represent the interests of the majority under their charge. This happened in SNT, when the Ejido Commission began selling off parts of the ejido. Witnessing the windfall gains reaped by these elected representatives, other ejidatarios have been inclined to do the same (Rivera 1987, 102–103). Often the situation gets out of hand; that is, ejidatarios may control the illicit sale of land during the initial stages of irregular settlement, but they sometimes lose this control as the settlement process quickens and "overflows." This is exactly what happened in certain places within Los Belvederes. The results are often highly conflictual. Factions of ejidatarios end up fighting each other, creating a situation in which colonos often find themselves negotiating with and paying money to one particular leader or group of ejidatarios only to be forcefully evicted by another.[2]

The most important point to make about SNT and about ejidos in general concerns the opposing interests of ejidatarios and colonos and the manner in which these interests have unequal protection under the law. The legacy of Agrarian Reform has given the ejidatarios a certain kind of leverage. The existence of government agencies responsible for agrarian reform and for political organizations of campesinos has created a situation inside the state apparatus whereby the interests of ejidatarios are better represented than the interests of colonos (Rivera 1987, 104). This does not mean that the interests of colonos and ejidatarios conflict at all times. Nor does it mean that ejidatarios always emerge from conflicts as beneficiaries. But it does mean that ejidatarios have an advantage.

By law, an agrarian community does not lose its land ownership even when the land has been illicitly sold off to urban settlers. Hence, when regularization becomes imminent the land must be expropriated by a decree signed by Mexico's president. One of the consequences is that occupiers who had already paid ejidatarios for their land must pay again. The initial transactions are not legally recognized, and the agrarian community has to be compensated for the expropriation.

Popular Organizations

The motive to settle Bosques was born out of our desperation for a decent place to live. Many of us came from the city's center expelled by real estate agents and big business, through the construction of highways, the extreme elevation of rents, the low minimum wage and unemployment, etc. When we arrived to this place it was uninhabitable . . . there were pits, ravines, steep cliffs, unstable rock, and no services whatsoever. The necessity for a place to live, the courage and determination of the colonos, was turned towards converting this place into a more or less livable settlement. Everybody, women and children included, worked collectively—despite conditions of hunger, cold, thirst and sickness.

—*UnoMásUno,* July 15, 1978

The preceding quote is from a grassroots activist living in Bosques del Pedregal, the first established settlement of the thirteen that now make up Los Belvederes in Ajusco Medio. In many ways his words illustrate a typical story: irregular settlement made possible by high degrees of popular mobilization and collective self-provisioning under conditions of great sacrifice and hardship. In this respect, there is little novelty to the story of irregular settlement in Ajusco. However, what is novel is the state's use of ecological arguments on a broad scale to prohibit access to land in the city's periphery (the politics of containment), and the countervailing forces, also incorporating ecological arguments, that were spearheaded by popular groups.[3]

To set popular mobilization in Ajusco in a broader context, it is useful to summarize several observations made by Julio Calderón (1986, 19–55) in his study that describes conflicts over land in the metropolitan area of Mexico City, including Ajusco, from 1980–1984.

1. During the five-year period from 1980 through 1984, there were 120 conflicts over land (involving state, private, and agrarian property regimes). The 120 conflicts comprised 439 events (e.g. a public protest, a press conference, forced occupation of a government office, a meeting between community leaders and authorities, participation in solidarity movements).

2. Of the 120 conflicts, 84 (70 percent) were *movimientos espontáneos,* characterized by (a) their shortness in duration—each movement was made up of only one or two events, most often a denunciation vis-à-vis a press conference or a one-time only protest or demand-making group demonstration, and (b) their confinement to official channels.

3. Of the 120 conflicts, 36 (30 percent) were *movimientos organizados,* characterized by (a) their medium- and long-term periods of duration— each movement encompassed between three and forty events over a period of one to three years; (b) a strong local leadership with organic connections to the grassroots; (c) the support they have from extralocal institutions or organizations including political parties and church groups; and (d) their articulation and coordination with other *movimientos barriales* or political fronts to strengthen their struggle through solidarity.

In light of the preceding definitions, the popular mobilization in Los Belvederes (between 1979 to 1990) could be classified as *movimientos organizados.* The zone of Ajusco has a certain notoriety for political opposition. Indeed, political opposition in the zone has legendary roots that extend beyond contemporary urban struggles. During the Mexican Revolution, the Sierra del Ajusco was a stronghold of Zapata. In 1918 Ajusco was the site of several battles between Zapatistas and the army loyal to Carranza. Going back even further, to the colonial period, Benítez Badillo (1986, 13) relays an interesting "imperial anecdote" that I reproduce here:

"The Emperor Maximiliano and Empress Carlota de Habsburgo had as their place of residence in New Spain the Castle of Chapultepec, from where one could observe the entire Valley of Mexico. One night the Empress, with her view settled on the peaks of Ajusco, noticed a series of lights that called her attention so she asked her servant: what are those lights? The response was: they are the bonfires of the rebels that oppose the new empire." The contemporary mobilization of Ajusco's urban poor has been characterized by its autonomy from the state and by its high degree of organization over the long term.

The mechanisms brought into play by popular groups in the competition for Ajusco's land have included all the different events noted by Calderón (e.g., public protest, press conference, forced occupation of government buildings) as well as strategies to strengthen solidarity—for instance, through the articulation and coordination with other grassroots movements and political fronts. In the late 1980s, this type of articulation and coordination led to the formation of the Coordinator of Colonias and Pueblos of the South (Coordinadora de Colonias y Pueblos del Sur), and the Popular and Independent Coalition (Coalición Popular y Independiente).

Bosques del Pedregal, Dos de Octubre, and Belvedere were the most organized and politically active settlements in the zone during the late 1970s and 1980s (see figure 8.1). They have also been the site of the greatest violence. To a significant extent state policy toward these settlements has shaped state policy toward all of Los Belvederes. Also, activists from each of these settlements in conjunction with external agents generated ecological projects that reflected the growing sophistication of urban social movements. Popular groups in Bosques del Pedregal went the furthest with ecological projects. Figure 8.2 outlines the popular groups, including the multisettlement organizations, that will be discussed in the next two chapters.

In terms of the production of irregular settlements, a distinction can be made between strategies of development (i.e., filling up the space quickly, establishing a popular elementary school as a shield) and strategies of making demands and claims. The latter set of strategies are where much innovation has occurred. Popular groups have become increasingly sophisticated in their use of the press, external agency (local, regional, national, and international), and capacity to assert technical arguments on par with the

Figure 8.1
"Support the struggle of Ajusco/Reject the evictions or relocations/Respect possession and colonos' rights." Note: This poster—two feet by three feet in size—was plastered all over Ajusco Medio in 1984.

ecological arguments of the state. The sophistication of popular groups in Ajusco, together with the fragmented and reactive nature of state action, led to the movement of the ecological line. The state's justification for moving the line incorporated some of the points raised by popular opposition. Government officials adopted these points as if they were their own. Yet, there has been no significant action on the part of the authorities to actually take "social capital" into account and foster an alternative approach to development.

The population out of which the popular groups listed in figure 8.2 emerged has both urban and rural origins. Some parts settled in Los Belvederes after migrating to the capital from rural areas of Mexico (e.g., Oaxaca, Puebla). But most inhabitants of Los Belvederes arrived from other parts of the Federal District, in some cases due to evictions caused in the wake of public works such as the construction of highways *(ejes viales)*. According to one study completed in 1985, 62 percent of the families in Los Belvederes earned less than the minimum salary, 15 percent earned the minimum, and 15 percent earned 1.5 times the minimum (Gotica Villegas, Pulido Vega, and Avila Bravo 1985). Of all those in the economically active population, 36 percent were employed in the secondary sector; 60 percent worked in the service (tertiary) sector (see table 8.1).

A striking aspect of table 8.1 is the low levels of un- and underemployment. This has to be clarified. First of all, it clearly shows that the population at the time was not marginal. The workforce was actively engaged and firmly integrated into the economy. However, stability of employment was, and still is, a serious problem. One survey reveals that three out of every four workers in the secondary sector have unstable employment (Arroyo Acosta, Correa Garcia, Gonzalez Villegas, Mosta Díaz, and Mora Alva, 1987, 82). This means a high degree of uncertainty as laborers continually search for new jobs. To make ends meet, many colonos have to work two or three jobs.

Over the years, popular organizations in Los Belvederes experienced dynamic rises and falls. This is common, but what is not so common is the primacy of ecopolitics and the way it has produced novel arrangements in the struggle over land. On this point, the role of external agency has been significant.

BOSQUES DEL PEDREGAL

Mesa Directiva
1977–1978
 Transition Group
 1978
  General Council of Representatives
  1979–1984
 Cooperative
 1983–present
DOS DE OCTUBRE

Unión of Settlers of San Nicolás
Totolalpan, 1977–1986
    Association of Independent
    Settlers of the
    Settlement Dos de
    Octubre, 1987–present
BELVEDERE
  Independent Promotional
  Commission, 1981–1982

   Association of Settlements of Ajusco,
   Casa del Pueblo, 1982–1990

MULTISETTLEMENT POPULAR ORGANIZATIONS

   Popular Front for the
   Defense of the Settlements of
   Ajusco, 1984–85

    Coalition, 1986–1990
    Coordinator, 1987–1990

| 1976 | 1979 | 1982 | 1985 | 1988 | 1990 |

Figure 8.2
Popular organizations in Los Belvederes, 1977–1990. Source: Interview data compiled by author.

Table 8.1
Employment, by sector, of the economically active population in Bosques del
Pedregal, Dos de Octubre, and Belvedere, 1985

Sector	Bosques del Pedregal Number of workers	%	Dos de Octubre Number of workers	%	Belvedere Number of workers	%
Secondary						
Unskilled labor	311	19	114	17	92	15
Certified labor	69	4	61	9	28	5
Carpenter	20	1	13	2	12	2
Mason	211	13	83	12	78	13
Subtotal	611	37	271	40	210	35
Tertiary						
Merchant	81	5	35	5	42	6
Public employee	162	10	115	17	71	12
Private employee	192	12	67	10	82	13
Domestic employee	137	8	58	9	27	4
Professional	—		—		12	2
Driver (bus, taxi)	—		—		25	4
Other	413	25	106	16	122	20
Subtotal	985	60	381	57	381	61
Unemployed or underemployed	53	3	18	3	23	4
Total	1649		670		614	

Source: Gotica Villegas, Pulido Vega, and Avila Bravo (1985).

External agents

It is often the case that popular groups do not become collective actors
spontaneously. Popular mobilization is often encouraged and supported
by agents external to the barrio. As Friedmann (1989, 508) points out:
"External agents, sometimes referred to as intermediate organizations,
provide an important conduit of ideas, information, resources, and moral
support for barrio residents. Without them, very little mobilization would
occur. Yet, it is also true that external agents cannot replace the political
will of barrio residents to join in common practice."

Two classes of external agents have been important in Los Belvederes:
university students and professionals. The part played by university stu-

dents (most notably from the departments of architecture, sociology, and anthropology at the Universidad Nacional Autónoma de México [UNAM]) has a tradition that goes back to the political radicalism engendered during the turbulent 1960s. After the 1968 student movement, many students opted to "use the squatter communities as a ground on which to build a new form of autonomous political organization" (Castells 1983, 195). These efforts contributed to the emergence of a new radical Left, and grassroots activists gained experience leading urban struggles. Related to this, a growing number of nonprofit research and consultant agencies began to work with grassroots organizations. The most important agency of this type in Los Belvederes was the Alternative Technology Group (Grupo de Tecnología Alternativa; GTA).

Developers

Two basic types of developers have been operating in Ajusco: those who have engaged in (or who have lobbied for) upper-class development and those who have illicitly subdivided the land for low-income settlements. These are not mutually exclusive activities; some developers do both. Either type of developer can be set apart from the other actors by their activity's direct tie to the deliberate commodification of land and realization of its exchange value. In Ajusco developers have principally been interested in reaping profit from land transactions, whether or not they were legally sanctioned to do so and whether or not they actually owned the land. Their claim to the land has been argued on the basis of their supposed ownership, with reference to the sanctity of private property and to principles of the market economy.

This is a diverse category of actors that does not express a uniform attitude toward the land problems in Ajusco. One view, common among the upper-class elites involved, is that poor people are the problem: unless the poor are kept out of Ajusco they will destroy an important ecological resource for the city. Furthermore, they argue that the area should be occupied only by wealthy people who can afford low-density and ecologically sensitive development. Another perspective has nothing to do with environmental concerns. It is a view that combines social Darwinism with raw profit motive. Competition for the land is viewed as a power game in the urban jungle: those who best play the game make the most money and control land development. Developers central to irregular settlement in Ajusco include the Teresa family, the Belvedere Association, and Rena and Associates. The first two of these developers have an adversarial relationship with the third. The manner in which the developers have operated is fundamentally the same. They write up a contract and collect an *enganche* (deposit) followed by installments including interest. Colonos who do not comply with the contract face expulsion.

During the initial settlement of Dos de Octubre and Belvedere, developers periodically prompted the state to carry out violent, large-scale evictions (desalojos). Among the factors leading to such desalojos were (1) the independent mobilization of colonos and their collective refusal to continue payments, and (2) the infilling of nonpaying settlers in the form of an *invasión hormiga*. *Invasión hormiga* translates literally as "ant invasion." It implies a relatively quiet, gradual process of encroachment until some threshold is reached and it becomes apparent that major inroads have been made. Sometimes the developers evicted settlers in order to clear a lot and sell it again. To cite a popular group publication, "clandestine subdividers and supposed owners typically carry out their operations for speculative purposes by resorting to different mechanisms and legal subterfuge, such as the creation of phantom civil associations, titles, wills or business contracts that are equally invalid" (Belvedere, Casa del Pueblo 1985, 3). Developers often have support from government officials and political functionaries.

The Teresa Family and the Belvedere Association

The Teresa family claims a large portion of Los Belvederes referred to as the Ex-Hacienda de Eslava. Their claim goes back to 1865, when the family originally purchased the Hacienda de Eslava (Schteingart 1987, 459). According to an agreement signed by authorities of the Federal District Department and by Fernando de Teresa (the executor of the estate) ownership was passed down through the family by means of will and testament.[4] When I was doing field research in 1987, Fernando de Teresa was an engineer in charge of the family business. The operation was divided into three functions: (1) legal affairs run by a group of lawyers, (2) technical affairs including the drafting of plans and the determination of land prices and payment schedules, and (3) a branch in charge of collections.

By the late 1960s the Teresa family had sold 80 percent of the land they claimed in the zone. They kept 20 percent of the land for themselves for strategic purposes. The Teresa family calculated that when the time came for land tenure regularization they would be the largest property owner in the zone, thereby giving them an upper hand in negotiations regarding sales. The Teresa family began to sell land in the zone during the early 1950s. In the beginning very large sections were purchased by several individuals. These individuals then resold the land in the form of lots measuring 1,000 square meters each (twenty meters by fifty meters). These lots were purchased by a group of nearly 1,000 middle- and upper-class professionals, many of whom were university professors and federal employees. On April 16, 1968, this group formed the Belvedere Association with plans to develop the site into a "model community."[5]

The bylaws of the Belvedere Association, dated 1967, state that the principal aim of the association is to develop an exclusive, upper-class residential development modeled after Swiss chalets. Interestingly, ecological measures were to be incorporated into the project's overall design in order to preserve the area's natural beauty. Emphasis was placed on the prestige of the place. It was to become the most beautiful and most desirable project in Mexico City, with an international reputation. A high stone wall was to be built around the site with "the objective of maintaining its exclusivity, beauty and autonomy." It was projected that a commercial zone would be built to meet the needs of residents and elite travelers, including a Hilton Hotel that would rival the best in the world for its view and luxury.

The "model community" project never materialized. There were insufficient funds and unlike what happened in other upper-class development projects in the zone, there was inadequate support from the state to proceed.[6] The important point here is that there have been designs for upper-class development on the zone for a long time. Such designs have made colonos in the low-income settlements wary that the government's ecological policies have been more concerned with residential segregation than with the ecological virtues of the zone.

Rena and Associates

Rena and Associates (henceforth Rena) are notorious for their fraudulent subdivision of land in the zone of Ajusco. Legally constituted as a civil organization on August 11, 1971, Rena has been prolific in their authorship of amparos[7] (at least six between 1984 and 1987) and their claim to land in Ajusco involves a long, convoluted story. Rena is the driving force in the push of irregular settlement beyond Los Belvederes, into El Seminario. In an amparo filed by Rena on November 21, 1986, Rena traces their claim to the land in Ajusco back to 1941 when the thread and fabric factory Fama Montañesa declared bankruptcy. When Fama Montañesa went bankrupt the owners handed over the estate to Jésus Martínez Orozco, at that time the general secretary of the Union of Workers and campesinos of the Fama Montañesa (Unión de Obreros y Campesinos de la Fama Montañesa). This transfer was part of a settlement giving land to the union in lieu of wages. Rena claims that by way of a barter agreement *(contrato de permuta)*, Orozco later transferred the property to them on August 3, 1972. Government officials have declared that Rena's claims are fraudulent.[8]

The paradox I wish to highlight is that nearly all actors in the zone (the state, popular groups, ejidatarios, and other developers) characterize Rena as a villain. Nonetheless, Rena's successes continue and the wealth they have accumulated has been tremendous. Besides superlative skills in the art of conning people, Rena has been very clever in exploiting divisions within the state and maneuvering through legal loopholes.

In view of Rena's capacity to circumvent zoning laws and subdivision regulations, some have gone as far to say that Rena could subdivide the Zócolo! The drawing in figure 8.3 illustrates the general sentiment toward Rena. The drawing illustrates a well-disguised man who melds into the background. This illusory figure just sold yet another chunk of Ajusco to an impoverished fellow who had to empty his pockets to pay for it. What we have here is capital accumulation through illicit land transactions. The amparo process, together with government connections, has been effectively used by Rena to push irregular settlement into the Lomas del Seminario part of Ajusco Medio (see figure 8.4).[9]

Figure 8.3
Caricature of Rena, a clandestine real estate developer active in Ajusco. Note: The top reads, "Until when?," the bottom "Rena Real Estate Agency." Source: *El Nacional*, May 16, 1986.

Figure 8.4
Land in the process of illegal subdivision, Lomas del Seminario, 1988. Note: The white line in the lower right-hand corner is a boundary of lime laid down by the *fraccionador* (illegal developer).

Agencies of the State

The state is set apart from the other actors in Ajusco by its claim that it acts in the public interest. This is the basis upon which the state's action regarding land and irregular settlements has been legitimized. Directly and indirectly, numerous government agencies have had a role in determining the politics of land allocation in Ajusco. These include the Tlalpan Ward, the Subdelegación of Tlalpan in Ajusco, the DDF, DART, CORETT, SRA, and various local and regional planning institutions. The subward of Ajusco is the government's "front-line" state agency. According to its charter, the mandate of this front-line agency is to "respond institutionally to the complex and intensifying social, political, juridical and ecological problems of the zone" (Subdelegación del Ajusco 1985c)

The Overall Balance of Forces

The relation of forces in the struggle for control over land in Ajusco Medio has changed over time. In Los Belvederes the relative weight of the state has tended to increase over time regardless of the property regime. During initial stages of settlement, on both ejidal and private property, the state's presence was minimal. It became more significant as conflicts arose and as demands were placed on it. Eventually, through political mediation and the initiation of land tenure regularization, the state became the most significant actor in the process of land development—most significant in terms of capacity to "govern" in the sense defined by Foucault. This sequence is not unusual. The state in Mexico has generally been characterized as a reactionary force in settlement planning. The politics of containment aim to reduce the lag in the state's capacity to control urban development at the city's outer limits.

The relative weight of popular groups in Los Belvederes has tended to rise and fall regardless of the property regime in question. During the initial stages of settlement in Los Belvederes, first on ejidal land and then on private property and on land in litigation, the relative weight of popular groups grew. During the period of consolidation they reached a peak. Following this peak, once the social and spatial structure of the settlements were assured of becoming firmly integrated into the formal legal and institutional terrain, the capacity of popular groups to organize the action of others and to overcome the resistance of vested interests tended to dissipate.

On ejido land, ejidatarios were initially directing settlement, while on private property and on land in litigation the developers were doing so. Both groups fell in importance during the period of settlement consolidation until the time came to open formal negotiations with the state over the terms of land tenure regularization.

The rise and fall in the relative weight of actors in the struggle for control over land is not the outcome of some chaotic clashing of competing forces. There is a certain logic to it—logic in the sense that there is an identifiable paralegal framework within which actors define their legitimacy. Here I draw on the work of Antonio Azuela de la Cueva (1987, 524), who suggests that "in every process of formation of a low-income settlement there is a series of social representations related to the control of land." Azuela de la Cueva (1987, 524) refers to these social representations as "forms of legitimation" and argues that

Control of land, that specific combination of economic and political power at the local level which allows someone to command the distribution of land between new settlers, never appears as an unnamed power. Instead, it is always expressed in such terms as to make it recognizable and legitimate. Regardless of the actual social origin of land control, those who exert it represent themselves as holding some kind of right over the land. In turn, the legitimacy of the settlers' possession depends on this right.

Azuela's insight into forms of legitimation allows us to see how illegal settlement processes are conditioned by the law. Far from being an anarchic or chaotic process at the margins of society, illegal land development is dynamically rooted in the legal and institutional terrain. This is clear when we examine the activity of ejidatarios, developers, and popular groups. Within each of these categories, the capacity for advancing the interests of the parties involved hinges on the legal form by which they identify themselves for legitimacy (e.g., ejidal organization, civil association, cooperative organization).

In Ajusco, forms of legitimation promoted by dominant social agents in the urban process have aimed at convincing settlers and authorities that their control over land is legitimate from a legal perspective. In the development of Los Belvederes, ejidatarios, developers, and popular groups have each incorporated certain contents of legal rules into their practices. In this way, "forms of legitimation are thus a compromise between the actual power of the social agents who control the urbanization process, on the one hand, and the legal rules of the form of land ownership involved" (Azuela de la Cueva 1987, 524). Thus, it is crucial to critically examine the means by which juridical and institutional arrangements have conditioning effects on the social relations of property (Azuela de la Cueva 1982, 83).

Any study of illegal settlement in Ajusco should be quick to discover that a central juridical strategy (besides legally constituting one's particular group or organization) is the amparo process. Discussed earlier, the amparo, known in English as a temporary restraining order or court injunction, is one of the key mechanisms brought into play in the power relations between actors. The amparo process is increasingly used by ejidatarios, popular groups, and developers.[10] Notwithstanding the stepped-up recourse to amparos, most conflicts over illicit land transactions (e.g., disputes caused by a single lot sold twice over) have not been dealt with in the court system. Outside the courts, battles over land become purely political and government functionaries intervene as *mediators* (not as judges) who enjoy wide discretionary powers. The consequence is that the rule of law gives way to ad hoc conflict resolution, and the outcome of state intervention depends on the respective ability of settlers and landowners to pressure authorities (Azuela de la Cueva 1987, 526).

Where the state intervenes as mediator, the contesting actors involved do not merely appear as buyers and sellers in conflict. Developers, popular groups, and ejidatarios have distinct juridical identities that influence their actions—first by implying different sets of rights and duties, and second, by compelling an obligatory relationship with specific government agencies. This setup has important implications in Ajusco, where much of the as yet undeveloped land pertains to agrarian communities. In situations of conflict, as Azuela de la Cueva (1987, 539) points out, "government agencies may enjoy wide discretionary powers to recognize certain kinds of social organizations such as settlers associations, but unless major political changes occur, they cannot modify the status of other organizations such as agrarian communities." During the early 1990s, such changes were accomlished by way of ejido reform. Yet, experts have been quick to note that these reforms (specifically, changes to Article 27 of the Constitution to allow the privatization of ejido land) are not likely to slow irregular settlement. The situation in the mid-1990s shows this to be an accurate critique. Much depends on the overall balance of forces. On this point, the degree of popular mobilization has been, and continues to be, key.

Popular Mobilization in the Formation of Los Belvederes: Origins, Structures, and Agendas

Crossing the Railroad Tracks

A threshold was crossed when the growth of irregular settlements in Ajusco Bajo expanded south across the railroad tracks around 1976. The railroad tracks were a barrier; once they were crossed the expansion of irregular settlements into Ajusco Medio was under way. The railroad tracks form one of the defining boundaries of Los Belvederes. The first place that irregular settlement crossed the tracks was in an area later called Sector 17, followed by the settlement of Bosques del Pedregal, Dos de Octubre, and Belvedere. But Sector 17 has a distinct history from the rest of Los Belvederes: it was included in the expropriation decree of San Nicolás Totalpán on May 13, 1980. Thus it was brought into the legally designated urban area during the same period as lower Ajusco.

Profiles of the Early Settlements

Together Bosques del Pedregal, Dos de Octubre, and Belvedere have been the three most politically active colonias in Los Belvederes. Tables 9.1 and 9.2 describe certain characteristics in these settlements. This chapter focuses primarily on Bosques del Pedregal for three reasons: (1) it was the first irregular settlement to become established in Los Belvederes, (2) during the 1980s, popular activists in Bosques provided barriowide leadership throughout Los Belvederes, and (3) Bosques went the furthest with its urban-ecological projects and proposals for alternative development. Families first began to establish the colonia Bosques del Pedregal in the summer of 1976. Around this time there were four attempts to eradicate the settlement. Despite evictions, colonos returned. In 1984, after eight years of persistently threatening to eradicate the settlement, the government finally extended the officially designated urban area to include Bosques and the rest of Los Belvederes. The government initiated a land tenure regularization program in 1985. In the next section I recount the history of Bosques up until 1983, which marked a major turning point. I also discuss Dos de Octubre and Belvedere for comparative purposes and to indicate their alignment with Bosques in multisettlement movements.

Sector 17

The origin of Bosques can be traced to the initial settlement that occurred from 1975 to 1977 in Sector 17, a colonia divided into seventeen blocks (manzanas), which adjoins Bosques on its northern border. The way land was initially allocated in Sector 17 set the stage for the emergence of popular groups that spearheaded the settlement of Bosques. Sector 17, which in the beginning was illegally situated on ejido land, is distinct from the rest of the colonias of Los Belvederes in that its land tenure had been legalized early on. Apparently, there were plans to incorporate Sector 17 into the officially designated urban area even before it became an established colonia. On November 1, 1974, the Treasury of the Federal District published a reference manual on urban land values for taxation purposes, and Sector 17 was included within "Región 55." On July 19, 1977, CORETT initiated the process to regularize the settlements in Región 55 by filing a request to expropriate a large part of the ejido San Nicolás Totolalpán. The request was approved May 13, 1980.

According to several organizers of the first popular groups that settled Bosques, the initial developers of Sector 17 were operating with inside information and government contacts. Specifically, leaders from Bosques argued that a cabal of ejidatarios from San Nicolás Totolalpán were illegally subdividing land for sale with the help of Alicia Torruco, at that time a deputy of PRI and an employee of the Agrarian Reform. The organizers suggest that Torruca and other leaders, such as Silva Flores Neri (a member of the PRI), knew that Sector 17 was going to be regularized. Flores Neri, a lawyer, filed an amparo to protect a group of families in the seventeen *manzanas*. He sold copies of the amparo for a fee (even though by law only those names recorded on the amparo would be covered). Also the president of the Ejidal Commission, Arcadio Camacho Zúñiga, from

Table 9.1

Characteristics of Bosques del Pedregal, Dos de Octubre, and Belvedere, 1983

	Bosques del Pedregal	Dos de Octubre	Belvedere
Land area (acres)	60	55	50
Population	2,783	2,199	4,968
Property regime	Ejidal	Ejidal and private	Private
Zoning[a]	Within buffer zone	Within buffer zone	Within buffer zone
Land use[b]	Urban residential	Urban residential	Urban residential
Land tenure status	Irregular; expropriation begun with CORETT but process is suspended	Irregular; boundary disputes among ejidatorios and developers, and among rival developers	Irregular; boundary disputes among developers

Source: Subdelegación del Ajusco (1985a) and field research interviews conducted by the author.

a.   Land uses permitted in the buffer zone included habitation for rural settlements, agriculture, cattle ranching, parks, and limited services (e.g., facilities for recreation and tourism).

b.   In 1983, approximately 65 percent of the total land area in Los Belvederes was divided into lots for one-room housing units. The rest of the land was divided among services, infrastructure, commercial use, and open space.

whom many of the settlers purchased their lots, allocated land titles that colonos used to resist eviction. Based on paperwork issued to them by the ejidatarios, colonos were able to argue that they were *avecindados*— legitimate residents living within an urban ejido zone. Even though many settlers of Sector 17 paid the Ejidal Commission for their lots, Rivera Lona (1987, 82) points out that settlement there was presented in the interior of the ejido as an invasion.

Origins of Bosques del Pedregal, 1976–1977

Around 1977 several local factors led to a spilling over of settlement from Sector 17 into Bosques: (1) there were several desalojos in Ajusco Bajo and

Table 9.2
Characteristics of housing units in Los Belvederes, 1983

	Lots	%
Total	4,803	100
Occupied	4,560	95
Vacant	243	5
	Housing units	%
Total	4,560	100
Consolidated	1,824	40
Precarious	2,736	60
	Precarious housing units	%
Total	2,736	100
With one room	1,450	53
With average of three rooms	1,286	47

Source: Gotica Villegas, Pulido Vega, and Avila Bravo (1985).

new arrivals poured into Sector 17, (2) the Pristas (members of the PRI), ejidatarios, and colonos began to fight over the price and terms of land sales, and (3) independent leadership with experience from previous struggles arrived to the area (e.g., organizers from the *colonia* Ajusco).

On the southern edge of Sector 17, a zone of transition emerged. Many of the people who first entered the transition zone paid the ejidatarios for the land, often making a deposit followed by monthly payments. The demand for land intensified, and soon more and more people took lots without paying anything. In this way the origin of Bosques is best characterized as a mix of illegal sales and small scale "invasions." People staked their claims by erecting small dwellings made of rocks, branches, and cardboard. On three separate occasions during the summer of 1977 ejidatarios initiated police action against the invaders. Hundreds of dwellings were torn down and burned. After each desalojo some people dropped out never to return. Others returned again and again, consolidating themselves as a group with common interests. A fourth attempt to eradicate the settlement was resisted. Colonos who were there recorded the incident: "The police, led by the chief (jefe) of colonias of Tlalpan, moved in to throw us out, but by then we were more organized. The police could see that the situation was going to get ugly, with them on the one side ready to enter and destroy, and the colonos on the other side standing their ground, refusing to be burned out" (from the grassroots pamphlet *Boletín de Colonos*, April 1986).

The confrontation ended in a standoff, with the police giving the colonos seventy-two hours to clear out peacefully. The colonos used the time to gather support for their struggle; they organized brigades to solicit support from Universidad National Autónoma de México (UNAM) students and from syndicates. Students helped the colonos make banners declaring relations of solidarity, which the colonos then placed throughout Bosques. They persuaded other experienced popular groups (notably, Campamento Dos de Octubre—Bloque Urbano de Colonias Populares) to give them banners.

They also sent letters and telegrams to the regente and to the president. When the police returned, on June 6, 1977, the colonos had a strategy: (1) They argued that they would not vacate the premises until they received a response to their letters, (2) They used banners showing the support they had from students and unions, and (3) they gave signals—holding "sticks, machetes and rocks"—that made clear their willingness to defend their claims (*UnoMásUno*, n.d., 1982). The colonos were successful. The police left saying that they would return on another day to dislodge them.

The colonos had formed a *mesa directiva* and were now in a position to consolidate their organization. Rather than directly confront the Pristas in Sector 17, the leaders of the Mesa and their followers met in an area later to become Manzana 18. The rise and fall of the Mesa, which occurred in less than a year, marked the beginning of ten years of struggle and five major transformations in the composition of the popular organizations directing events in the zone. In the following section I trace the development of these organizations; the focus is on their strategies and methods aimed at securing land.

Popular Organizations in Bosques del Pedregal, 1977–1983

The Mesa Directiva, 1977

The first task of the Mesa Directiva, a group of five or six activists, was to organize four commissions: (1) planning and subdivision (*lineamiento*), (2) press and propaganda, (3) security and vigilance, and (4) land allocation (*reacomodaciones*). Members of the Mesa were not elected; they were self-appointed, "natural leaders," their power determined by their capacity to allocate land. The first major decision of the Mesa was to define what would be the boundaries of Bosques. To maximize their control and enhance their chances of establishing the new settlement, they observed the following territorial boundaries: (1) to the north, the street Encinos, which marked the outer edge of Sector 17 controlled by the *pristas*, (2) to the west, the natural barrier presented by a river-ravine, (3) to the south, a dirt access road, and (4) to the east, land that was still under cultivation (milpas) (see figure 9.1). This boundary setting is clearly an instance of popular planning. The space was strategically defined; it was not merely filled in some chaotic or anarchic fashion. Once the territory to be occupied was defined the leaders set aside land for streets, services, and open space. With some help from architecture students, leaders marked off street lanes with lime and branches, following the trajectory of the streets downhill.[1] At the same time they marked off *manzanas* and allocated lots. Initially each lot was approximately 500 square meters, but they decided to divide the lots into two approximately 250-square-meter lots so that more people could be brought in right away. The organization of the colonia depended in part on the accomplishment of this territorial division.

At first, the objective was to populate the settlement as quickly as possible. However, land was not allocated indiscriminately; there were two basic conditions—set by the Mesa. First, each incoming family had to set up a dwelling unit right away and occupy it. This meant that they had to bring all their belongings and begin living in the colonia, despite the considerable risks and hardships involved.[2] If someone set up a dwelling unit but did not occupy it full-time, the Mesa would reallocate the land. Brigades actually checked on occupancy. The idea was that if a family was really needy than they would be willing to set up right away despite the insecurity. Until the settlement had a significant number of full-time resident families, the risk of eviction would be high.

The second condition a family who wanted land had to meet was financial. Families had to pay a weekly fee *(cuota)* to the Mesa. The fee was intended to cover expenses to (1) support the leaders to work as activists (2) promote an amparo (including money for transportation, photocopies,

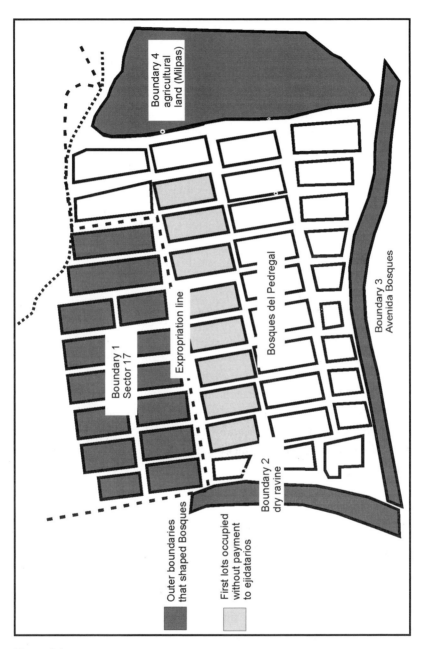

Figure 9.1
Four boundaries *colonos* used to define the settlement of Bosques del Pedregal.
Source: Interview data compiled by author.

etc.), and (3) fund the construction of a school. According to the oral histories of two leaders (Pablo and Chavo) who were on the Mesa but later defected from it, the Mesa became exploitative. It lost its function of serving and protecting the colonos. Instead, the leaders took advantage of the situation in a way that primarily benefited themselves. In particular Narciso, a leader who took over the commission of land allocation by arguing that he had "pull" *(palanca)* within the delegation, and other leaders in the Mesa were accused of selling land and charging fees to the colonos that never got used for the colonos' benefit. Pablo and Chavo claim that "political manipulation was common—they brought us like sheep to the acts of the PRI" *(Boletín de Colonos,* 1986, p. 11). Furthermore, the two informants attribute no concrete benefits to the Mesa (e.g., progress on the school was not forthcoming).

Transition Organization, 1978
Over time discontent with the Mesa grew. People began to refuse to pay the weekly fee. Groups outside the control of the Mesa began to engage in collective projects, for example the construction of an elementary school. Dissident members of the Mesa and others—Pablo, Salvador, Wulfrano, Marcilino—began to consolidate an alternative organization that openly opposed the Mesa. An early attempt to negotiate failed. Dissidents of the Mesa argued that the only way to gain security of occupancy in the settlement would be to put into action a Council of Representatives (Consejo de Representantes)—a new popular organization based on the democratic participation of as many colonos as possible. But according to one of the dissidents, the Mesa could not accept the idea of a Council. Instead, they claimed the need for a strong leader, someone with connections and who would command.

The alternative group set up its own commissions (basically the same as those of the Mesa) and began to allocate land without charge, thereby gaining popular support. It was about this time that the Mesa sought advice and support from Francisco de la Cruz, the leader of the Campamento 2 de Octubre in Ixtaclaco.[3] At the time, the popular movement based in the *campamento* was the most radical and influential in the city (Castells 1983, 196). De la Cruz sent 80 to 100 people to Bosques to claim land and to support the Mesa. Things became volatile: conflict between the Mesa and the emergent rival organization intensified. Dissidents from the Mesa publicly denounced de la Cruz for his extortion and manipulation of the colonos: "To combat the caciquismo and style of leadership that always conducts corruption, and does not promote the participation of our community in the solution of our problems, we denounce the militaristic manner of Francisco de La Cruz" (*UnoMásUno*, July 15, 1978). They complained that he instigated group infighting and gangsterism, which he ultimately controlled to his own advantage. In a final showdown, the leaders of the Mesa, Manuel Reyes and Jesús Rivera, were forced to leave Bosques. They moved on to lead the formation of a newer settlement to the east: Dos de Octubre (see figure 9.2).

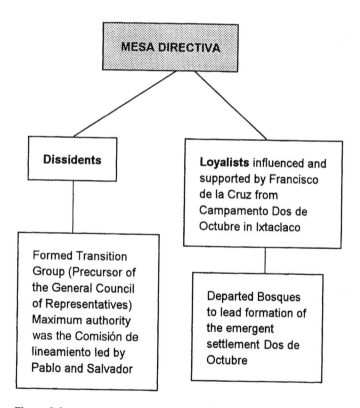

Figure 9.2
Origin of the transition organization, 1978. Source: Interview data compiled by author.

General Council of Representatives

What began as a dissident group within the Mesa developed into the General Council of Representatives (hereafter referred to as the Council). In a meeting on June 24, 1979, a group of thirty (out of a total of forty) leaders—each of whom had been elected to represent his or her respective *manzana*—voted to approve the bylaws and articles of incorporation of the Council. At the same time, they elected the representatives that would serve in the Council. During a general assembly (asamblea general) on August 11, 1979, the decisions were ratified and, after filing proof of the events with a notary public, the General Council of Representatives, Civil Association became a legal entity.

The organizational structure of the Council was modeled on the Workers Union of UNAM (STUNAM), where grassroots activists in Bosques had gained experience (see table 9.3). The objective of those who founded the Council was to create a form of legal representation that would encourage the maximum participation of people in matters pertaining to the settlement. At about the same time the Council in Bosques del Pedregal was established, colonos of Dos de Octubre legally constituted their own organization: the Union of Settlers of (Unión de Colonos de) San Nicolás Totolalpán. This organization used the Bosques Council as a model for its own organization. In 1982, colonos of Belvedere formed the Association of Settlers of Ajusco, Casa del Pueblo, Civil Assoication (Asociación de Colonos del Ajusco, Casa del Pueblo, A.C.). The popular groups in all three settlements mentioned sought juridical status for their grassroots organizations. They believed such status would serve as a form of legitimation. Forms of legitimation arising from popular social practices substitute for state law as far as they fulfill the role of a legal system in "normal" situations (Azuela de la Cueva 1989a). Popular groups often solicit juridical recognition for their organizations in order to convince settlers and authorities that their control over land is legitimate. Here we see an inherent contradiction: colonos organize to challenge dominant social and political relations of power, but in so doing they take on an identity defined by the very system they oppose. This has important implications: struggles tend to be defined by the legal and institutional terrain; they become institutionalized through political mediation.

Table 9.3
Structure of the General Council of Representatives, Bosques del Pedregal, 1984

Body	Composition	Power/Function
General Assembly	All members	Maximum authority; ratifies decisions
Executive Committee	Five representatives elected during general assembly	Represents the settlement in all negotiations
Assembly of Block Representatives	One representative (plus one alternate) from each block (thirty-one total)	Acts as main decision-making body
Assembly of Sector Coordinators	One representative from each sector (eleven total), plus heads of any special commissions	Serves check-and-balance function against the activities of the Assembly of Block Representatives
Sector Assembly	All members of a sector[a]	Serves to mobilize people for political purposes or tasks requiring collective labor
Block Assembly	All members of a block[b]	Has the immediate authority to resolve problems internal to the block, subject to the rules of the association

Source: Field Research and interview data gathered by the author.
a. The settlement is divided into eleven sectors, each composed of two, four, or five blocks.
b. The settlement is divided into thirty-one blocks, each composed of approximately fifteen to twenty individual lots.

To be a member of the Council in Bosques, a settler had to be in possession of a lot. If there were two, three, or four families on one lot, only one of the families could vote—a built-in bias against renters. The association's main objective, as originally stated in its articles of incorporation, was "to promote the measures, and to complete the obligations necessary in order to bring about the inscription and legalization of the lots in the Colonia Bosques del Pedregal." Toward this end the council mobilized thousands of people on marches to local government offices and sometimes to Los Pinos—the residence of the president.[4] The Council aimed to do what was necessary to upgrade the colonia both in physical and social terms. For example, the objectives of acquiring basic infrastructure (e.g., a drinking water supply) and basic services (e.g., an elementary school) were listed as priorities. To improve living conditions and to create jobs, the association aimed to promote cooperatives. Other more explicitly political strategies were the fostering of solidarity between their group and other groups of colonos, workers, and students.

Box 9.1 summarizes the division of labor within the Council, including the functions and obligations of its six principal entities. It also describes the duties and responsibilities of the individual members within each of these divisions. The Council functioned with a high degree of democratic participation. During the period of 1983 to 1984 I did extensive participant observation at each level of the organization. From what I could see, the decision-making process genuinely involved open discussion and debate. The Block Assembly (Asemblea de Manzanas) was the most intimate meeting, where immediate neighbors gathered to discuss current events. The Assembly of Block Representatives (Asemblea de Representantes de Manzanas) was usually the most dynamic meeting. The General Assembly (Asemblea General) was usually well attended, often with over 1,000 participants.

Box 9.1
Functions of the General Council of Representatives, Bosques del Pedregal, 1983–1987

General Assembly

The General Assembly is the maximum authority in the colonia. No distinction is made between the entire colonia and the organization. The General Assembly is a convocation of all members of the association and their families. In theory only one titular per lot is recognized as having the right to vote. But votes during a general assembly are taken by a show of hands and do not verify possession. General assemblies can be "ordinary" and "extraordinary."

Ordinary assemblies are held at the end of every month, whereas extraordinary assemblies can be convoked at any time by the Council of Representatives to resolve urgent matters (according to Article 2,675 of the Código Civil).

Executive Committee

The Executive Committee's five members are empowered to represent the association on any issue or negotiation, whether of a juridical, administrative, or political nature.

Assembly of Block Representatives

A member of the assembly of elected block representatives has the following responsibilities:

1. Complete tasks agreed upon in the General Council of Representatives (CGR).
2. Convoke regular meetings in his or her *manzana*
3. Inform the assembly of the positions and agreements made by the CGR and lead discussion
4. Analyze the problems of the *manzana* and colonia and propose solutions that can be considered in the CGR
5. Participate without excuse in the meetings of the CGR
6. Present in written form the agreements made by the assembly to the CGR
7. Form the commissions assigned by the CGR

Representatives will be chosen in a free and secret election by 50 percent plus one of the titulares and can be removed by the same. The position lasts six months with no limit on reelection.

Council of Representatives

The council is the main decision-making body of the association. It is composed of the representatives of each block and their alternates. Only the representative has voting power. For the agreements and decisions of the Council to be valid, there must be a quorum of 50 percent plus one of the block representatives present for the vote. Meetings are held every month. The Council sets up special commissions and is a forum wherein the following issues and topics are put forward for deliberation and possible resolution:

1. The report that will be made by the Coordinating Commission
2. Reports of the different bodies of the Council and of the commissions
3. Proposals for modification of the bylaws
4. Evaluation of the function of the organization, particularly the Coordinating Commission
5. Definition of the general political strategy and position of the group until the next General Assembly
6. Reviewing the proposals made by the *manzana* assemblies

7. Diffusion of information, consider linkages with other colonias
8. Modification of the fees imposed by the *Assembly of Sector Coordinators*

Coordinating Commission (CC)

The CC is composed of the coordinators of each of the sectors of the organization. (There are eleven sectors, divided into four, five, or two *manzanas*). Also included are the coordinators of the various commissions within the organization. Duties and functions of the CC include

1. Coordinate the commissions
2. Act as a check and balance against the activities of the CGR
3. Propose the plan of work for the CGR
4. Convoke the meeting of the CGR and General Assembly

Block Assembly

At the level of the block *(manzana)* exists the immediate authority to resolve problems internal to the block, subject to the rules of the association. In a meeting at the block level, one person plus an alternate are elected by simple majority (50 percent plus one) to represent the *manzana,* and the interests of the members therein, within the Council. Representatives elected to represent the block can be removed from their position at any time by the people who voted for them.

Source: I compiled this information from interviews, direct observations, and archival research (including reference to a legal document prepared by Lic. Jesús Castro Figueroa, a notary public who legally registered the association).

The Council's structure was not without faults. For instance, there was bias against renters. There was also grossly unequal gender representation in leadership positions—the Executive Committee was composed entirely of men. Aside from these serious limitations, the organization was remarkable in its accomplishments. The Council acted to a large extent as a government within a government. Compensating for the failure of the court system to deal with conflicts over illicit land transactions (for example, a lot being sold twice), the Council provided both judicial and police functions and acted as the cooperative manager for land allocation and development. It also had a tremendous capacity to mobilize people and did so by bringing into action a sophisticated network of people, facilities for printing bulletins and newsletters, and strategically placed loudspeakers. People were mobilized to collectively organize the basic infrastructure (streets, schools, meeting places, and water and electricity distribution) and to press their claims against the state. The Council was one of the popular organizations that spearheaded the colonia ecológica productiva movements. Yet, the height of the Council's capacity to mobilize people and accomplish tasks occurred during a specific historical juncture (1979–1985), after which the Council declined and eventually disbanded.

The Council was most active when the threat that all the settlements in Los Belvederes would be eradicated was high. Accordingly people were concerned and attentive to calls for mobilization. However, by 1985 several factors had undermined the effectiveness of the Council. First, the Council lost its sense of urgency as the colonia transformed from an incipient, highly insecure, and low-density residential outpost to a relatively well-consolidated settlement with a much higher density, between 100 and 200 people per hectare (Gotica Villegas, Pulido Vega, and Avila Bravo 1985). The focus of struggle turned from the strategic infilling of the land under popular control, to consolidation activities and demands for land tenure regularization. As the settlement became consolidated, the prospect that it would be completely eradicated became less likely, therefore less fearsome. Second, as time went on living conditions in the colonia became less desperate: faced with popular pressure the government committed itself to regular deliveries of water, the secretary of public education (Secretaria de Educación Pública; SEP) began to staff the schools, and bus service was introduced. Under these conditions—even though the threat of eradication still existed—fewer and fewer people participated in the constant meetings and mobilizations called for by the Council. People grew tired. This form of decline in popular organizations has been well documented.

An important issue is the decline of the Council in the context of political mediation, even before the state initiated land tenure regularization. The Council met its demise as it gradually lost control over available land, community functions, and resources. For example, once the SEP agreed to staff the school the Council lost that site to the state. The members of the Council wanted service from the SEP, but by winning institutional commitment the association lost control of an important rallying point (constituting an instance of what Max Weber described as expropriation of the means of administration). In the beginning, children in the settlement received their education in a makeshift popular school set up by the Council and staffed by professional and nonprofessional volunteers. When the SEP arrived, the state assumed responsibility for the children's education. In the same manner the Council lost control of a health center and local store.

Another set of factors that led to the demise of the Council stem from the divisive tactics of the state. In 1980, PRI set out to "double its efforts to increase its incorporation into the organic structures of the popular sector by fostering new forms of participation and mobilization" (Garcia Pérez 1985, 169). This was the year in which the government restructured the Junta de Vecinos, a citizen body of elected grassroots representatives meant to serve as a consultative council to authorities of the Department of the Federal District. In Bosques a damaging rift developed within the council leadership over the issue of whether or not to participate (and in what manner to participate) in the Junta de Vecinos.[5] Leaders of the council became divided as rivalries developed. Efforts were made to transform the structure of the council several times, but participation waned. A splinter group formed a cooperative.

While there may have been a temporary *reflujo* (ebb or downturn) in the level of popular organization in Bosques during 1983, popular groups in Dos de Octubre and Belvedere continued to be very active. In Bosques the most violent conflicts and desalojos occurred before 1979, whereas in Dos de Octubre and Belvedere the greatest stress began to build after 1979. I turn here to broadly compare and contrast the settlement history of Bosques with that of Dos de Octubre and Belvedere.

Dos de Octubre and Belvedere

Unlike Bosques, which is entirely on ejido land, the mix of property regimes in Dos de Octubre and Belvedere is more complex and hotly contested. Figure 9.3 shows the overlapping claims involved. According to the Teresa family and the Belvedere Association, less than 9 percent of Dos de Octubre is ejido land. In contrast, ejidatarios claim that at least 75 percent of Dos de Octubre is on ejido land. The state asserts that approximately 20 percent of the colonia pertains to the ejido of SNT and that the remaining 80 percent is private property. Within the parts of Dos de Octubre and Belve-

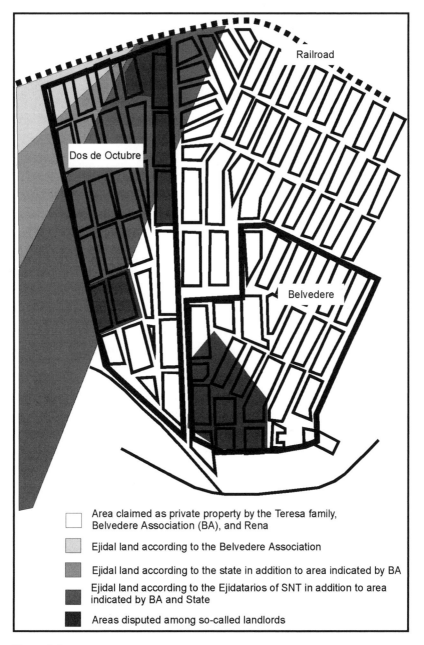

Figure 9.3
Land tenure disputes in Dos de Octubre and Belvedere, 1987. Source: Archival research and interview data compiled by author.

dere that have been generally acknowledged as private property, the Teresa family (in alliance with the Belvedere Association) and Rena have made conflicting claims. This tangle has led to violent conflicts and multiple deaths. Without going into the amount of depth devoted to Bosques I turn here to highlight key historical events in the formation of popular groups in Belvedere and Dos de Octubre.

Dos de Octubre

The first settlers of Dos de Octubre erected provisional dwellings as early as 1976. Some of these earliest colonos purchased documents from the Belvedere Association—photo ID cards that identified them as *colonos avecindados indocumentados* (see figure 9.4). These ID cards were blatantly fraudulent: a colono avecindado *Indocumentado* (undocumented) is pure juridical fiction. Besides, only a *comisariado ejidal* has the right to grant *colono avecindado* status; developers do not. Nevertheless, developers issue the false documents as a form of legitimation.

On December 7, 1976, a new settler of Dos de Octubre purchased an ID card in order to occupy Manzana 38, Lot 2. The card was purchased from Ignacio de Teresa with the Belvedere Association as the executor. Sometime later, the same colono—under the threat of forceful eviction— had to purchase the lot again, only this time from Rena. The colono showed me a contract that he signed on March 30, 1979, with Rena. He also showed me receipts for monthly installments paid to Rena over the period of 1979 to 1981 (see figure 9.4). Most recently the state has declared that transactions with Rena would not be recognized. The colono's money was lost. This is a typical situation.

Just as in Bosques, while some paid for the land in Dos de Octubre others squatted. The first large-scale occupation of land ("invasion") in Dos de Octubre occurred in 1977, when 800 families staked out claims. Under the threat that they would be eradicated by the ejidatarios of SNT, colonos joined together in 1978 to form the Union of Settlers (Unión de Colonos) of San Nicolás Totolalpán. The union attempted to secure from the state the promise of land tenure regularization and an amparo to protect themselves from desalojos. But this effort was unsuccessful at first.

On September 19, 1979, approximately 800 families were eradicated as their dwellings were demolished or burned. Of these 800 families, approxi-

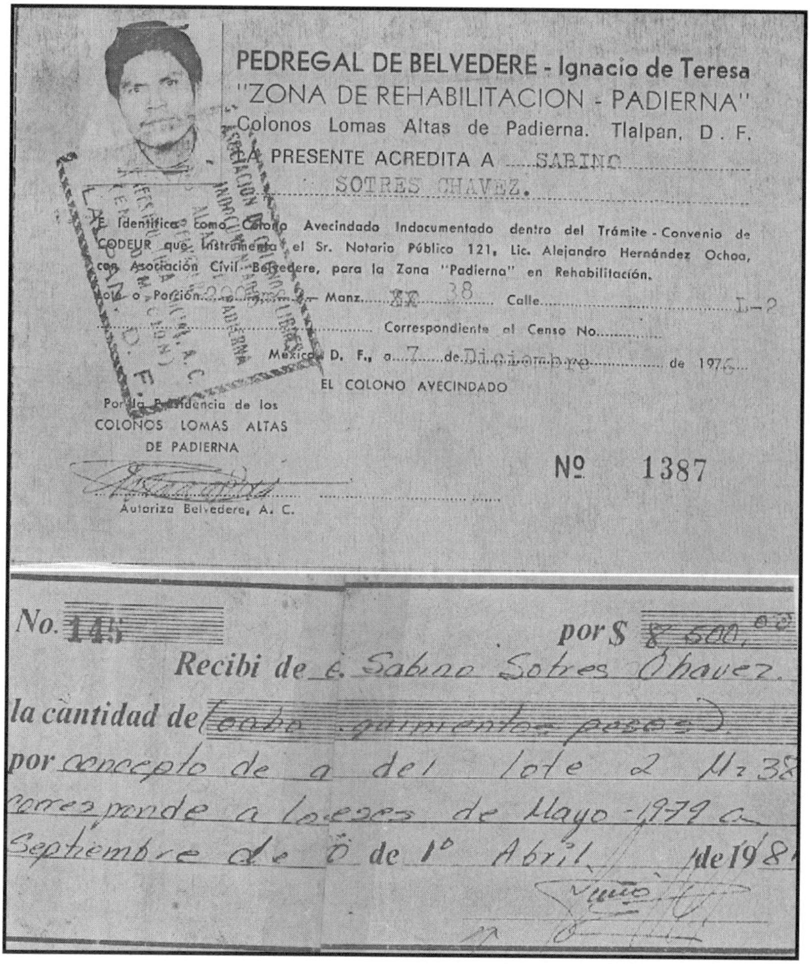

PEDREGAL DE BELVEDERE - Ignacio de Teresa
"ZONA DE REHABILITACION - PADIERNA"
Colonos Lomas Altas de Padierna. Tlalpan, D. F.
LA PRESENTE ACREDITA A SABINO
SOTRES CHAVEZ.
E Identifica como Colono Avecindado Indocumentado dentro del Trámite - Convenio de CODEUR que Instrumenta el Sr. Notario Público 121, Lic. Alejandro Hernández Ochoa, con Asociación Civil Belvedere, para la Zona "Padierna" en Rehabilitación.
Lote o Porción ... Manz. 38 Calle ... L-2
... Correspondiente al Censo No. ...
México D. F., a 7 de Diciembre de 1976
EL COLONO AVECINDADO
Por la Presidencia de los
COLONOS LOMAS ALTAS
DE PADIERNA
Autoriza Belvedere, A. C.
Nº 1387

No. 145 por $ 8 500.00
Recibi de C. Sabino Sotres Chavez.
la cantidad de (ocho quinientos pesos).
por concepto de a del lote 2 Mz 38
corresponde a los meses de Mayo 1979 a
Septiembre de ... de 1º Abril de 198

Figure 9.4
Fraudulent ID card issued by the Teresa family to a *colono* occupying a lot inside Dos de Octubre.

mately 150 remained in the area along the railroad tracks at the colonia's northern perimeter.[6] In the early part of 1981—under the protection of an amparo—the colonos finally moved back to the lots from which they had been evicted two years earlier. But the amparo they had was not respected and on October 2, 1981, the colonos suffered another massive desalojo. At least 200 dwellings were burned down. The action was initiated by the Ejido of Commission of SNT with support from government authorities and police.

Colonos regrouped and consolidated their organization. A turning point in the fate of the settlement came with the successful establishment of an elementary school. This was also the case in Bosques. Indeed it is a key strategy of popular groups to make the installation of an elementary school a priority. There are two reasons for this. First, parents consider education as vital and they do not want their children to miss any school time. Second, the elementary school serves as a shield *(escudo)*. In a manner of speaking the state's hands are tied when it comes to the prospect of using force against children. The school becomes a safe place.[7]

From 1982 to 1983 colonos of Dos de Octubre made great strides toward upgrading the settlement. They secured the school as well as basic, even if rudimentary, infrastructure and services. But progress was not without infighting and conflicts. Political prisoners were taken on a number of occasions, and colonos had to maintain constant vigilance.

Belvedere

From its inception, the colonia Belvedere has been one of the most conflict-ridden in the entire metropolitan area. Most recently, it has been given the dubious distinction of being one of the Federal District's forty-four most impoverished human settlements (Joyner et al. 1996). Belvedere originated as a result of clandestine subdivision by the Teresa family, the Belvedere Association, and Rena. As in Dos de Octubre, both the Teresa family and Rena sold the land. Yet the colonia was entirely situated on land in litigation, so all sales were fraudulent. The illicit transactions generated a great deal of violence resulting in multiple deaths.

As in Dos de Octubre many colonos were compelled to pay a deposit and then installments to the so-called private owners. Eventually some of the colonos "began to have doubts . . . they began to ask themselves about the legal origin of the land they occupied, and about the supposed legitimacy of the fraccionadores."[8] In 1981 a group of grassroots activists formed the Independent Promotional Commission (Comisión Promotora Independiente) and initiated a rent strike. The commission took the position that no further payments would be made until all litigation had been settled.

The year 1982 was one of constant conflict. Developers moved against the "agitators" who refused to pay. Colonos responded with numerous mobilizations and protest marches. On May 12, 1982, the colonos won recognition from the delegado Ernesto González Aragón, who promised to prevent the developers from continuing to sell the land. But the promise was an empty one. The illicit transactions continued, and the state moved to repress the popular groups active in the colonia. On November 16, 1982, much of the colonia Belvedere was violently eradicated. The following quote is a historical account of the desalojo prepared by colonos who were present: "At ten in the morning around 300 men, including riot police, pistoleros, mounted police, personnel and functionaries from the Tlalpan Ward and from the Department of the DDF entered the upper part of the colonia. Through brute force 400 families were eradicated. In the process colonos were assaulted . . . and robbed . . . and a newborn baby of only six months was killed" (from the archives of the Sociedad Cooperativa, Bosques del Pedregal en Lucha).

The response of the colonos was intense mobilization: for three days they protested in front of the offices of the DDF in the city center. On the third day they were received by the authorities, who then promised them that there would be no more desalojos and that their possession of land in Belvedere would be respected. This turned out to be another empty promise. Colonos returned to the settlement, but they were dealt another massive desalojo on December 15, 1982. When this second desalojo occurred the colonos were better organized. Following the first desalojo one month earlier, the colonos had secured juridical recognition for their group. On November 22, 1982, they had formed the Association of Settlers of Ajusco, Casa del Pueblo, and on November 25, 1982, they had filed an amparo to protect themselves from further desalojos. Ironically, on December 15, 1982 (the day the second massive desalojo was executed) the amparo was granted. Procurement of the amparo enabled the Association of Settlers to use established legal channels to denounce the desalojo as a violation of their rights under the law.

As a result of the amparo—granted by the fifth district judge of the DDF—authorities of the Tlalpan ward were obligated to respond to the colonos' accusations; otherwise, these authorities would be prosecuted before the public ministry. This gave the colonos time. With an amparo in their possession, the colonos moved back to the colonia on December 24, 1982, and they once again took possession of their lots. This marked a turning point for the colonia. To quote a thesis written about Belvedere, state-community relations in the colonia shifted from an initial "repressive stage" to a second "stage of negotiation": "In this second stage, the mobilization of colonos began to exert force. Administrative authorities became obligated to seek negotiation with the colonos and to abandon repressive tactics including desalojos. The state sought to mediate the interests of all the agents present in the zone (from the supposed landowners and clandestine subdividers to the colonos) . . . yet no definitive solutions attended to the demands of the colonos.[9]

Expert Judgment and the Emergence of Ecopolitics

Toward the end of 1983 it appeared less and less likely that the state could eradicate (much less relocate) all families from Los Belvederes. The sheer magnitude of the settlements—more than 20,000 people occupying nearly 5,000 lots—made it practically impossible. Yet no resident could be sure, and the level of insecurity among colonos remained high. In August 1983, the threat of eradication reached a new peak. The state announced that "expert judgment" called for measures to eradicate Los Belvederes. The expert judgment was written in response to the question: "do the settlements of Los Belvederes have an adverse impact on the flora, fauna and ecology of the zone?"[10] The conclusion of the judgment can be paraphrased as follows: As a result of an investigation (done in consultation with the Subsecretariat of Ecology[11] and the Tlalpan ward), which took into account the dictates of official plans and programs, it is concluded that there should not be any human settlements in the zone. The settlements of Los Belvederes do not comply with the intent of the buffer zone, which is to conserve the ecological equilibrium of Mexico City. Furthermore, given that the Sierra del Ajusco is one of the last remaining areas that oxygenates the air of the Federal District, its ongoing destruction is particularly alarming. The balance of the ecosystem is progressively disturbed such that all inhabitants of the city are adversely affected.

This statement was written at a time when the politics of containment, which incorporated ecological arguments, were becoming firmly established in the creation of new institutions, laws, programs, and plans. A steady stream of official declarations denounced settlement in Los Belvederes. For instance, the director of DART, Hidalgo Cortes, argued that "the industry of invasion" must be stopped and that in the extreme case upwards of 10,000 families will have to be relocated out of the zone of Ajusco Medio" (*El Día*, October 2, 1983). At about the same time, a newly created interagency commission composed of representatives from the DDF, SRA, Secretariat of Agricultare and Hydroaulic Resources (Secretaría de Agricultura y Recursos Hidráulicos; SARH) and SEDUE recommended that 9,200 hectares of the zone of Ajusco should be expropriated for a park.[12]

As if to prepare public opinion for a massive eradication of settlers in Ajusco Medio, the press secretary of Tlalpan announced, on March 19, 1984, that a large portion of the population of the zone would have to be relocated to other parts of the city. The reasons given were environmental problems, the impossibility of installing public services, and the fact that people cannot prove that they hold legal titles to the land (*El Día*, March 20, 1984). Supporting this position, the president of Mexico City's Consultative Council (Consejo Consultivo) also announced that the settlers of Ajusco Medio should be relocated (*Ovaciones*, May 7, 1984). Popular groups in Los Belvederes now had to contend with a new line of argument, one that placed an unprecedented emphasis on ecological concerns. This spurred grassroots environmental activism and attempts to define a political ecology of human settlements—the subject of the last part of this book.

IV

Toward a Political Ecology of Human Settlements

10

Grassroots Environmental Action at the Frontier: The Colonia Ecológica Productiva Movement

Our Popular Front of Settlements, with Bosques, Dos de Octubre and Belvedere at the core, demands regularization by way of expropriation. At the same time, we insist that the government take into account our work, cooperation and collective efforts—especially our technical proposals to create settlements of a new type: ecological spaces in which harmonious relations within civil society and between humankind and the natural environment are possible.
—Frente Popular para Defender los Asentemientos Humanos del Ajusco, Los Asentamientos Humanos del Ajusco: En la Cara de la Represión

Faced with renewed threats of eviction, colonos of Los Belvederes mobilized to repudiate the government's plans. In January 1984, settlers of Los Belvederes banded together and formed a barriowide urban movement under the name Popular Front for the Defense of the Settlements of Ajusco (Frente Popular para Defender los Asentemientos Humanos del Ajusco). The front was spearheaded by popular groups from three settlements: Bosques del Pedregal, Dos de Octubre, and Belvedere. This was the high water mark of a short but remarkable period (mid-1983 to early 1985) during which the terms of struggle for land dramatically shifted. Activists of the movement merged conventional arguments and strategies demanding adequate housing and a secure life space (on the basis of Constitutional rights—art. 27) with more innovative arguments and strategies involving ecological proposals.

Members of the front proposed an integrated approach to urban development based on production that is "socially necessary, ecologically sound, and economically viable." Such a strategy, they argued, is in sharp contrast to the government's centralized sectoral approach to planning, which treats housing, economic and ecological problems as separate issues.

Activists of Los Belvederes recognized that it is not possible to solve these problems separately. Faced with chronic scarcities of income and resources, they saw no alternatives outside of promoting self-reliant forms of urbanization. Their approach called for the development of appropriate technology that can reduce the cost of generating employment and recover capital investment through the rational use and reuse of natural resources. The movements went beyond subordinated forms of protest; the popular groups pressed claims for participatory democracy and for expanded power in local development decisions.

It must be kept in mind that colonos saw the rural-to-urban transition of land use in the zone of Los Belvederes as inevitable. For them, there was never a question of whether the space should or should not be preserved for ecological conservation. Such a prospect was dismissed as an impossibility. Thus, from the beginning the question they deliberated was, Who will ultimately get to settle in the area: the wealthy or the poor? In this respect colonos never felt that they were violating a space that probably should be protected. On the contrary, they argued that only their alternative plans offered a viable approach to the zone's inextricably bound socioeconomic, political, and ecological problems.

During peak activity (1984–1985) the alternative projects promoted by activists in Los Belvederes were fostered by many external agents: university groups, newspaper reporters, independent research associations, a union of progressive lawyers, and to some extent political parties in opposition to the PRI.[1] There was even international support from groups in Austria, Germany, and the United States.

Ecopolitics, External Agents, and State-Community Relations

Popular groups in Los Belvederes joined forces with student activists from the Urban Sociology Department at UNAM (ENEP-Acatlán) to counter the state's arguments against the permanence of their settlements. In particular, colonos and students concentrated on delegitimizing the "expert judgment" announced in August 1983. They argued that the judgment was a foregone conclusion that actually had little to do with the ecological realities of Ajusco. They took issue, for example, with being blamed in the judgment for the extinction of certain species of flora and fauna that have disappeared throughout the entire Sierra del Ajusco, contending that this extinction process was already under way before the formation of Los Belvederes. It began as a result of multiple causes, of which, the settlements were only one.

For many years, the zone of Ajusco had been negatively affected by the widespread deforestation and pollution caused by Peña Pobre, the paper pulp factory, and by other industry in the area. Acid rain had also been a factor. A most glaring contributor to contamination in the zone had been the state itself through the operation of an open air waste dump site contiguous with Los Belvederes. Every day hundreds of tons of untreated wastes were deposited in this dump site at the very same time that the state cited colonos for their contamination of the pristine zone. In January 1984, a group of biologists (Interdisciplinary Group for Agrobiological Studies; Grupo Interdisciplinario de Estudios Agrobiológicos) from UNAM wrote up the results of a field study. The biologists, in support of the colonos of the zone, concluded that "the population of these colonias causes little harm to the environment, given that it has already been contaminated by industry, automobiles, and by the excessive falling of trees—without reforestation—by companies which exploit the forest in a way concerned only with profit in order to produce paper or furniture." Many of the arguments wielded by colonos and sociology students against the judgment drew from the biologists' report. The continued operation of the garbage dump was a blatant contradiction that underscored the lack of integrity in the state's ecological arguments about the Los Belvederes. Eventually the dump was shut down. Even so, colonos were convinced that the judgment was nothing more than a pretext to justify their eradication. They refused to let their settlements be singled out as a necessarily destructive force that upset the ecological equilibrium of the *entire metropolitan area!*[2]

The contradictions in state policy were angrily noted in grassroots bulletins circulated throughout Los Belvederes. One such bulletin voiced a scathing critique:

If we let them relocate us, in our place they are going to build upper-class residential developments, worth many thousands of millions of pesos—including luxury hotels, Swiss chalets, elegant offices, clubs for tennis and golf—all for the officials and politicians, corrupt thieves, who without shame have come to enjoy what they have robbed from the nation, from the people, from our colonia. . . . They want us to believe that we are contaminating the underground water supply, and that we are damaging the 'lungs' of the city by chopping down trees—when in reality it is the authorities who are contaminating the zone, for example, with the garbage dumps they have placed near our colonia. . . . Or how about the thousands of trees that the paper company Peña Pobre cut down in Ajusco? Or the thousands of trees plowed over to make room for Adventure Kingdom [the nearby "Disneyland" of Mexico], or the residential estates such as Jardines de la Montaña, Moyers Country Club and Forest Hill—Don't the rich pollute? Why have they singled us out? They act against us with false pretexts to clear the area and take advantage, for themselves, of all that we have worked so hard to earn.[3]

Given the mass opposition to threats of relocation and the sheer magnitude of the population involved, together with the skillful use of ecoarguments by grassroots activists, the state relented. On May 28, 1984, the regente of the DDF, Ramón Aguirre Velázquez, announced the Program for the Conservation of Ajusco (México 1984). The new positon was that sections of Los Belvederes would be incorporated into the legally designated urban area. Although the program declared that "there would not be a massive eradication" of families from the zone (6), it also stated that the area for residential use would be restricted and that it would have to be "densified through the relocation of families inside and outside of the zone."

The Program for the Conservation of Ajusco

The Program for the Conservation of Ajusco laid out an agenda for state intervention and improved state-community relations in Los Belvederes. Its main points are summarized in box 10.1.

Settlers viewed the program with skepticism. They feared that demographic densification through relocations was a euphemism for eviction (*El Día,* June 8, 1984). Promises to regularize land tenure in some areas and not in others, on a case-by-case basis, were viewed as divisive tactics that would cause confrontations among settlements (*UnoMásUno,* May 29, 1984; *El Día,* May 29, 1984). The state's overtures to help people return to their points of origin was rejected as absurd: "[F]irst they will have to address the problems in the countryside that compelled people to migrate to the city in the first place," stated the newspaper *El Día* (June 8, 1984). Regarding the arguments that a "normal" infrastructure was prohibitively expensive, colonos agreed that novel lower-cost arrangements would have to be made. Toward this end, popular groups offered

Box 10.1
The Program for the Conservation of Ajusco

---

1. It established limits between the ecological reserve zone—to be designated as a national park—and the zone for residential use. Land for residential use was to be expropriated by the DDF and *densified through the relocation of families inside and outside of the zone.* Parcels subject to regularization were to be determined on a case-by-case basis. Also, incentives (one-way bus tickets) were made available to rural migrants willing to return to their points of origin.

2. Based on the argument that "normal" infrastructure was cost prohibitive, it stipulated that only rudimentary public services would be provided in the settlements and that the free supply of water would be stopped.

3. Regarding public security, it outlined efforts to improve state-community relations. Popular groups were called upon to stop their violent provocation and to instead use institutional channels to voice their grievances. With indirect reference to opposition movements in Bosques, Belvedere, and Dos de Octubre, it was declared that pressures from minority groups would not be tolerated.

4. Finally, to assert the state's physical presence in the zone, the program called for the construction of a "front-line" government outpost.

Source: México (1984).

to contribute their labor as a means to lower costs. At the same time, colonos reacted sharply against the possibility of losing their free water supply (250 gallons per week per family delivered by truck). They viewed this measure as a scare tactic linked to the state's warning that pressures from minority groups would not be tolerated.

The Popular Front for the Defense of the Settlements of Ajusco protested against the government's conservation program with such well-worn strategies as marches and protests,[4] press releases,[5] and letter-writing campaigns.[6] Colonos also repudiated contents of the program by developing and acting on their own brand of ecopolitics. Popular groups critiqued the state's technical arguments and promoted alternative development plans within official forums *(convocatorias)* and through official channels for community participation in development decisions.

On August 13, 1984, settlers of Dos de Octubre and Belvedere, along with students from the departments of architecture and biology at UNAM, submitted to the state's Special Commission for the Sierra del Ajusco, an *Urban and Ecological Project*. It was intended as an alternative to the official plans.[7] Earlier in the year, with reference to the same project, a grassroots group from the Belvedere settlement published an open letter in the newspaper *UnoMásUno*. It was addressed to President de la Madrid, Mayor Ramón Aguirre Velázquez, and "all democratic and revolutionary organizations." The letter provided background that aimed at legitimating the group's claim to the land. The letter also provided an innovative response to language in the government's planning documents; in it the grassroots activists argued that

Confronted with ecological constraints we have designed a settlement suitable to a zone of rural-urban transition, where inhabitants take into account and coexist with the natural environment. The designs include plans for organic waste recycling systems as a way to avoid the high costs of drainage while preventing black water filtration into the underground water supply. Also we will prevent deforestation of the zone and the circulation of automobiles will be limited. We will realize all these plans on the basis of collective work. (*UnoMásUno*, March 9, 1984)

In August 30, 1984, Bosques del Pedregal participated in a public hearing conducted by the Special Coordinating Commission for the Conservation Program of the Sierra del Ajusco (Comisión Coordinadora Especial del Programa de Conservación de la Sierra del Ajusco). At the hearing, entitled "An Open Forum about the Ecology of the Southern Zone of Mexico City," grassroots leaders of Bosques presented a slide show about their settlement. They brought up arguments for turning the zone into a greenbelt for production as opposed to a greenbelt for consumption. More specifically, they presented a proposal to transform their settlement into a colonia ecológica productiva.

The colonia ecológica productiva model is ambitiously intended to integrate—within the territorial framework of the local community—the need for residential space, employment, food, and cultural and political self-determination. Activists of Bosques stated that the main objective of the project is "to promote an ecologically valid, economically viable, and socially necessary solution aimed at resolving the problems in the Sierra del Ajusco, and more specifically in Bosques del Pedregal." This was the most sophisticated proposal of its type in the zone.

The Colonia Ecológica Productiva Movement

On July 1, 1984, leaders of Bosques del Pedregal's General Council of Representatives met and agreed to promote a series of collaborative efforts calling for environmental protection and local community development. Their first initiatives included the following: (1) combat the plagues infesting the trees in the settlement, (2) protect all the vulnerable trees in the settlement with a ring of rocks around their base, (3) reforest the zone with 20,000 fruit trees and 20,000 'oxygen' trees to be supplied from the Tlalpan ward (at the time this decision was made 5,000 fruit trees had already been planted by the Bosques colonos), (4) establish two compost piles in Sector 1 of the settlement, and (5) form the Ajusco Cooperation Group (Grupo Cooperación Ajusco) to define an alternative approach to development that would transform Bosques into an ecologically productive settlement. The council promoted these measures as part of a barrio-wide effort to develop a positive image of Los Belvederes. Popular groups in Belvedere and in Dos de Octubre promoted similar measures. However, the Ajusco Cooperation Group and its colonia ecológica productiva initiative focused on Bosques del Pedregal.

The Ajusco Cooperation Group

The Ajusco Cooperation Group was composed of five organizations. Two of these were popular groups from Bosques: the General Council of Representatives and the Cooperative Society of Bosques del Pedregal in Struggle. The other three were external agents: the Group of Alternative Technology (GTA), the Center for Biological Investigation (Centro para la Investigación Biologico), and Agro-Industry Integrated (Agro-Industria Integrada). The most influential of the external agents was the GTA. In view of the central role the GTA played in the conceptualization and promotion of Bosques as a colonia ecológica productiva, it is worthwhile to briefly describe its agenda and the context out of which it emerged.

Alternative Technology Group (GTA)

The GTA was founded in 1977 under the direction of Josefina Mena, a charismatic architect. It is a small, multidisciplinary group of researchers committed to generating new technologies and to promoting alternative development (Mena 1987). The group has a history of working with low-income groups in both rural and urban settings to improve livelihood through the implementation of sustainable practices. The GTA has special expertise in alternative technology for the recycling of organic household waste.

The GTA views housing as a factor of production and as generator of resources that can be used to elevate public health and the quality of life. The group's work aims at transforming household wastes into primary materials or inputs for ecologically sustainable development. Describing her group's work, Mena (1987, 547) claims: "we aim at delimiting a school or scientific discipline that we have termed 'Productive Ecology.'" In the process GTA members have devised a technology to treat human waste effluent so that it may be reused—for instance, to irrigate urban agricultural plots. They have also devised a technology to transform organic garbage (together with human waste) into a high-grade, marketable fertilizer. These technologies (SIRDOs) were centerpieces in the CEP proposal brought before the state.

SIRDO-seco and SIRDO-húmedo

SIRDO is an acronym used to describe an integrated system for organic waste recycling (i.e., *sistema integrado de reciclamiento de desechos orgánicos*.[8] The terms *seco* (dry) and *húmedo* (wet) differentiate the two basic types of SIRDOs (Mena 1987; Schmink 1984).

The simpler of the two models, the SIRDO-seco, is a closed system that resembles an outhouse (see figure 10.1). In 1985 three SIRDO-secos were put into operation in Bosques. The GTA improved the design and operation of SIRDO-secos based on experiences in rural areas in the states of Quintana Roo, Yucatán, Campeche, and Chiapas. They also gained experience in Mérida and in other parts of Mexico City. Each unit can be utilized by several families simultaneously. Inputs may include urine, fecal matter, and organic matter of all sorts such as vegetable husks and peelings, discarded food, weeds, paper products, and even dead pets. Indeed, right after Bosques put its first SIRDO-seco into operation, a pet dog had died giving birth. The pet belonged to Polo Clave, a dedicated community leader who placed the dog into the SIRDO after cutting her up (to facilitate decomposition). Polo did this to make a point. When the SIRDO was

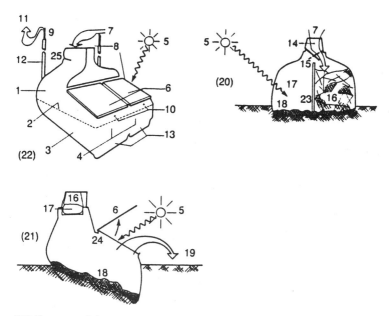

(20) Transversal view
(21) Longitudinal view
(22) Perspective

1. Material of low thermal conductivity
2. Ground level
3. Impermeabilisation
4. Two twin chambers for alternate use
5. Solar energy
6. Solar collectors of various types depending on local conditions
7. Entrance for faecal matter and organic household waste
8.
9. Flytrap
10. Opening for oxygen
11. Air flow
12. Vent (black)
13. Alkaline filter

14. Toilet
15. Dispenser plate to direct waste to one bin or the other
16. Phase one of the aerobic process
17. Phase two of the aerobic process
18. Sanitary organic fertiliser of excellent quality (60-80% organic carbon)
19. Extraction of fertiliser (alternately from each chamber every six months)
23. Wall separating compartments
24. Support angle
25. Step

Figure 10.1
Schematic illustration of SIRDO-seco. Source: This illustration was copied with permission from the Grupo de Tecnología Alternativa, S.C. It is registered under the Certificado de Invención no. 6758 M.R. 101432 SECOFI, México.

shown to local authorities and/or to possible funding agencies, colonos were quick to point out that not only did the unit turn garbage into a resource, it was also odorless; and they reasoned, if it can handle a rotting dog it can handle anything. Poetically, the dog ended up as fertilizer used to feed the Clave family's lemon tree and family garden.

SIRDO-secos are designed so that liquids evaporate and nothing needs to drain off or seep into the ground. The unit has two holding bins that are used alternately. Initially, one bin is kept empty while the other is filled. The first bin to be put into use is prepared with start-up materials including some foliage and lime. When the first bin is filled to capacity, it is sealed off; the alternate bin is then prepared and put into use. While the alternate bin is in use, the contents of the full bin are allowed to aerobically decompose. The material is fully decomposed when it has the form of a brown granular powder that can be safely removed and used or sold as fertilizer. The unit is designed to operate on a six-month cycle: by the time the bin in use becomes full, the other bin is ready to be emptied and made available (see figures 10.2 and 10.3).

The SIRDO-húmedo is a more complex system. In 1985, the construction of a SIRDO-húmedo was begun in Bosques with funding from an Austrian solidarity group (see figure 10.4). The solidarity group was started by Dr. Peter Baumgartner specifically to support projects in Bosques. I introduced Dr. Baumgartner to the people of Bosques in 1983. From 1983 to 1985 he managed to raise thousands of dollars in Austria for the CEP initiative including the construction of SIRDOS. The SIRDO-húmedo system involves a decentralization of urban infrastructure. The system can be designed to serve from 20 to 1,000 families. The larger system can provide access to a settlement area up to three to four kilometers in diameter. The unit in Bosques was designed to provide service for 120 families. The SIRDO-húmedo performs the same functions as a SIRDO-seco, but it is also designed to treat and recycle *agua negra* (sewage effluent) and *agua gris* (literally, "gray water"; soapy dish, bath, and laundry water) (see figures 10.5 and 10.6).

Each housing unit linked to the SIRDO-húmedo system is connected by two tubes: one for gray water and one for sewage effluent. The tube for gray water conducts fluid away from each house to a filtering device that enables 70 percent of the fluid to be recycled for irrigating vegetables, fruits, and all types of cultivation (assuming all the proper precautions are taken). The *aguas*

Figure 10.2
The SIRDO-seco, eco-outhouse composter, Bosques del Pedregal, 1985. Note: Dr. Peter Baumgartner inspects the contents of a SIRDO-seco holding bin that is full and in the process of decomposing. Inside the shed is a waterless toilet through which human and other organic waste is deposited into the bins. Stairs leading up to the enclosed toilet are in the back of the unit, out of view.

negras are conducted to a sedimentation tank in which anaerobic digestion takes place. The sludge from this initial process is then passed into another holding tank. In this tank the sludge is mixed with other solid organic household waste (deposited on a daily basis), and the contents undergo aerobic decomposition. The process is accelerated by design features that take advantage of solar energy and create optimum conditions for aerobic bacteria. The final product in solid form is an organic fertilizer of excellent quality.[9] In addition to the recycled gray water, the residual fluid from the treatment of *aguas negras* is clarified and can be used—after a tertiary biological treatment—for irrigation purposes or even for fish farming.

One of the key lessons that the GTA learned in developing and implementing the two SIRDOs is that the introduction of alternative technolog-

Figure 10.3
Fertilizer being extracted from the SIRDO-seco. Note: The woman shown is a member of the Cooperative Society Bosques del Pedregal in Struggle. She is holding a cupful of organic fertilizer in the form of a brownish granular powder. She just extracted it from the SIRDO-seco in the background. The final (dry) product from the SIRDO-húmedo has the same form.

ies depends as much on social and political processes as it does on technical processes. For this reason, they have worked closely with social change agents in the form of grassroots organizations, particularly cooperatives.

The initial contact between the GTA and Bosques was made by a settler of Bosques, Juan Barriga. At the time, Barriga was a grassroots leader active in the Cooperative Society of Bosques del Pedregal in Struggle; he was also a captain in the army, where he studied civil engineering. Barriga heard about the SIRDO technology and became interested in it as a possible answer to the state's arguments that settlers of the zone were contaminating the underground water supply. Around the same time, members of the cooperative in Bosques were visited by Jim Conrad, an environmentalist

Figure 10.4
SIRDO-Húmedo under construction, 1985.

from the United States. In discussions with Conrad, the idea emerged to promote Bosques as a colonia ecológica productiva.

The cooperative decided to invite the GTA into Bosques to help transform their settlement into a colonia ecológica productiva. To document the project and to determine appropriate strategies for its promotion and implementation, the Ajusco Cooperation Group was formed. Settlers of Bosques raised funds to cover expenses (e.g., copying costs and materials for presentations including audio tapes, poster boards, and photographs) from a number of sources. For instance, they sold donated clothing and other items collected from upper-class neighborhoods.

The Ajusco Cooperation Group produced a document (thirty-six pages, plus appendixes) with the title "Colonia Ecológica Productiva." The CEP document is not merely an interesting proposal; it represents a tightly argued countervailing strategy backed by popular mobilization. The proposal is an excellent example of grassroots initiative in the form of proactive, ecologically integrated planning. The CEP document was presented

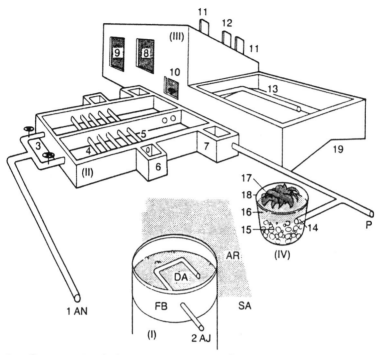

I. Separate networks for *aguas negras* and soapy water
II. Accelerated sedimentation tank (anaerobic)
III. Biological chamber (aerobic)
IV. Evapotranspiration beds
1. (AN) Entrance to the network for *aguas negras*
2. (AJ) Entrance to the network for soapy water
3. Gate valves to alternate use
4. Accelerated sedimentor
5. Sludge retainer
6. De-scummer
7. Clarification unit
8. Access door for organic waste material
9. Access door to the biological chamber
10. Gate valve to let sludge pass (every 24-48 hours)
11. Solar collector

12. Oxygen vents
13. Sludge dispenser into biological chamber
14. Perforated tube inside evapotransporation bed
15. Gravel
16. Sand
17. Earth with ornamental plants, flowers or plants to feed livestock
18. Impermeable evapotranspiration container
19. Unlevel biological chamber floor
P. Run-off to fish tanks
FB. Biological filter (slow)
DA. Aerobic decomposition of organic material in soapy water
SA. Area of communal horticulture
AR. Water suitable for irrigation

Figure 10.5
Schematic illustration of the SIRDO-húmedo. Source: This illustration was copied with permission from the Grupo de Tecnología Alternativa, S.C. It is registered under the Certificado de Invención no. 6758 M.R. 338568 SECOFI, México.

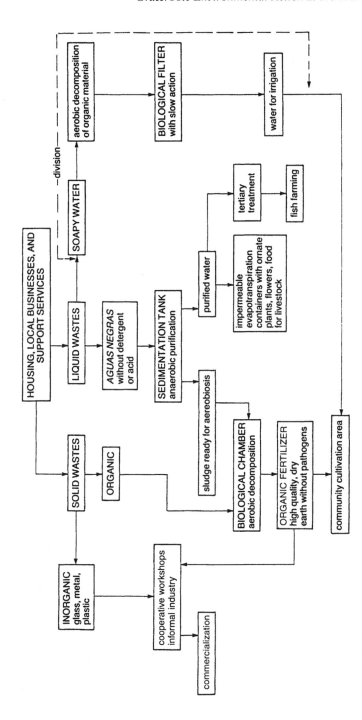

Figure 10.6
An integral barrio-based approach to recycling. Source: This flow chart appeared in the *Colonia Ecológica Productiva* document prepared by the Grupo de Cooperación Ajusco in 1984.

by grassroots leaders to government officials in a range of settings: public hearings; planning sessions organized by COPLADE; administrative offices of the DDF, SEDUE, and Gobernación (similar to the FBI in the United States); and the Tlalpan Ward. Box 10.2 outlines the main arguments and working hypotheses of the CEP initiative.

In addition to outlining the broad arguments and working hypotheses noted in box 10.2, the CEP document spelled out a wide range of specific proposals concerning (1) urban land use and tenure, (2) environmental control and ecological protection, (3) human ecology and public health, and (4) barrio-based pilot projects for production in the primary and secondary sectors of the economy. Central to each of these proposals was the call to build and put into operation the alternative technology referred to as SIRDO. Box 10.3 and table 10.1 contain paraphrased listings of the Ajusco Cooperation Group's objectives and specific proposals.

Box 10.2
Main Arguments and Working Hypotheses of the CEP

1. The current population of Bosques is not marginal; it is productive and can be an active agent in technological development. Furthermore, it is possible and necessary to cojoin productive development with ecological equilibrium and in this way generate a mode of urban development that does not provoke antagonism between these two factors.
2. The conservationist focus that the Tlalpan Ward has maintained conceives of Ajusco as a national park. But this conception disassociates the necessity of reforestation from the possibility of production. Given the demands on this space, it will be impossible to maintain it as a greenbelt for consumption. Instead, it should be developed as a greenbelt for production.
3. Through community pilot projects—involving the operation and maintenance of appropriate technologies as well as the generation and distribution of resources—the autoadministrative capacity of the community can be elevated; so can its productive capacity. In terms of production, it is estimated that with an initial investment of 100 million pesos, 340 jobs can be created.
4. Promoting social experimentation in Bosques to ameliorate the problems in Ajusco is in the national interest. If the colonia ecológica productiva model is successful, it could benefit others by way of example.
5. The project coincides with the political intent of the Mexican Constitution and government of the Federal District. It fits within the guidelines of the Development Plan for the Metropolitan Zone of Mexico City and the Central Region.

Source: Bosques del Pedregal (1984).

Box 10.3
Colonia Ecológica Productiva: Objectives and Proposals

Objectives

1. Promote an integral barrio-based urban development model to general jobs and resources and to foster the production of basic goods *(satisfactores)* in a way that is socially necessary, ecologically valid, and economically viable
2. Positively intervene—vis à vis the aforementioned urbanization model—in the transformation of daily life, generating new relations between humankind and the surrounding natural environment, combining production with an ethos of noncontamination and sustainability.
3. As part of a collective strategy to reduce the cost of generating jobs, employ alternate technologies *(sociotecnicas)* such as recovering capital investment through the rational use and reuse of natural resources (e.g., transform liquid and solid wastes into valuable derivatives that can go into production of basic goods)
4. Over the medium and long term, implement a conceptual transformation of residential space from being a place of consumption to a place of production
5. Put into operation an urban-rural buffer zone (a greenbelt for production) to mediate the contradictions between the city and the outlying rural land in production, recovering or reincorporating intrinsic rural values into urban lifestyles

Proposals

Urban Land Use and Tenure

1. Generate new forms of collective urban land tenure aimed at securing a satisfactory solution to problems associated with conventional forms of regularization (e.g., a cooperative land bank)
2. Establish productivity as the normative criterion of land use rights, and control speculation
3. Formulate and promote intersectoral actions that articulate the resources of the community, the Grupo de Cooperación Ajusco, CORETT, SARH, SEDUE, and the Tlalpan ward

Environmental Control and Ecological Protection

1. Systematize and develop the reforestation campaign already initiated throughout Los Belvederes. Community groups can continue to provide the labor to plant the trees. But the Tlalpan ward and SARH must provide a sufficient number and variety of trees as well as adequate tools and inputs for the task (e.g., organic fertilizer, farming equipment, etc.).

2. Systematize and develop efforts to control plagues. This goal requires financial and technical resources to start community education programs and to design and implement techniques for maintaining the health of the woods. Along these lines, support the research of the Centro de Investigación Biológica.

3. Install two pilot units (SIRDO-húmedos) for the integral recycling of organic wastes in order to provide a part of the colonia (300 families) drainage service and garbage collection and treatment. A total investment of 18 million pesos is required. Half of this amount corresponds to labor costs. Solicit funds to cover these costs from the Regional Employment Program (Programa Regional de Empleo; PRE) of the Secretariat of Programming and Budgets (Secretaría de Programación y Presupuesto; SPP). To pay for expenses other than labor (e.g., materials, training, the right to use patents) solicit credit from either National Bank of Public Works and Services (Banco Nacional de Obras y Servicios Públicos BANOBRAS) or the Trusteeship Fund of Popular Housing (Fideicomiso Fondo de Habitaciones Populares, FONAHAPO). To monitor the operation of the SIRDO units, solicit the participation of the GTA and the General Directorate of Hydraulic Construction and Operation (Dirección General de Construcción y Operación Hidráulica; DGCOH).

Human Ecology and Public Health

1. Consolidate the work of the Cooperative Society's (Sociedad Cooperative's) Health Committee, which has been promoting activities to improve environmental conditions, health, and nutrition. Solicit educational materials and medical supplies.

2. Form a working relationship with the National Center for Traditional and Herbal Medicine (Cuernavaca, Morelos). Solicit funds from SARH to buy seeds, materials, and information.

3. Solicit resources and/or grants to organize grassroots groups dedicated to ameliorating problems of alcoholism, drug addiction, and juvenile delinquency and to stimulate positive community activities.

Production in the Primary Sector: Four Pilot Projects

1. *Horticulture and fruit growing*   Do an analysis of the soil and water; get seeds and farming equipment from the DDF and SARH. Implement plans to stimulate urban agriculture and terrace gardening. Aim to generate eight jobs per sector in the *colonia* (eighty-eight jobs in all).

2. *Rabbits*   Secure financial resources and/or materials from the PRE or SARH to construct cages and expand the production of rabbits for their meat and pelts. Set up several modules of twenty to thirty rabbits each to generate up to six jobs.

3. *Mushrooms*   Secure resources from the Tlalpan Ward and SARH (e.g., funds, mushroom spores, materials for beds and boxes) to generate six jobs in the production and marketing of mushrooms.

4. *Piscicultura*   When the SIRDO-húmedos are built and functioning properly (i.e., providing a supply of recycled water) and when the main water supply to the *colonia* is regularized, make arrangements to raise fish. To create up to six jobs in fish production in conjunction with the SIRDO system, funds must be secured from BANOBRAS and FONAHPO to build the holding tanks and the facility necessary for the tertiary treatment of SIRDOs effluent.

Production in the Secondary Sector: Inorganic Recycling and Community Workshops

1. Through the already established Sociedad Cooperativa de Producción SIRDOTEC, train a group of *colonos* to manufacture SIRDO-secos and SIRDO-húmedos. The proposal calls for resources from the PRE and FONHAPO to establish a materials park and to generate twelve jobs.

2. Parallel to the action of environmental improvement and the installation of SIRDOs, initiate community workshops for recycling plastic, glass, and metal. Solicit grants from CREA and SEP for technical training and to incorporate young people into the activities.

3. Over the medium term—once the productive processes in the primary and secondary sector pilot projects are stabilized—begin processing products generated in the community workshops for microregional commercialization (dehydrate mushrooms and medicinal herbs with solar energy, prepare preserves, etc.).

Source: Grupo de Technología Alternativa (1984).

In many ways, the CEP approach is consistent with Habitat II's 1996 program for poverty reduction. Among the initiatives called for by Habitat II are recommendations for poverty reduction at the neighborhood and community levels. These recommendations are outlined as a series of subprograms with such strategies as (1) generation and support of livelihood opportunities, (2) formal and informal sector local economic development, (3) environmental infrastructure reticulation—networking water, sanitation, drainage, solid waste recycling, and disposal—via community contracting, (4) community management involving action planning, formation of cooperatives, advocacy and participation, security of tenure, and access to credit (Habitat 1996c).

Table 10.1
Community pilot projects

Community pilot projects		Productive capacity	
Project	Jobs generated per project	Material needed (cost in pesos)	Capital generated per year (income in pesos)
1. Reforestation	10 people/ sector	Trees and sticks, fertilizer (20,000)	
2. Rabbits	2–3 people/ project	18m² area/project; frames and cages	20–30 litters/ month; 800–1000 kg of meat/litter
3. Infestation Control	5–7 people for initial demo		
4. Mushrooms	3 people/project; 10 people/ sector	18m² area; wood, bituminized cardboard, nails, knives, and garden tools	150 kg/month
5. SIRDO (system for organic waste recycling)	4 people/sector (permanent); 20 people/sector (temporary)	Manufactured parts from community workshop (4 million/sector)	(1.3 million/ sector/year; 14 million/total settlement/ year)
6. Health	10 people/ sector	Graphic educational material	
7. Horticulture and fruit growing	8 people/sector	Instruments for soil and water analysis, seeds	(1 million/sector/year; 10 million/unit/ year)
8. Fish farming	5 people/sector	(3 million/ sector)	(0.5 million/ sector/year)
9. Community workshops		100m² area/ sector	(5.5 million/ unit/year)
Total for units involving production	30 people/sector; 340 people/ unit	(approx. 100 million)	(39.8 million/ unit/year)

Source: Grupo Technología Alternativa (1984).

Habitat II's recommendations for poverty reduction at the local level share many similarities with the principles and strategies advocated by the CEP enthusiasts more than a decade ago. Both approach human settlements problems as problems of environment and development simultaneously, and both emphasize the importance of action at the local level. Such an approach suggests key steps toward implementing principles of productive ecology in the production and reproduction of human settlements. In terms of recognizing the importance of local-level action, there is a parallel with *Agenda 21*. Atkinson (1994) did an analysis of *Agenda 21* (the forty-chapter document signed by heads of state at the UN Conference on the Environment and Development, UNCED, in 1992). He found that more than two-thirds of all actions identified in *Agenda 21* as necessary to engender sustainable development will have to be achieved at the local level (98). Of course, local-level activity is embedded in a broader regional, national, and international context. Success of local-level initiatives, such as the subprograms specified by Habitat II or the CEP initiative, will ultimately depend on a multilevel range of interacting variables. In the case of the CEP initiative, many important factors were in place, such as local social and cultural capital in the form of self-organized community structure, community access to technical knowledge for social experimentation and social learning, and environmental resources including productive land and water. But is also the case that many factors were not in place. In particular, there was little real support for the CEP from the state.

Outcome of the CEP and Other Ecological Initiatives (1984–1990)

The colonia ecológica productiva initiative was most actively promoted and used as a basis for popular mobilization during 1984 and 1985. The settlers and external agents that strived to realize its objectives knew it was an ambitious challenge. When optimism in Bosques was high, Josefina Mena—director of the GTA and a member of the Ajusco Cooperation Group—explained that "The social dynamic in Bosques del Pedregal has acquired a very different rhythm. This is reinforced by the CEP project— a project that expresses the distinct character of the settlers, including their priorities and their 'productive ecology' consciousness. What is happening in Bosques represents a fundamental questioning of conventional planning concepts. Concepts of zoning, urban spatial categories, the conception of housing and the role of infrastructure have been transformed" (Mena 1987, 554).

At the same time, enthusiastic support was expressed by independent researchers. For instance, Martha Schteingart (1986, 30) of the Colegio de México argued that the CEP project "is an example of creativity and of the search for alternatives to improve the community, demonstrating at the same time, the importance of the independent organization of settlers and of their urban struggle to offer new solutions to the urban and ecological problematic." These observations were made at a time when the CEP project in Bosques, as well as similar initiatives of the groups in Belvedere and Dos de Octubre, were in full swing. The projects were supported by large numbers of settlers, and the initiatives got a great deal of positive coverage from the press.[10] Public protests, marches, and other forms of popular mobilization made ecological arguments and the call for sustainable development their rallying point.

The intensity of grassroots ecopolitics reached its peak in late 1984, when the state announced that it would remove the provisions in the Program for the Conservation of Ajusco that had called for a massive relocation of families out of the zone. The new official strategy was announced at the end of 1984 during a government forum referred to as the Ajusco Reunion (Reunion Ajusco). In short, officials announced that all of Los Belvederes would be included in the legally designated urban area, and each family could acquire security of tenure on an individual basis. Officials announced that the government agreed to "make an exception . . . given [the settlements] antiquity and magnitude" and that "the Federal District administration will begin the juridical regularization of the zone in favor of the residents" (DDF 1985, 3). The regularization was to be carried out under ten conditions (see box 10.4).

The shift in state policy dramatically changed the terms of struggle. "Winning" the right to have their settlements incorporated into the legally designated urban area reduced the settlers' sense of insecurity and urgency. A "victory" of this sort is a result of political mediation. It tends to remove the impetus for further mobilization; struggle becomes institutionalized. From the standpoint of popular mobilization, this pattern of development is typical and presents a major dilemma: "Periods of strong mobilization correspond to the struggle to satisfy permanent needs (e.g., access to legally

Box 10.4
State Stipulations regarding Land Tenure Regularization in Los Belvederes, 1984

1. The 431 acres of Los Belvederes will continue to be considered a special area of ecological protection ("colonias con una vocación predominantemente ecológica").
2. Standard infrastructure will not be installed, only rustic forms that will not disturb the ecology.
3. A strip of land along the access road will be expropriated, and *colonos* must travel to the city through northern connections.
4. The bus terminal will be relocated.
5. Lot layouts will generally be respected, but no more subdivision will be permitted.
6. Each household will receive a *constancia* (certificate) for their lot as proof of legal possession (only one lot per household will be allowed). But the *constancia* does not grant ownership, and occupants will not be allowed to rent, sell, or mortgage their property for ten years.
7. Lots will be allocated only to those current occupants identified by the government's census.
8. Provisions will be made to develop three urban subcenters, each with a capacity to meet the needs of 11,000 people in terms of health, education, culture, and commerce.
9. Land will be set aside for green space, recreational areas, reforestation, and agricultural production.
10. The Tlalpan Ward will provide orientation and information to avoid confusion and misinterpretations.

Source: México (1984, 3) Programa de Conservación del Ajusco.

secure land on which to live, running water, a school) or to unexpected crises such as transportation fee increases, grave accidents, etc., but as soon as the critical juncture is passed demands resume on general terms and organization disappears" (MRP 1983, 6).

The popular groups of Los Belvederes were no exception. Popular mobilization in Bosques and throughout Los Belvederes declined after mid 1985, but it did not subside altogether. Winning the right to have their settlements incorporated into the legally designated urban area highlighted problems outside the realm of ecological discourse. New battles had to be waged to secure the best terms for regularization and to resolve litigation regarding fraudulent land transactions and boundary disputes. To this end, settlements throughout the zone joined forces and created the Popular and Independent Coalition.

Although settlers no longer had to worry about forced evictions by the state, for many a new threat took its place: the prospect of being uprooted by economic forces. Many settlers on ejido land and on private property feared that although they may have gained the legal right to (re)purchase the lot they occupied, they would not be able to afford to do so. They also worried about the prospect of high taxes and fees for services and infrastructure.

The rise of new concerns and priorities was paralleled by a fall in concern with the CEP and similar initiatives. A crude measure of this shift is reflected in the content analysis shown in table 10.2. In 1984, there were twenty-four newspaper articles about popular groups in Los Belvederes rallying under ecological banners. This constituted 21 percent of all articles on urban development in the zone. In 1985 the percentage of articles covering this subject dropped to 10 percent of the total and by 1986 it was zero. Of course, one could argue that the press is not a reliable source and that it merely featured fewer articles on such events. But the shift in media coverage is important in its own right. Besides, no one familiar with events in the zone would doubt that the axis of popular mobilization actually did shift. It moved away from ecological arguments toward juridically defined contests regarding tenure regularization and the terms of contracts between the buyers and sellers of land. Part of the reason for this shift was the lack of state support for the CEP-type initiatives. Indeed, in many ways authorities undermined the initiatives and repressed the activists promoting them. Many of the authorities responsible for overseeing the transition of Los Belvederes into the legal urban land market were in collusion with the Teresa family, RENA, and other (supposed) land owners. I use the word *collusion* because there was an obvious drive to reap huge profits from the commercialization of the land. The collective organization of colonos opposed this drive, hence one of the reasons for lack of state support, especially for the calls for alternative forms of collective tenure. For instance, in Belvedere the grassroots organization Casa del Pueblo, and the university groups working with the settlers bitterly denounced the state's support of "groups that instigate divisions in the community and hinder the projects to create an ecological settlement" (*El Día*, March 31, 1985). All during 1985, Belvedere was embroiled in violent conflicts stemming from disputes over land ownership; there were multiple deaths. On May

29, 1985, university groups from UNAM announced that the profound conflicts in Belvedere "made it impossible to implement any new project" (*La Jornada,* May 30, 1985).

During a press conference coordinators from the university regretfully announced that they had to sever their relationship with Belvedere and that the projects they worked on over the past eighteen months had to be aborted (*La Jornada,* May 30, 1985). The fate of the ecological projects in Bosques was not so terminal. On May 5, 1985, the first eco-outhouse was put into operation. Not much later, three similar units were in operation, and construction began on the SIRDO-húmedo for 120 families. State support for these specific projects was promised, but it never materialized. On November 15, 1985, approximately 1,500 settlers of Bosques took over the administrative offices of the Tlalpan Ward for one hour. To no avail, they protested the lack of support and demanded to know what happened to the supposed 1,700 million pesos (approx. $US850,000) that were promised for financing services and ecological projects (*El Día,* November 16, 1985).

Funding had to be raised from other sources. Early on, money was collected from clothing drives and other community-based fund-raising events. But most of the funds came from two outside groups: the MISEROR Foundation (a Catholic philanthropic agency based in Germany) and the Austrian Solidarity Network (a group of "green" Austrian professionals sympathetic to the ecology movement). Due to galloping inflation during the late 1980s, the funds raised were only enough to cover the cost of four eco-outhouses. Enough money was garnered to begin, but not to complete, the SIRDO-húmedo system. Consequently, after absorbing a great deal of community effort and resources, the SIRDO-húmedo technology sat idle for a long time, creating the appearance that all the effort had been lost.

The CEP's Demise

From 1987 to 1990 the CEP initiative in Bosques del Pedregal was essentially dead. In 1989, the GTA secured funding from the Canadian International Center of Investigation for Development to improve the design of the SIRDO-húmedo and to conduct further research on the use of bioferti-

Table 10.2
Content analysis of newspaper clippings about Los Belvederes, 1982–1990

Main topic of newspaper clipping (including feature articles, brief reports, press releases, commentary, and letters to the editor)	Number of newspaper clippings and percent of total per year							
	1982	1983	1984	1985	1986	1987	1988–1989	1990
Popular mobilization[a]: demands or claims regarding land tenure, services, and political action	14	14	16	15	39	56[b]	21	7
	45%	40%	14%	22%	39%	37%	25%	32%
Ecological arguments and projects calling for alternative (sustainable) development	0	0	24	7	0	1	0	0
	0%	0%	21%	10%	0%	1%	0%	0%
Diverse topics (e.g., poverty, residential segregation) covered from a procolono perspective	9	4	21	13	13	14	15	4
	29%	11%	18%	19%	13%	9%	19%	18%
Official declarations and politically conservative commentary concerning the politics of containment[c]	8	17	56	32	47	80	45	11
	26%	49%	48%	48%	47%	53%	56%	50%
N	31	35	117	67	99	151	81	22

Source: Between 1983 and 1987 there were extended periods of time (totaling twenty months) during which I clipped all articles about the zone in question from four daily newspapers: El Día, UnoMásUno, El Nacional, and La Jornada (after 1985). I also collected articles from two other sources: (1) archives of popular groups and students and (2) Servicios Informativos Procesados (SIPRO, A.C.). SIPRO provides a comprehensive data bank of clippings from eleven newspapers, including the four just listed. Data for 1988 to 1990 are from Miguel Díaz Barriga and Karen Kleiber (1995), an unpublished manuscript titled, "The Press and Urban Conflict in Mexico City."
a. Clippings refer to marches, protests (e.g., the forced occupation of an administrative office), meetings between popular groups and authorities, plantones (sit-ins), the formation or activity of cooperatives and civil associations, and so on.
b. This figure includes twenty-two clippings with details about the participation of popular groups from Los Belvederes in metropolitanwide urban movements (e.g., the Metropolitan Front [Frente Metropolitana], in which the Ajusco-based Coordinator of Colonias and Pueblos of the South [Coordinadora de Colonias y Pueblos del Sur] participated.
c. Descriptive reports of desalojos are included in these figures. From 1982 to 1986 sixteen clippings reported desalojos in Los Belvederes. In 1987, twenty-eight clippings reported desalojos in El Seminario.

lizer in urban agriculture and on the recycling of inorganic wastes.[11] But by 1989, the GTA no longer had a good working relationship with the community. Up until that point the SIRDO-húmedo core plant had been completed, but the tubing network necessary to connect the 120 families to the system existed only on paper plans. Activists in the community had ceased to view the SIRDO project as a priority. Instead, their attention focused on the process and terms of land tenure regularization. But then official arguments about the contamination of Ajusco's aquifer took on renewed life, and progress with the land tenure regularization process was again interrupted—events that breathed new life into the SIRDO-húmedo. Once again settlers mobilized in support of the SIRDO project. On December 21, 1990, they convinced the delegado of Tlalpan to agree to purchase the materials necessary to connect the SIRDO to 120 families on 101 lots, provided that the citizens donated all the labor. In January, 1991, construction of the 2,274 meters of plastic sewage-pipe network began. But it soon stopped again due to organizational and financial problems.

The relationship between the GTA and the General Council of Representatives became strained. The GTA faulted the leadership of Bosques for allowing the government to define the terms of crisis in the zone. According to the GTA, the leadership of Bosques during 1984 was proactive; i.e., it offered alternatives and demanded space for social experimentation. But by 1990, the leaders were employing tactics out of step with the political times *(coyuntura)* in Mexico (Mena Abraham 1990). The council, for its part, faulted the GTA for raising expectations that could not be fulfilled. Aside from the successful operation of the four eco-outhouses, the SIRDO-húmedo and other ecological initiatives failed to make headway from 1985 to 1990. The leadership of Bosques argued that the CEP lost its capacity *(convocatoria política)* as a rallying point; although the CEP sounds good in principle, the reality is that it proves to be immensely difficult to translate into action.

Coulomb (1991) found grounds to be critical of the GTA's handling of another situation in which the SIRDO technology ran into problems. With reference to the case of El Molino, Coulomb argues that the approach of the GTA was disempowering. He notes that external technical agents (such as the GTA) sometimes use colonias as experimental testing grounds without full disclosure and without the consent of the colonias involved. In the case of the SIRDO's failure to work in El Molino, Coulomb asserts that it was not yet a proven technology and certain factors associated with its function clashed with social norms and expectations. The group in El Molino is now experimenting with an alternative to the SIRDO (Coulomb 1991; Merffert 1993).[12]

The GTA was well intentioned, I am certain. It had intended to work closely with the 120 families who would have been directly involved with the SIRDO-húmedo once the hookup was completed. Connecting the dwelling units to the system was planned to occur in stages over the course of several months. Concurrently, there was to be a period of social experimentation. The GTA received a two-year research grant from the Canadian International Center of Investigation for Development to design and monitor methods for community operation of the SIRDO-húmedo. Mena Abraham (1990) notes that "The primary objective of the investigation is to determine the technical, social, and economic conditions that will enable the autonomous operation of the SIRDO." The SIRDO is now dead. Despite what the state said about not hooking up ordinary services, it has begun hooking up a traditional drainage system at enormous cost.

Although the state provided no support for the grassroots ecology initiatives, it did incorporate aspects of the proposals into plans for the zone. For example, in the Ajusco Reunion it was announced that rather than install "regular" infrastructure, the state would need to use "rural and rustic" forms of infrastructure. What was meant by "rural and rustic" was not immediately specified but the idea was to avoid disturbing ecological equilibrium, and with this aim the state promised to support the initiatives of popular groups.

The Subdirectorate of Ajusco: The Front-Line State Agency

The Subdirectorate of Ajusco is a satellite office of the Tlalpan Ward inside the territory of Los Belvederes. To use language taken from the agency's original charter, the subdirectorate was established in 1985, "with the objective of responding institutionally to the increasingly complex social, political, juridical and ecological problems of the region."[13] In 1987, this front-line agency had three departments: (1) Urban Development and Ecology, (2) the Program to Contain Urban Expansion, and (3) Land Tenure.

The subdirectorate is subordinate to the Tlalpan Ward. In one of the few studies that have focused primarily on the ward as a political structure, Agustín García Pérez (1985, i) points out that the "Ward presents itself as the first political entity facing the inhabitants of the DDF, that is to say, when confronted with an urban problem, citizens will generally go first to the Ward Office to resolve their problems."

The present division of the DDF into sixteen Wards was determined in 1970 when President Echeverría Alvarez reformed the Organic Law of the Federal District (Ley Organica del Distrito Federal). The reform was part of a restructuring process referred to as the *desconcentración administrativa*. Echeverría imagined that the capital would be composed of sixteen cities and that "each one of these 16 cities would have, as a center of action, a great administrative building, headed by a genuine Delegado" (García Pérez 1985, 59). The delegado's authority is just beneath the *jefe*—an appointed mayor, also known as the regente. The regente is the DDF's highest authority appointed to office by the president.[14]

As a result of the administrative decentralization the ward assumed greater powers and autonomy, even though it continues to be subordinate to the DDF. It has explicit functions of political control exercised in coordination with delegational committees of PRI (García Pérez 1985). The Subdirectorate of Ajusco has been the state's front-line agency with regard to urban land use conflict and the institutionalization of struggle.

Political Mediation of the "Problematic"

When the Programa de Conservación del Ajusco was announced on May 23, 1984 (and later revised in December of the same year), the groundwork was laid for the creation of the subdirectorate. In May 1985, under the direction of (1) the regente of the DDF, (2) the general secretary of the government of the DDF, and (3) the delegado of Tlalpan, concrete plans were drawn up to begin the land tenure regularization program of Los Belvederes and to construct a front-line state agency for its implementation. The physical structure of the agency was composed of two detached single-story office buildings covering approximately 1,500 square feet each. At this time, the Special Coordinating Commission for the Conservation Program of the Sierra del Ajusco (Comisión Coordinadora Especial del Programa de Conservación del Sierra del Ajusco) was set up. The sub-directorate was created as the local base from which to execute strategies specified by the special commission in their "Plan for Immediate Actions Plano para Acciones Inmediates" (see figure 10.7). According to the subdir-ectorate's 1985 Work Program (Programa de Trabajo), the Plan for Im-mediate Actions was to be carried out in two stages. The first stage called for the establishment of the subdirectorate in Los Belvederes in order to "penetrate and politically stabilize the zone" by asserting the "physical presence of the state" (1).[15]

The first stage was concluded by July 1985. It involved the state's direct contact with 5,000 colonos in a massive undertaking in which 4,480 *con-stancias de ocupación* were distributed. *Constancias* are provisional land titles that demonstrate an occupant's legal possession of a lot without granting that occupant legal ownership. Also during stage 1, three police command posts were put into operation, a government subsidized grocery store (CONASUPO) was opened, and the state began to make infrastruc-tural improvements (e.g., several miles of roadway were paved). Stage 2 of the Plan for Immediate Action was to be undertaken once the zone was politically stabilized. In legal and institutional terms, stage 2 called for the authorities to establish social control.[16] This was the task of the subdirect-orate of Ajusco.

The Plan for Immediate Actions

To politically stabilize Los Belvederes and to bring its social organization "within the state of law" the subdirectorate focused its resources on two programs: Community Development (Desarrollo de la Comundiad) and Land Tenure Regularization (Regularización de la Tenencia de la Tierra). Both programs were carried out in conjunction with other agencies. When the subdirectorate first opened its doors to the public in 1985, it was a bustling base camp with twenty vehicles and seventy full-time employees (each of whom was supplied with lunch and dinner). By 1988, when the situation in Los Belvederes had become less urgent, the number of employ-ees at the subdirectorate dropped to thirteen; and the number of vehicles to three. Over time, the point of greatest political conflict in the zone shifted from Los Belvederes to newly emergent irregular settlements nearby in El Seminario. As this happened, authorities of the subdirectorate's Program

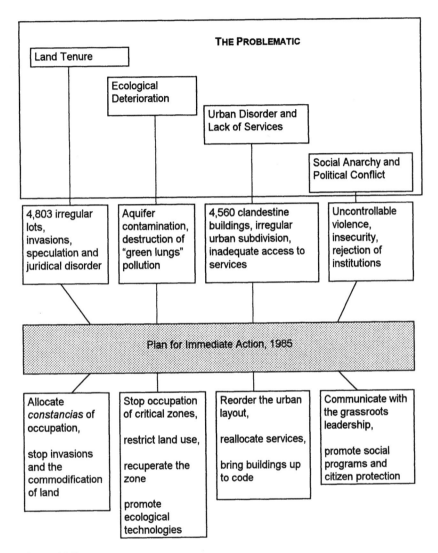

Figure 10.7
The "problematic" and the "plan for immediate action" as illustrated by the Special Commission for the Conservation of Ajusco. Source: Adapted from a working document of the Subdelegación del Ajusco, titled "1985 Programa de Trabajo: Lineamientos para su Ejecución" (1985c).

to Contain Urban Expansion complained about the reduction in their staff and vehicles. They argued that although the gravity of conflict had diminished in Los Belvederes, it had begun to flare up in El Seminario. Yet, resources allocated to the subdirectorate continued to decline. During this time the institutional terrain was shifting. While the subdirectorate was in decline another institution, COCODER, was on the rise. COCODER was home base to the newly created force of *ecoguardias*. *Ecoguardias* have a quasimilitary approach to rural development in the Federal District: agricultural land is viewed as under attack and it is the job of ecology troops to protect it.

Initial efforts at the subdirectorate were tied up with the determination and distribution of *constancias*. Issued free of charge under the direction of the Tlalpan Ward and CORETT, these constancias were intended as a measure to end violent land disputes and to stop real estate speculation. In effect, the procedure created a one-to-one relationship between the state and residents. Each *constancia* stated that the lot's occupant would be prohibited from selling, renting, or mortgaging their lot for a ten-year period beginning from the date the *constancia* was issued. All pretense of enforcing this restriction was dropped in 1986, when the authorities decided to have the occupants purchase the land directly from supposed landowners (including the Teresa family and members of the Belvedere Association). Attention was thus focused first on private property and on property in litigation. Colonias that were undeniably on ejido land were put on hold.

The subdirectorate said it would mediate boundary disputes and put an official stamp of approval on the contracts made between colonos and landowners. This procedure generated popular opposition. A barriowide coalition of popular groups (the Popular and Independent Coalition) formed to reject the terms of regularization. One of the main points of contention concerned the Community Development Program.

The Community Development Program

As specified in the subdirectorate's 1985 Work Program (Programa de Trabajo), the Community Development Program had two components: (1) the Program of Community Organization for Social Integration (Programa del Desarrollo de la Comudad y Organización Comunitaria para la Integración Social), and (2) the Program of Social and Cultural Action (Pro-

Box 10.5
Activities of the "Social Integration" Program

---

1. Compile data on all existing popular organizations—including details about each group's membership, geographical sphere of influence, relation with other organizations, social objectives, and disposition.
2. Put together an inventory of the organizational requirements necessary to execute the programs of the subdirectorate.
3. Set up a program to evaluate existing popular organizations. Design strategies and make arrangements, as deemed necessary, to either support or disintegrate these groups.
4. Create an individual file for each of the leaders in the colonias. Include in this file the individual's personal data, experiences, complaints, successes, and institutional disposition *(comportamiento institucional)*.

Source: Subdelegación del Ajusco (1985c)

grama de Acción Social y Cultural). The second element amounted to very little; its highlight was an occasional state-sponsored basketball game or beauty pageant. The former component was more significant. At its core were strategies of surveillance and social control. A document that listed the activities of the Program for Social Integration specified four such strategies (see box 10.5).

In an interview I conducted with the director of the Community Development Program, I asked if the subdirectorate actually engaged in strategies aimed to repress or "disintegrate" certain popular groups. The response was affirmative:

We are an administrative institution, but we also have a political nature, one-hundred percent. So we have friends and we have enemies; the Coalición is our enemy because they are not with the PRI. As I was telling you, the Coalición is disorganized—in part, our institution contributes to this. It is not a mere accident that I don't take it upon myself to always hand out documents to the same person in the Coalición. I give out a copy here and a copy there. This gives us the advantage. It is one of our methods, a form of contra-política. . . . While these strategies of disintegration exist, and they are well implemented, you won't find them written down anywhere. For example, you see the list of [grassroots] representatives in the meetings with COPLADE, its only those affiliated with the PRI that get invited; we don't invite the Coalición. In this way we selectively support some groups and push others down; this is our business here.[17]

In another interview with the director of the Ajusco satellite office, I asked if the subdirectorate actually kept individual files on each of the colonias leaders. Again the answer was affirmative, and to my surprise I was handed two files as proof. Each file contained a two-page "political profile" of a community leader prepared by an *oreja* (translates literally as "ear"—the title given to an undercover agent working for the state). One of the profiles described Jerómino Martínez, a leading member of the Association of Settlers of Ajusco, Casa del Pueblo, a popular organization based in the colonia Belvedere. The other profile described Ernesto Bravo,[18] a leading activist from the colonia Bosques del Pedregal.

The file on Jerónimo Martínez illustrated a fair degree of background research.[19] The profile was addressed to the director of the subdirectorate. It characterized Martínez as a dangerous threat, and it concluded with the ominous statement "waiting for your instructions." This report was dated July 31, 1987. Less than one month later, on August 15, 1987, Martínez was arrested by the police while he was en route by bus from Oaxaca to Mexico City. Martínez was arrested on the charge that he was smuggling drugs (supposedly fourteen kilograms of marijuana), and he was imprisoned. Colonos of the colonia Belvedere argued that the charges were false and that Martínez was actually a political prisoner, because he had led a protest movement against the terms of the land tenure regularization program. I asked the director if the allegations were true; he said yes, Martínez was in fact a political prisoner.[20]

The political profile on Ernesto Bravo, also dated July 31, 1987, was even more damning. It included the observation that "Apparently he does not belong to a political party, but nonetheless—of all the leaders in the zone of Ajusco Medio—Bravo is the toughest. It can be said that he is the base of the political group the Coalición Popular Independente . . . he has become an acrid critic of the system and a tireless instigator of disrespectful demonstrations."

The conclusion to the profile on Bravo noted that "the advance, slow but sure, of the System in the zone has caused the popularity of Mr. Bravo to decline—that is, the decisive intervention of our Delegado in the regularization of land tenure in the zone has been a determining factor in [Bravo's] decline." At the end of this profile there also appeared the ominous statement "waiting for your instructions." Bravo was not detained by the police. His brother, on the other hand, is on Mexico's list of desaparecidos (missing persons) (*La Jornada*, October 25, 1986). Although Bravo has managed to avoid becoming a political prisoner, he has been repeatedly harassed and this has taken a heavy toll on his family.

It is worth noting here what I mean by exacting a toll. Bravo has a wife and four daughters; threats against him deeply affect them all. This became all too clear to me one day when I visited his small, one-room house. I went there that day to continue a series of interviews with Bravo, but he had not yet returned from work. During the preceeding months, I had become quite close to his family; so his wife, Gabriela Bravo, invited me in to await his return with her and the four children. While waiting Mrs. Bravo began to describe to me how her nine-year-old daughter suffered recurring nightmares. Every night for two weeks the little girl woke up screaming, having dreamt that her father was being tortured and put to death by "people in the government." Describing the night previous to my visit, Gabriela told me how the nightmares had reached a frightful climax.

Because the father was late in getting home, the daughter, fully awake, convinced herself that the nightmares she was having had become real. From 9:30 p.m. until 12:30 a.m. the girl was in a state of panic, crying hysterically: "We must go out and look for daddy, they're going to kill him!" The girl feared for her own life as well: "Mommy, Mommy, I don't want to die." The mother, while telling me all this, broke down in tears. Sobbing, she described to me how the daughter wrung her hands in desperation, how she hyperventilated, how nothing could be done to calm her. The mother actually feared that her girl's heart would stop. She feared that her daughter's torment was so great that she may forever suffer emotional problems. That night, when the father finally got home, the girl fell silent, emotionless. She withdrew. The child, the mother told me, "doesn't have mental problems." "But, what will happen?" she added, "if this keeps up."

This may seem to be an exaggerated case. However, around the time that the little girl was having her nightmares, the father was one of the principal activists of the popular movement called the Coordination of Colonias and Pueblos of the South. This group was protesting the nearby construction of an office complex for occupation by Gobernación. Popular groups mobilized to stop construction of the office, arguing that they had proof its design included a torture chamber in the basement as well as ovens to cremate dissidents to make them dissappear (*La Jornada,* June 21, 1987; *El Día,* July 16, 1987).

In the final analysis, the Community Development Program did little community development. Its primary objective was social control. In addition to overseeing the Community Development Program, the subdirectorate ran a program for ecological preservation. Almost nothing, however, was achieved along these lines. Most effort was channeled toward land tenure regularization. The distribution of provisional land titles was carried out in 1985 by twenty public notaries and twenty brigades of social workers. This first stage of the Plan for Immediate Actions was implemented without much trouble (*Excelsior,* May 10, 1985). Popular opposition developed once the authorities decided to orchestrate the private sale of the land (as opposed to its expropriation).

It was arranged that colonos submit their entire payment directly to the Teresa family. The Teresa family (henceforth called Teresa for short) then kept 35 percent of the payments and distributed the remaining 65 percent to members of the Belvedere Association. The 65 percent was paid out to compensate those who had purchased 1,000-square-meter lots from Teresa in the late 1960s. This thirty-five/sixty-five deal had been worked out between Teresa and the other major landholders in the zone.[21] The local government endorsed this strategy and actively promoted the transactions. It gave legitimacy to the deals by channeling the procedure through "tripartite contracts," official documents signed by the three actors involved: the developer, the state, and the colono. These tripartite contracts were pivotal in the process of political mediation.

Tripartite Contracts

Dos de Octubre was the first colonia in Los Belvederes where a tripartite contract was put into effect. It was signed on August 5, 1986, by the delegado of Tlalpan; the director of the General Directorate of Territorial Resources (Dirección General de Recursas Territoriales; DGRT), the *"propietario"* (Landowner), Fernando de Teresa; the Belvedere Association; and a colono, Jesús Rivera, who represented the thirty-five families involved. Rivera played an important part in initiating the proceedings. In a dramatic political reversal, Rivera, who was previously considered by the government as one of the most "acrid critics of the system," became a sectional representative of PRI.

The tripartite contract, or *convenio de regularización,* was an agreement between all parties concerned specifying the terms of sale and regularization. The contract signed in Dos de Octubre was a nine-page typewritten document beginning with separate declarations by each party. These declarations were followed by ten clauses.

Under the rubric "Declaraciones," the state agreed to help resolve all problems associated with land tenure and titles. The state also declared that once individual contracts were finalized, the area would be legalized to the full extent of the law. The *propietario* declared that the land in question was legally theirs and that it was owned free of any liens or other liabilities. The Belvedere Association declared that they were involved as the duly designated representative of the Teresa family and the other landowners of the property referred to as the Ex-Hacienda de San Nicolás de Eslava. For their part, colonos had to declare that they had occupied the lots of the *propietario* "without authorization of the DDF or of the propietario, thereby violating the Urban Planning Law of the Federal District and the Urban Development Law."

Following the declarations, one of the tripartite contract's main clauses stated that "upon the celebration of the present accord, the landowner and colono are obligated to individually negotiate their own separate buy-sell contracts [contratos de compra venta]." This obligation had to be met within twenty days from the date the *convenio* was signed. The price would be approximately 3,000 pesos per square meter (approx $US3.00 per square meter).

The thirty-five families that entered into this agreement did so fearing that should they delay, the price of the land would climb beyond their reach. Their decision to buy into the deal created additional pressures for others to follow, and the move in this direction generated great friction among the popular organizations of the zone as well. The groups most adamantly opposed to the signing of the tripartite contracts joined together and formed the multisettlement coalition noted earlier. The coalition opposed the terms of regularization on three grounds.

First, the coalition argued that most of the colonos in the zone had already paid for the land once, sometimes twice over. Over the years, payments were made to either developers or ejidatarios, and many colonos had papers to prove it. Besides, the coalition argued, even if the colonos had not paid anything for the land, it was the colonos' labor that made the land valuable. To ask 3,000 pesos per square meter was considered criminal. Why, the coalition asked, should the landlords reap such wealth when it was the colonos that labored the earth to make it habitable? Rejecting the option to purchase the land under these terms, the coalition pushed instead for the government to expropriate the land so that it could be transferred to the colonos for a nominal fee.[22] Anything near the asking price of 3,000 pesos, the colonos contested, was tantamount to a *desalojo por vía económica*: "they couldn't clear us out with the police force so they intend to do it now with market forces!"[23]

Second, the coalition opposed the signing of the tripartite contracts because it was not satisfied that Teresa actually owned the land. The coalition demanded to see Teresa's titles, or at least some proof that Teresa had been paying property takes on the land. No such proof was delivered (*La Jornada*, August 23, 1986).

The third point of contention that motivated the coalition's opposition related to collective relations and solidarity. There was concern that the legalization of the zone—by way of institutionalizing private property on an individual basis—would undermine the strength of popular organization and thereby erode the potential for collective approaches to the zone's problems. More specifically, legalization individualizes the settlement problem, dividing up the land in a way that creates a specific relationship between each colono and the state, thus undermining the basis for the land's collective administation. In this way, the coalition's argument goes, the efficacy of autonomous popular groups and popular mobilization is lessened. Popular organizations become fragmented; they lose their internal solidarity and are pushed toward integration with the state. These are observations that Castells (1983) also made in his study on popular struggle in Monterrey, Mexico, in the late 1970s. The coalition thus raised the question: Does effective security of occupancy require full legal title as government agencies insist? Their answer was a resounding *no!*

Some observers agreed with the coalition's analysis. For example, Ward (1986) argues that the creation of a battery of institutions to provide full legal title to *colonos* reflected governmental needs more than those of low-income residents. Regularization became an "end in itself . . . the key ingredient around which state-community mediation was channelled" (66).

Popular Opposition to the Tripartite Contracts and the State's Response

Colonos that opposed the signing of the tripartite contracts were denounced by the state. In September 1986, the delegado of Tlalpan, Ramos Galindo, called such opposition the work of "minority groups" trying to impede the regularization process. Galindo argued that these groups do not want to see regularization advance because "during many years they have been able to take advantage of disorganization; their force is derived from the existence of problems and difficulties" (*El Nacional,* September 29, 1986). At about the same time, Veloz Bañuelos, director of DGRT, stated that expropriation of private property is not feasible in the short term: "Viewing it as a Federal Executive mechanism of [land tenure] resolution, we're not afraid of expropriation; however, to bring it about would take at least 10 to 15 years, involving elaborate negotiations"(*El Día,* August 25, 1986).

The official federation of colonias populares (the CNOP of the Federal District), made its views known by publishing a letter to the regente in *Excelsior.* The federation was in favor of "ordinary" regularization, so it denounced those who sought to block the process by demanding expropriation. The *junta de vecinos* from the Tlalpan Ward made its views known in a letter to the regente (Ramón Aguirre Velázquez) published in the newpaper *El Universal.* Written by the president of the *junta* in Tlalpan, the organization supported the private, market-oriented approach to regularization. The letter argues that "Dissident groups are trying to disorient the local community and public opinion. The conduct of these extreme groups, based on lies and verbal agression, aims at undermining the ongoing regularization process with political motives and the objective to continue living in disorder and anarchy. We would like to see 12 year prison sentences for those who are getting in the way of regularization" (*El Universal,* November 10, 1986).

In December 1986, Galindo made the same arguments in an extensive interview with the daily newspaper *El Día.* Galindo claimed that the government, given conditions of austerity, did not have the resources to carry out an expropriation of the zone (*El Día,* December 21, 1986). Concerning Bosques del Pedregal, Galindo stated, "since they were not invited into the process of ordinary regularization, they are the ones voicing the most attacks, believing that we are out to divide them, instead of giving us our due thanks considering that we are also going to pull them upwards through a different process" (*El Día*, December 22, 1986). Regarding the colonos' contention that the supposed landowners of the zone are fraudulent, Galindo stated "The Public Registry of Property establishes who are the landowners, how many there are, and what it is that they have. I have no reason to doubt the Public Registry of Property or the Office of Territorial Regularization because they are the entities created to deal with this subject. Yet, colonos have lost confidence because they believe that they are more papist than the Pope. It's just that the colonos don't want to recognize these landowners" (*El Día*, December 22, 1986).

On December 12, 1986, the government ceremoniously granted the first 200 land titles to colonos from Dos de Octubre (*El Día*, December 13, 1986). By March 21, 1987, the DGRT reported that 1,505 households in Los Belvederes had signed contracts with the landlords.[24] The coalition aimed to stop this process by organizing public demonstrations, marches, and sit-ins *(plantones)*,[25] and, more aggressively, by forcefully occupying the Tlalpan Ward office (*La Jornada*, May 4, 1987). Figure 10.8 shows a scene from one of the rallies led by the coalition in Bosques del Pedregal. The stuffed effigies hanging by their necks in front of the defunct SIRDO-húmedo project were intended to represent what were cast as the coalition's villanous enemies. These included a couple of real estate agents, a few government officials, and a symbolic character dubbed *carestía*. *Carestía de la vida* refers to the high cost of living. In a gruesome end to this particular rally, the effigies were thrown into a heap, covered with gas, and ignited. Childern were instructed by their parents to throw rocks at the flaming pile.

None of the coalition's demonstrations or tactics had much effect. Then, in an unusual move during May 1987, the coalition approached the ejidatarios of San Nicolás Totolalpán to form a united front (*El Día*, May 4, 1987; *La Jornada*, May 4, 1987). The ejidatarios of San Nicolás Totolalpán, who also disputed the regularization process, agreed to collaborate. They argued that much of what Teresa claimed as private property was actually ejido land. Specifically, the ejidatarios claimed all of Dos de Octubre and all of Bosques del Pedregal, as well as parts of Belvedere, Mirador Uno, Mirador Dos, and Lomas de Cuiltepec.

Figure 10.8
Effigies hanging by their necks, Bosques del Pedregal, 1987.

The Tlalpan Ward, DGRT, and CORETT agreed to suspend land transactions in the face of the pressures exerted by the coalition in conjunction with the Ejido Commission of San Nicolás Totolalpán (*El Día,* May 21, 1987). At one point, 500 members of the coalition threatened to storm the delegation; ensuing negotiations took more than eight hours. The coalition and ejidatarios asserted that the Teresa family's contracts were fraudulent and that, in many cases, the contracts were signed by colonos under pressure. The protestors demanded that the perpetrators' bank accounts be frozen in order to guarantee that monies due (plus interest) would be returned to those who were defrauded. Also, the ejidatarios expressed their discontent that they had still not been fully indemnified for an expropriation of 339 hectares of their land (they were still due 20 percent at the time) (*La Jornada,* May 21, 1987).

These protests halted the regularization process for about one month. On June 26, 1987, Rodolfo Veloz Bānuelos (director of DGRT) announced that the line of contention had been settled and that the land tenure regular ization process would continue (*La Jornada,* June 27, 1987). Although the ejidatarios did not completely agree with the new line, they did agree to stop protesting. Thus, the ejidatarios ended their association with the coalition. Member groups of the coalition had trouble winning support for their ecological projects. The calls for alternative, collective tenure arrangements that were a part of the colonia ecológica productiva project were denounced as subversive. The next chapter explains these developments under the rubric of "ecology as politics."

11
Ecology as Politics

Given the sheer magnitude of the population, the sophistication of popular resistance, and the state's inconsistency, plans to eradicate Los Belvederes were never realized. By 1993, much of the tension that had characterized Los Belvederes in the 1980s had subsided in the wake of land tenure regularization. The government had expropriated 440 hectares and finished up the business of land tenure regularization in Ajusco Medio; 23,000 land titles were allocated. Emphasis on regularization was national in scope. President Salinas reported that between 1989 and 1991, his administration had granted 1.2 million title deeds to legalize 1.2 million lots in *colonias populares* throughout the country (México 1991b).

The Pronasol program put money to work in sixty-four *colonias*—habitat for 400,000 people (Serrano Gutiérrez 1993a). Most of the colonias were within the Tlalpan Ward's southern fringe, which included Ajusco Medio. The amount spent in 1992 totaled 220 billion pesos (approximately US$70 million). Table 11.1 shows that more than half the money went to install drainage pipes. Approximately 2 percent of the funds were allocated to establish a new ecological park in Ajusco.

Negotiating Life at the Limits: Cultural Embodiment

In 1985, when the government extended the officially designated urban area to include Los Belvederes it asserted, once again, that absolutely no new settlements would be permitted in the zone of Ajusco. The buffer zone was eliminated by adding it to the area for ecological conservation, and new limits between the areas for urban use and ecological conservation

Table 11.1
Allocation of Pronasol Funds to sixty-four *colonias populares*, Tlalpan, 1992

	Millions of pesos	Thousands of US$[a]
1. Drainage for sewage and rain	121,680	38,753
2. Potable water	48,959	15,593
3. Schools (maintenance)	12,470	3,972
4. Pavement	10,907	3,472
5. Parque ecológico	4,000	1,274
6. Potable water treatment plants	3,300	1,051
7. Youth center	2,600	828
8. Station for the transfer of refuse	2,500	796
9. Tlalpan Ejido Park	750	239
10. Health program	120	39
Total	220,000	70,064

Source: Serrano Gutiérrez, Alicia (1993b).
a. Calculated at 1992 exchange rate of 3.142 pesos per US dollar.

were clearly marked with numerous signposts and *mojoneras* (boundary markers). Yet beyond these *mojoneras,* irregular settlements as well as upper-class residential estates continued to penetrate the so-called ecological reserve. It's an ongoing drama. This contradiction can be is mediated through political culture as well as politics. Figures 11.1 to 11.9 capture the process of political mediation in action. The photos show activity at one of the state's lines of containment. The line in view is made up of government forces cutting off access to the street. But it is also symbolic of the urban-ecological line in Ajusco. Activity at the line is the cultural embodiment of state-community negotiations and power relations at the frontier. At such a frontier, ecology is politics and limits are negotiated in terms of power and social control.

In the photos, demonstrators are shown approaching the residence of Mexico's president. Once they reach the police blockade, negotiation begins; it's routine. The negotiations concern the security of the marchers' irregular housing inside the ecological reserve. The blockade is successfully penetrated. Since the squatters ultimately won the right to be included inside the city's legally designated urban limits, their passing of the blockade can be viewed as a metaphor for their passing of the actual urban-ecological line at the city's periphery.

Figure 11.1
"No government office in Contreras." Note: The banner is carried by two boys at the front of the march to Los Pinos. The banner illustrates an office building of the Mexican government (Gobernación) that was under construction in the ward of Magdelena Contreras at the time of the march. The protesters believed it was a political prison equipped with a crematorium to kill their leaders. Source: Photos taken by author, September 1987.

Urban-Environmental Politics as Preemptive Reform

At the outset of his term, President Zedillo announced his commitment to extending efforts aimed at promoting ecological sustainability in the Valley of Mexico.[1] During a speech before an environmental conference, Zedillo outlined his broader views on Mexico's urban prospects under the rubric, "Ecology: The Guarantor of Sustainable Development":

Cities are indispensable for sustainable development. We should not lose site of all that which urban spaces offer us for the creation of civilized environments, economic opportunity and social well-being. So, the question today is how to achieve a sustainable city; that is to say, how to make ecologically viable the re-

Figure 11.2
Speaker van leads march. Note: Participants in the march included member gropus of the Popular Coordinator of Colonias and Pueblos of the South (one of the five major networks or umbrella organizations of NGOs that joined together in a metropolitanwide front with a national agenda for social change). Source: Photos taken by author.

source utilization and waste generation of our metropolis, which is one of the greatest cities in the world. A first response resides in successfully satisfying the necessities of the inhabitants of the metropolitan area of Mexico City with resource utilization and waste generation compatible with ecological equilibrium, without loosing site of the national effect and our commitment facing our planet. (Zedillo 1994, 84)

Further into the speech, Zedillo argued that current difficulties make it imperative to encourage the participation of the different sectors of society in the diagnosis, investigation, and solution of urban-ecological problems. He announced his support for a "new ecological culture that stimulates the participation of the children and the families of the metropolitan region in the creation and caretaking of green zones" (97). In his first annual address to the Mexican Congress (Zedillo 1995), he stressed his intention to apply new efforts to consolidate a greenbelt for Mexico City, including

Figure 11.3
"Mothers defend the security of your children!" Source: Photos taken by author.

stricter prohibition against any irregular settlements in the Federal District's most sensitive areas.[2] Zedillo warned that there will be an "energetic punishment for violators" who trespass in the greenbelt zone of Ajusco or in any other of the city's ecological reserves. Along similar lines, the Program for the Urban Development of the Federal District (1995–2000) outlines how "firm steps must be taken in order to preserve and restore natural conditions and to avoid at all costs the occupation—with urban uses—of the ecological conservation zone" (cited in Urrutia 1995).

During April 1996 the Assembly of Representatives of the Federal District (ARDF) passed a new environmental law. This new law imposes severe penalties (up to 20,000 pesos in fines) (approx. $US2,660) for various "ecological crimes." The most severe penalties are slated for infractions labeled as "ecocide"—including the illegal occupation of an area designated for ecological conservation (Valasis 1996). Paradoxically, even as the teeth in the politics of containment get sharper, other developments are taking place that fuel encroachment into the so-called for-ecological-

Figure 11.4
"The coalition of colonias demands expropriation of Ajusco Medio and services."
Source: Photos taken by author.

conservation area. Proposals are being reviewed by the government that would allow the construction of hotels and other tourist attractions within the special zones of controlled development (ZEDECS) (Zedillo 1995; Olayo 1995; Pacheco 1995). The ZEDECS are inside the ecological reserve.

The Integral Program for Transport and Highways (1995–2000) includes plans to construct a twenty-two-kilometer highway (La Venta—Colegio Militar) that will cut across the wards of Cuajimalpa, Alvero Obregón, Magdalena Contreras, and Tlalpan—including the zone of Ajusco. Critics argue that this highway will greatly increase the illegal traffic and speculation in land holdings within the ecological reserve. The government has responded to this critique by arguing that the road will provide a new physical limit to growth. A coalition of local *comuneros,* ejidatarios, and colonos disagree. The coalition banded together to oppose the new road on the grounds that it will (1) destroy streams and springs and (2) foment a "furious speculation of ejido land" (Greenpeace México

Figure 11.5
Popular groups selecting representatives, who are then allowed to pass through the blockade. Source: Photos taken by author.

1995). The main concern of the multisector coalition is that the highway will set in motion dynamics that undermine rural livelihoods and displace lower-income families. There is evidence that—far from being a barrier to growth—the proposed highway will open access to higher-income groups in search of housing outside Mexico City's already built-up urban area (Greenpeace 1995).

The blatant contradictions and polarization inherent in such developments certainly do not escape the attention of politicians. As environmental problems have worsened, political parties outside the government's dominant political party (the PRI) have begun to wave ecological banners in their oppositional campaigns. The widely supported center-left-nationalist presidential campaign of Cárdenas in 1988 had a platform calling for job creation via soil enrichment programs, watershed restoration, fisheries, and reforestation projects (Reding 1988). The injection of ecological initiatives into the political arena has helped spur legal-institutional changes.

Figure 11.6
More representatives passing through the blockade. Source: Photos taken by
author.

Under the rubric: "strategies for social participation in environmental
management," President Zedillo's National Development Plan (1995–
2000) puts a premium on civil society's input. The plan declares that the
authorities should promote social participation in the formulation, application, and vigilance of environmental policies and that this should be
done by creating and fortifying consultative councils to encourage the contributions of citizens and NGOs in the diverse tasks of public administration (Zedillo 1995, sec. 2.3).

To elicit input from civil society's for the purpose of resolving housing
and urban-environmental problems, SEDESOL (together with other commissions from the Mexican government's House of Representatives and
Senate) convened a series of public forums in 1995. As Pradilla Cobos
(1995b) points out, this top-down–bottom-up collaboration among the
executive and legislative branches of government is novel. Rather than a
simple acceptance of the plan endorsed by the executive branch, there

Figure 11.7
"Masses" making themselves heard by banging rocks on a metal pole. Source: Photos taken by author.

was active legislative input into the plan itself. There was also input from organizations outside the government. This may be a promising development, but critics are wary: they view the initiatives as the latest in a long series of efforts on the part of government to secure regime legitimacy while failing to actually deal with the root causes of urban-environmental problems (Pradilla Cobos 1995b).

In view of the political context in which it takes place, Mexican environmental policy over the past decade, according to Mumme (1992), can best be understood as *preemptive reform.* Mumme (1992) conceptualizes the Mexican government's environmental initiatives as "a situated and reactive response to the threat of unconstrained environmental mobilization and as a part of a basic strategy of containing political mobilization at a time of declining system legitimacy and economic crisis" (126). This policy has had a divisive impact on grassroots organizations and popular move-

Figure 11.8
Protesters holding onto a rope at the tail end of the march as insurance against getting secretly nabbed by the police. Source: Photos taken by author.

ments. As Mumme describes it: "Preemptive reform challenges the independence and development of the environmental movement in Mexico and has been partly effective in dividing the environmental movement and neutralizing its effectiveness in the Mexican political arena" (126).

With respect to housing conditions, there is rising criticism that SEDE-SOL's social program to eradicate poverty (Pronasol) has not lived up to its promise. Researchers at CIDAC (1991, 62) maintain that "despite the recent increases in the production of housing and of financing made available by government institutions, the housing problem presents itself with ever greater force as a cumulative deficit that is now at the point of exploding." In part, this is because most government support for housing has favored the middle class, which has a greater capacity to repay credits. Javier Hidalgo, the director of the mass-based NGO Assembly of Barrios, notes that although Pronasol funds have been channeled to provide some infrastructure and services for low-income settlements, the impact has

Figure 11.9
The grassroots commission passing through the final line of blockade on its way
to the president's residence. Source: Photos taken by author.

done little to diminish the growing backlog of unmet need. To the extent
that support for new construction was given during the early 1990s, it
went mostly to upscale "private initiatives to construct five-star hotels and
triple A offices in anticipation of NAFTA" (Hildalgo 1995).

Critics argue that Pronasol, which is supposed to be nonpartisan, has
been used to bolster lagging support for PRI. Barry (1992, 101) argue that
as "a manifestation of the government's neocorporatist strategy, Pronasol
has bypassed the old corporatist network and established new and innovative forms of PRI patronage and clientelism." Projects funded through
Pronasol are managed by community committees *(comités de solidaridad)*,
which supervise local skilled labor and the purchases of material. Communities that receive Pronasol funds must contribute a minimum of 20 percent
of project costs (usually in-kind contributions; i.e., labor and materials).
As a result of this arrangement, a direct link is established between Pronasol
and civil society. Along these lines, Fox and Hernández write,

Pronasol projects are accompanied by a powerful official discourse of community participation and "co-responsibility" between the state and the citizen, and they require the formation of local solidarity committees, which in turn choose from a set menu of possible community improvement projects, such as electrification, paving roads, fixing schools, and so on. The political goal is to promote a direct link between PRONASOL (the symbol of the president) and the local community often bypassing both local authorities and traditional political bosses. (1992, 191)

Critics draw attention to the rapid liquidation of state-owned enterprises that have been sold to fund Pronasol activities. These procedures raise doubt about the ability of the government to keep up the pace of its accelerated public spending; a major source of its funding is nearly depleted. Reporting on the four year period of 1990 to 1994, the World Bank and the International Bank for Reconstruction and Development note that Pronasol has funded over 75,000 projects throughout Mexico at an average cost of US$11,000 each (1994, 74). Recognizing the positive side of these investments, the World Bank notes that studies have found that municipal fund projects often cost one-half to two-thirds as much as similar projects managed by state or federal agencies (74).

Does Pronasol set out to tackle the historically rooted problems that reproduce the skewed distribution of income and resources in Mexico and that give rise urban-ecological problems in the first place? Or, is Pronasol's main concern one of social control—where the task at hand is to pacify and provide temporary relief by targeting politically contested areas? From a critical perspective, the simple responses to these questions are *no* to the former and *yes* to the latter. But such questions and answers don't delve very deep. The state has long demonstrated the capacity to implement preemptive reforms, or as Ward (1989, 320) puts it, to institutionalize "change to effect no change." In view of the new constraints as well as opportunities facing civil society in Mexico, Fox and Hernández raise the crucial issue: "Both grassroots movements and NGOs face the classic challenge of Mexican politics: What kinds of new styles and institutions make it possible to change the system even more than one is changed by it?" (1992, 193). An effective response to this inquiry—in theory and practice—becomes all the more imperative as the politics of containment engender an increasingly costly industry of destruction in the 1990s.

Irregular Settlements and the Industry of Destruction in the 1990s

The politics of containment are more than a failed set of policy initiatives. They have given rise to an *industry of destruction:* a cycle of irregular settlement–eradication–irregular settlement, in which certain individuals, public and private, reap huge profits through illicit land transactions. The approach engenders violent confrontations and loss. Meanwhile, encroachment into the ecological reserve continues (see figure 11.10).[3] In an article titled "Inexorable, the Advance of Urban Sprawl over the Areas of Ecological Conservation," Sosa notes (1991) that from 1988 to 1991 the number of irregular settlements in the ecological reserve rose from 88 to 400. By 1992 the number had risen to 494 (Mejía 1992). These irregular settlements—with a total of more than 38,000 families distributed among nine

Figure 11.10
Irregular settlements encroaching beyond the outer limit of the legally designated urban area, Ajusco Medio, 1993. Note: The roadway constitutes the boundary between the legally designated area (on the right side, where part of Los Belvederes is pictured) and the area for ecological conservation (on the left side). The irregular settlements have spread across the road into the ecological reserve. Source: Photos taken by author.

wards—contain up to 600 housing units each; they occupy gorges, ravines, hillsides, and woodlands of the ecological reserve.

According to figures released by the COCODER, 1,778 hectares of the 85,554 hectares that make up the ecological conservation area (ECA) were occupied by 299 irregular settlements in 1994. The number of families living in these settlements was estimated as 10,716. Of the total area involved, an estimated 600 hectares were occupied by way of land invasions. The rest involve other forms of illegal and semilegal forms of subdivision.

COCODER acknowledges that indeed there were 494 irregular settlements in 1992 but state that from July 1992 to May 1994, they reduced this number to 299. During this twenty-two-month period, a total of 24,166 families were evicted from 195 irregular settlements. Table 11.2 provides land area and demographic details about the irregular settlements that occupied the ECA in 1992 and 1994.

According to Sergio Reyes Osorio, the spokesperson *(vocal ejecutivo)* of COCODER, from January 1994 to September 1994 COCODER had to use public force more than 100 times to recover illegally settled land.[4] Reyes Osorio notes that one of the most persistent cases is in Ajusco Medio. In late 1994, a group of 150 families took over an empty government building and its surrounding land in that zone. The building, referred to as Ecoguardas, was set up for evironmental education and conservation programs inside the ecological reserve.[5] Regarding this situation, Reyes Osorio exclaims:

The problem of invasions is a constant struggle and the case of Ecoguardas illustrates it very well. In four months, they inserted themselves there five times and five times we conversed with them so that they would get themselves out without requiring the use of public force. The last time they arrived more aggressively, including with machetes—according to them for clearing the land—and now they don't even allege to have purchased in good faith. This leaves us no other option but the desalojo. They insert themselves, we evict them, they insert, we evict. It is a story that never ends.[6]

Indeed! Over the course of a year (1994–1995), the number of irregular settlements mushroomed from 299 to 500 (Balinas 1995). This leap occurred despite the fact that COCODER uprooted, on a daily basis, an average of sixty to seventy dwelling units from the ecological conservation area. During May 1996, a group of 400 people from eleven of the newest

Table 11.2
Irregular settlements in the nine wards of Mexico City's Federal District that make
up the ecological conservation area, 1992, 1994

Ward	No. of irregular settlements 1992	1994	No. of hectares May 1992	No. of families May 1994
Alvaro Obregón	12	0	0	0
Cuajimalpa	45	39	398	1,465
Gustavo A. Madero	27	6	12	793
Iztapalapa	71	12	12	875
Magdalena Contreras	19	11	362	987
Milpa Alta	47	47	71	1,039
Tláhuac	36	27	93	479
Tlalpan	124	105	529	3,554
Xochimilco	113	52	301	2,524
Total	494	299	1,778	11,716

Source: Invaden Zonas Ecológicas: Afectan Asentamientos mil 778 hectáreas
(1994).

irregular settlements in Ajusco Medio stormed the main administrative
office of the Tlalpan Ward. They demanded that their petition for land be
attended to. The spokesperson for the Tlalpan Ward said that the protesters were illegally settled in the ecological reserve, hence it would be difficult
to attend to their petition (Jiménez and Luisa Pérez 1996).

Of course, it is not just in Ajusco that the irregular settlement problematic seems to be a never-ending story. On a daily basis, the desalojo machine
is doing its work in many parts of the metropolitan area. True, some of
the most serious cases persist in Ajusco Medio—especially the highly conflictive formation of Zacatón, Los Zorros, and Tierra Colorada—a group
of irregular settlements contiguous with Los Belvederes (Quadri de la Torre
1993b). But there are a number of other hot spots as well.[7] The government
has been attempting to construct a thirty-two-kilometer fence to protect
an ecological conservation area in the Sierra de Guadalupe, but the fence
keeps getting knocked down and encroachment continues (Páramo and
Medina 1995). Perimeter fences marking the urban-ecological line have
also been torn down in the Sierra Santa Catarina (Pillsbury 1994). This

last case has some parallels to the case of Los Belvederes in Ajusco. It has involved grassroots environmental activism as a countervailing strategy against the government's threat of eviction.

In 1992, the Inter-American Development Bank approved ten loans totaling US$1 billion for projects designed to resolve urban and rural environmental problems and to manage renewable natural resources in a sustainable manner (Environmental Committee of the IDB 1993). One of these loans went to Mexico. It was a $100 million loan that included a foreign debt buyback mechanism. The money was targeted to establish two ecological reserves: one measuring 8,500 hectares in Sierra de Guadalupe (in the northern part of the MAMC); and the other measuring 2,900 hectares in Sierra Santa Catarina (in the Southeast of the Federal District). This project was intended to be one of the showcases of the Mexican government's Comprehensive Program for Mexico City Pollution Control (Programa Comprehensivo para el control de la Polución dentro de la Ciudad de México), a $4 billion effort to control pollution levels, in which several donor countries and international lending institutions have participated. Part of the funds were allocated for planting 27.5 million trees and 33.8 million shrubs and other species (Environmental Committee of the IDB 1993). Plans for the reserve have impacted forty-seven settlements, with occupants totaling by 336,000 people.[8]

When the people living in the irregular settlements of Sierra Santa Catarina got wind of the government's plans to eradicate them, they banded together in protest. They formed the Front in Defense of Sierra Santa Catarina (Frente para la Detensa de la Sierra Santa Catarina; Pillsbury 1994). The movement produced a counter proposal (a thirty-page document) much like the CEP plan advocated by the coalition in Los Belvederes during the mid 1980s. Two of the front's principal objections to the project were that it ignored the potential for community-based productive ecology as a basis for sustainability and that it shut out input from the people affected. The front argued that it is wishful thinking to set the area aside as a pristine reserve. Indeed, the fence that the government had earlier erected to stop urban encroachment up the mountainside was torn down; the front argued that as soon as it goes up again it will be torn down again.

Confrontation between ongoing forces and countervailing forces of the sort locking horns on the Sierra Santa Catarina slopes clearly gives rise to profound urban-environmental problems. Yet, there seems to be a consensus among Mexicanist scholars that things are not likely to change in the near future. Gilbert and Varley (1991) outline three reasons why they think the politics of containment will probably fail to limit irregular settlement: (1) the government continues to encourage home ownership, (2) the PRI is under electoral threat and can scarcely afford to tighten the urban poor's access to land, and (3) land and settlement management presupposes capacity to deliver infrastructure and services, but the government doesn't have it. Clearly, an alternative approach is necessary. But to what extent can social experimentation and grassroots environmental action—such as the colonia ecológica productiva movement—really provide solutions? To what extent might popular initiatives also be ecology as politics in the pejorative sense?

CEPs: The Biggest Ruse of the 1990s?

Gabriel Quadri de la Torre, the president of Mexico's National Institute of Ecology, argues that CEP movements are the biggest ruse of the 1990s. Quadri (1993b, 1994b) raises some worthwhile points along these lines, but he also misses something fundamental. It is instructive, nonetheless, to note Quadri's arguments in some detail. Grassroots environmental action can be, and often is, a proactive and potentially progressive force. However, to embrace such dynamics in an uncritical manner would be naive.

Quadri (1993b, 1994b) states that those who organize invasions under the pretext of promoting an alternative type of ecological human settlement are actually clandestine, money-grubbing developers. He asserts that these clandestine developers are getting rich marketing the productive ecology argument in the sphere of irregular human settlement expansion. The idea of colonias ecológicas populares is in effect sold to the colonos as a strategy to avert pending eviction and to justify the permanence of irregular settlements inside the Federal District's ecological conservation area. Quadri (1993b, 37) believes that grassroots environmental action of this sort is nothing more than a sophisticated *mercadotécnica* (market technique) whereby diverse groups offer services of representation and political mediation for a price—a high price that the new colonos pay. The situation does sometimes match Quadri's assessment. But this is not the only way events take shape. The CEP movement in Los Belvederes is a case in point.

If there were nothing more to Quadri's critique other than the argument that CEPs are a ruse perpetuated by money-grubbing land sharks who lead unknowing masses for a price, his viewpoint would hardly be worth noting. But its complexity warrants consideration. Quadri (1993b) argues that even in those cases where grassroots environmental activism may be deemed genuine, CEP advocates lack a well-defined praxis. CEP ideas, Quadri claims, are at their best are the work of academics (from Universidad Autónama Metropolitana (UAM), UNAM and Universidad Veracruzana) who advance *ecotécnicas* on the basis of wishful thinking. He notes that there may be a few limited successes—notably in solar energy production and the recycling of sewage water and waste—but such successes have occurred mostly among those with greater resources (upper-income groups); very rarely have such projects taken place in entire blocks or colonias. There are indeed problems with some of the appropriate technology that holds a central place in CEP praxis. Quadri (1993b) is quick to pick up on this weakness and conclude that the SIRDO-húmedos have failed miserably. While arguing that the SIRDO-húmedos are technically and socially deficient in design, Quadri acknowledges that the SIRDO-secos have had a bit more success. Some are operational in colonias of Mexico City as well as in Veracruz and in the urban periphery of Oaxaca.

Quadri's main concern is that CEP arguments will bring nothing but endorsements of urban encroachment into Mexico City's ecological reserve. He argues that the implications of such a process are disastrous: (1) it condones those who sidestep or ignore public plans; (2) it presupposes the impossibility of stopping urban sprawl; (3) it perpetuates the notion that the state can only manage by crisis in response to political expediency; (4) it subsidizes those who break the law, requiring the state to subsequently justify its own contradictory action; (5) it creates sprawling detached dwelling units at great environmental cost; (6) it discredits the law, advance proof that it is OK to violate it, and (7) it suggests there are no real limits: the entire terrain is grist for the urban mill (Quadri 1993b, 38).

If CEP arguments amount to nothing more than clever countervailing forces to circumvent mainstream policy (without engendering real alternatives), then Quadri's concerns are well founded. Also well founded are Quadri's arguments that "the application of an ecotécnica in housing, whether for production or service allocation, requires prolonged studies, training, observation, education and organization of the people involved" (1993b, 38) and that under current conditions (social, cultural, market) the prospects for CEPs are slim.

Human Agency in Context

Quadri ultimately arrives at the conclusion that Mexico City should aim to be more like compact European cities and less like the sprawling cities of the United States. However, he also asserts that the changes necessary to increase density, as opposed to sprawl, will be resisted by the rich and poor alike. So, where does this leave his analysis? From where will come the human agency to resolve Mexico City's urban-ecological problems? It is with respect to this question that Quadri's critique of CEPs misses something fundamental. Quadri has a controlling-layer perspective of the problem; that is, he defines the problem in terms of the inadequate control and enforcement of preestablished norms and regulations. In Quadri's view the government needs to do a better job at the politics of containment. He rejects CEP arguments out of hand, as if they only relate to illicit land acquisition schemes. He does not recognize cases in which there is real potential for empowerment and social change in the urban milieu. Yet, such cases do occur. Necessity is the motivating factor; it generates deep and genuine support for the ideals and praxis of an alternative development. In a growing number of places—not just in Mexico but throughout cities in Asia, Africa, and Latin America—people join in struggle under banners of productive ecology, not just to get land, but out of necessity for securing decent livelihoods.

Grassroots Activism and Empowerment in Comparative Perspective

Solutions to many urban-ecological problems depend on effective social organization and collective self-provisioning at the grassroots level. In Asia, Africa, and Latin America there is a rich tradition of community-based urban development efforts (Mitlin and Thompson 1995). The mobilization of civil society at the grassroots and beyond generates experiences of social experimentation, social learning, and *empowerment* that are well-springs of praxis and opportunity. Mexico City, like many other cities around the world, harbors an increasingly mobilized civil society. In an article titled "Ecopolitics and Environmental Nongovernmental Organizations in Latin America," Price (1994, 42) notes that there are more than 500 environmental NGOs active in Latin America. In addition, there are up to several thousand community-based organizations (CBOs). Mexico alone has more than forty environmental NGOs at the national or regional level—more than half of which have their headquarters in the Mexico City (42).

From across the political spectrum, environmental as well as other types of NGOs are viewed with high hopes. For instance, United Nations Boutros Boutros-Ghali (1995), former secretary-general of the United Nations, expresses hope that the new networks linking individuals and groups across boundaries of nations promises "the emergence of transnational democratic politics" (3–11). Similarly, Rosenau (1995) suggests that the twenty-first century may witness the articulation of NGOs and social movements as new forms of institutionalized control mechanisms in the sphere of global governance. There now exist a number of extensive international networks that specifically support urban environmental planning and community-based development initiatives.[9] Brecher and Costello's (1994) work offers insights into these developments from the perspective of political ecology.

From Downward to Upward Leveling: The Global Reality

Brecher and Costello (1994) suggest that transnational networks (including alliances of environmentalists and labor unions; farmers and public health activists; advocates for human rights, women's rights, and Third World development, among others) may become an effective countervailing force to the "downward leveling" that is occurring in the wake of global economic restructuring. Downward leveling is a process whereby countries reduce labor, social, and environmental costs of production in order to attract corporate investment. The result is a "disastrous *race to the bottom* in which conditions for all tend to fall toward those of the poorest and most desperate" (Brecher and Costello 1994, 4).

In opposition to downward leveling Brecher and Costello (1994) make a case for *upward leveling*. The basic idea is to raise the standards of those at the bottom so that their downward pull on everybody else can be reduced. On this point, the authors argue that "Upward leveling does not mean that everyone can or should live like the wealthiest citizens of the wealthiest countries—its goal is ecologically sustainable well-being for all, not conspicuous consumption and unsustainable waste" (9). In terms of strategy: "Upward leveling requires grassroots rebellions against downward leveling, local coalition-building, transnational networking, and creating or reforming international institutions. . . . Only by combining their efforts can those resisting the effects of globalization in Chicago and Warsaw, Chiapas and Bangalore begin to bring the New World Economy under control" (Brecher and Costello 1994, 9). Ironically, communication and networking among NGOs in a north-south direction may be stronger than that occurring in a south-south direction. Martínez-Alier and Thrupp (1992, 115) argue that the lack of south-south communication stands in the way of a combining of efforts, "but all over the Third World a political ecology with a populist orientation is developing."

A number of factors are encouraging the rise of NGOs and CBOs. One significant factor lies in new sources of international funding and support for people's participation in development.[10] Mitlin and Thompson (1995) relate that international aid donors have discovered participation as one of the priorities of development: "the activities of the institutions of civil society—NGOs, residents associations (including those formed by squatters and by tenants), self-help housing associations and cooperatives—have therefore attracted the interest of donors for their role as development organizations" (235). Other factors that have prompted the rise of NGOs and CBOs are their attractiveness as vehicles of social change, the failure of governments to deal with environmental problems, and the focus on sustainable development (Price 1994; Carley and Christie 1993).

Upon reviewing a wide range of cases wherein the use of participatory strategies were formally sanctioned, Mitlin and Thompson (1995) conclude that they can amount to a double-edged sword: "On the one side, participation can bring about increased access to, and control over, vital resources and decision-making processes by local people, cutting away bureaucratic red tape and institutional constraints as it proceeds. On the

other hand, it can be used by governments and donors to justify and reinforce inequitable social relations of power. For this reason, the key question remains: who participates in whose project?" (232).

When it comes to mediating popular participation, Fox and Hernández
(1992) note that a hallmark of the Mexican government has been its ability
to thwart social change through a skillful combination of carrots and
sticks. Consequently, Mexican NGOs do not yet have the same level of
influence as do their counterparts, for instance, in Chile, Brazil, or Peru
(188). However, civil society in Mexico is becoming more and more autonomous from the state.[11]

From Reactive to Proactive: Cases from Urban Mexico

Fox and Hernández (1992) argue that the most significant change in the
activities of Mexican NGOs and CBOs is the transition from a reactive
stance emphasizing protest toward a proactive stance emphasizing efforts
to build concrete social and political alternatives (167): "Since the 1970s,
Mexico's social movements for reform and democracy have been making
the transition from confrontational opposition *(contestación)* to the construction of positive policy alternatives *(proposición)*. In the process, a
new sense of citizenship has emerged, combining community-based self-
organization for socioeconomic development with a political push for
accountable government. These two streams of social change followed
separate paths in the past, but they began to come together in the late
1980s" (168).

The case of Ajusco illustrates the shift from *contestación* to *proposición*.
But the Ajusco movement's attempt to transform irregular settlements into
colonias ecológicas productivas as a barrio-based approach to socioeconomic development did not pan out. The Ajusco case does not stand alone;
a significant number of related cases in other parts of Mexico City have
been recorded. These include, among others, studies about Chalco—one
of the fastest-growing areas of irregular settlement in the metropolitan
area (Hiernaux 1991)—Tepito (Redclift 1987), San Miguel Teotongo
(Coulomb 1991), and Sierra Santa Catarina (Pillsbury 1994). Indeed, a
growing number of case studies now document linkages among urban
processes, poverty, and the environment in Mexico City.[12] Initiatives advo-

cating barrio-based productive ecology have taken place throughout urban Mexico.

The case of El Molino is one of the most noteworthy examples in Mexico City of community groups initiating productive ecology initiatives. El Molino is a self-help settlement with more than 500 housing units on fifty hectares of land in the southeastern periphery of Mexico City (Suárez Pareyón 1987). The site was developed by four cooperatives who collectively acquired the land from the government. Meffert (1993) notes that one of the cooperatives (UCISV—Libertad; Union of Settlers, Renters, and Homeless—Liberty) combined collective self-help with concepts of *popular* (alternative) urbanization and local democracy at the neighborhood level (324). One of the alternative technologies that settlers of El Molino experimented with was the SIRDO, a technology for organic waste recycling with a central role in the struggle for land in Ajusco, discussed in chapter 10. The situation in El Molino is an ongoing learning experience. One of the greatest concerns is that the eventual privatization of land tenure in El Molino will undermine the organization's collective foundation (Meffert 1993; Coulomb 1991).

Outside the MAMC, other attempts at bottom-up productive ecology in Mexico can be found in the cities of Durango, León, San Luis Potosí, and Jalapa, among others. The most noted case is that of Durango, where an urban social movement called the Popular Defense Committee of Durango (CDP) established its own environmental protection NGO. Success of the CDP is attributed to its broad base of support, including its own constituency of urban poor as well as allies among government reformers and local notables (Fox and Hernández 1992). Acting under the direction of its Ecological Defense and Preservation Committee formed in 1988, the CDP initially constituted a protest movement (*contestación*) against the ill effects of the city's water pollution. But as Fox and Hernández (1992) point out, the CDP soon shifted its position from a reactive stance to a proactive one. By way of a broad-based consultative process the CDP articulated alternative policy proposals. The group began as a local, community-based emergency program to reclaim and protect seventeen local springs. From there it scaled up its outreach and initiatives to develop regional water management proposals (183).

Drawing attention to new productive experiences at the grassroots, Schteingart and d' Andrea (1991) document ways in which urban services and community participation in the production of those services have become important issues. Along lines similar to Schteingart and d' Andrea (1991) and Coulomb (1991), this book argues that citizen mobilization for services cannot be understood merely as a reactionary force. Many of these struggles involve a qualitative shift in urban politics. Investigators have not paid close enough attention to the role of popular groups in the design, development, and management of urban services. There is untapped potential that goes beyond "saving resources." A contradiction expresses itself in this regard: "On the one hand, popular participation—understood in terms of resource savings, and strategy aimed at alleviating the weight of the crisis in the development of the city—if it is deemed appropriate by the state and technical bureaucracy—is promoted. On the other hand, participation implies the planting of seeds of autonomous demands, especially for democracy, that emerges as a death threat to the current system of urban management" (Coulomb 1991, 270).

Despite potential benefits associated with genuine popular "participation," and in spite of all the official discourse that pretends to promote such participation, there is no one really promoting a genuine bottom-up popular control in the development of urban services. Similar to Coulomb's findings, the Ajusco case illustrates a *mobilization of bias* against the state's support for grassroots initiatives outside the norm. There is an interesting study that provides insight into the mobilization of bias while suggesting ways to deal with it. Documented by Hiernaux (1991), the study was part of an interdisciplinary group effort at the UAM, Unidad Xochimilco. It focused on three aspects of peripheral urban expansion: (1) socioeconomic formation, (2) spatial structure, and (3) environmental conditions. The study focused on the extension of urban services to the city's periphery, the relation of this process with social dynamics (in particular, state-community relations), and how all the processes affect the environment. The UAM research team stressed that "the periphery is not only a form of settlement, rather it signifies a form of subsistence. The proliferation of small businesses (family owned or not) is a phenomenon parallel to the consolidation of the settlement" (286). The businesses are of three basic types: services, material depots, and workshops.

The significance of the UAM research is that it examines the importance of urban services in a new light. As Hiernaux (1991, 283) describes it, "access to urban services now forms part of an ensemble of specific demands that interrelate diverse facets of social, economic and urban life. To acquire services is an urban demand, this is certain, but it is intimately related with the necessity for employment, conditions of transportation, education, health and environmental conditions." However, official discourse continues to focus on the importance of water, drainage, and electricity in terms of health standards and the well-being of the population. Attention to factors of health is certainly important. However, such a focus tends to obscure the importance of urban services for the productive plant to be found in the developing peripheral settlements (Hiernaux 1991). The key is to link the provision and development of urban services with economic development that is socially necessary and ecologically sustainable.

Latin America, Africa, and Asia

The Cooperative Housing Foundation (CHF) is an example of a nonprofit international NGO that advocates grassroots productive ecology in human settlements of the Third World. Since 1962 the CHF has worked with grassroots groups in eighty different countries. In one of their publications, *Partnership for a Livable Environment* (1992, 19), the CHF describes a wide variety of community-based environmental initiatives taking place in urban squatter settlements throughout Africa, Asia, and Latin America. Initiatives they describe include (1) using low-cost, indigenous construction materials so as not to deplete wood resources; (2) recycling solid and liquid waste to prevent the pollution of land and water; (3) using appropriate technologies, such as improved cookstoves, to reduce air pollution and deforestation; (4) planting slopes with hedge grasses, gardens, or fruit-bearing trees to prevent soil erosion and airborne dust pollution. These initiatives are much the same as the ones advocated by the participants in Ajusco's colonia ecológica productiva movement. According to the CHF, success stories are found in those cases that identify "a technically appropriate intervention that is acceptable to and can be implemented and maintained by the community" (19). As it turns out, this did not happen in Ajusco.

Around the world, research on grassroots activism has documented a wide range of initiatives and opportunities (Friedmann and Rangan 1993; Pye-Smith, Borrini Feyerabend, and Sandbrook 1994; Fuglesang and Chandler 1993; Atkinson and Vorratnchaiphan 1994). In a study titled "Urban Agriculture for Sustainable Cities: Using Wastes and Idle Land and Water Bodies as Resources," Smit and Nasr (1992) report on eighteen cities they visited in Asia, Africa, and Latin America. They found that community-based urban agriculture is a large and growing practice, much more pervasive than is commonly realized. Other studies support their findings (Wade 1987; Lynch 1994).

Kumar and Murck (1992) document the growth of cottage industries in Cairo, Egypt, based on the informal recycling of municipal waste. Along similar lines, Furedy (1992) documents the practice of nonconventional options for community-based waste management in Asian cities. Furedy describes projects in Bangalore, Manila, Madras, Jakarta, and Kathmandu. He notes that each of these projects went beyond community participation in waste collection and incorporated other social and ecological goals. Furedy contends "the significance of these approaches is that they combine social, economic, and environmental motivations for recovery and recycling and thus have the potential (as yet unrealized because of their small scale) to develop a broad base of cooperation for environmental improvement in Third World cities" (42).

In a speech delivered at the International Summit on the Environment, Janice Pearlman (1990) spoke about "Mega-Cities: Innovations for Sustainable Cities of the 21st Century." She drew attention to initiatives in Mexico and throughout Asia, Africa, and Latin America, where community-based organizations have undertaken projects aiming to (1) regenerate environments by promoting circular systems for water, sanitation, garbage, food, and energy; (2) alleviate poverty by generating income through informal sector activities; (3) engender greater local participation in planning, service delivery, resource allocation, and urban management; and (4) empower women with greater choice, access, and voice. Gaye (1992) draws attention to such initiatives in African cities. He suggests that they constitute a widespread—yet disorganized—form of popular urban ecology defined as "the activities of low-income communities to improve their local environment and living conditions" (1992, 101).

On the theme of empowerment and political ecology in Asian cities, Mike Douglass (1989a, 1989b, 1992, 1996) has conducted some of the best comparative case studies available. His work links concerns about livelihood opportunities, empowerment, and the availability of income with concerns about ecological sustainability. Douglass documents how many low-income urban communities throughout Asia engage in environmental management as a part of collective survival strategies. Based on research completed in Bangkok, Thailand, and other Asian cities, Douglass and Zoghlin (1994) articulate a set of four themes that can fruitfully guide comparative urban research in Asia as well as in Africa and Latin America. The four themes are listed in box 11.1.

Douglas and Zoghlin (1994) advance a much needed holistic perspective. Their insights, (1) that collective environmental management is also economic-cum-welfare management and (2) that households are best understood as units embedded in social networks and particular political-economic-cultural constellations of power, are significant advances in the political ecology of human settlements. Grassroots environmental activism certainly does not embody the solution to all of any city's urban-

Box 11.1
Political Ecology and Human Settlements: A Guide for Comparative Research

> 1. Environmental management can be seen as an integral part of household-level economic-cum-welfare strategies.
> 2. Strategies involve decisions about household allocation of resources to income-producing, habitat-managing, and socionetwork-sustaining activities.
> 3. The capacity for households to collectively engage in improving environmental resource management is contingent upon particular sets of circumstances and political, economic, and cultural configurations: the state and its interventions, the structure of the economy and labor systems, and certain social and cultural institutions that foster reciprocal and redistributive relations in communities.
> 4. Combine (1), (2), and (3) to suggest that certain constellations of factors—including allowance for human agency through leadership, the dedication and commitment of outside as well as internal organizations, and other often unexpected sources of empowerment—are more conducive to community-based initiatives than others.

Source: Douglass and Zoghlin (1994, 174).

ecological ills. Far from it. Yet, from the perspective of political ecology, popular groups are key. Mexico City's urban-ecological problems are rooted in history, power, and an environment characterized by uneven development. Ultimately, solutions to these problems require the effective mobilization of civil society and empowerment of the poor. Grassroots environmental action has proven to be a fertile field in the search for alternatives.

12

Conclusion

Human Settlements and Planning for Ecological Sustainability

The opening pages of this book drew attention to a paradox. An increasing number of government officials, policymakers, and planners are now quick to declare "The world cannot afford to approach human settlements development through *management by crisis responses.*"[1] Yet, management by crisis persists in nearly all the fast-growing cities of Asia, Africa, and Latin America; Mexico City being one of the most noted examples. This paradox raises some critical questions. Why does management by crisis continue to be so pervasive? Are worthwhile alternatives so difficult to locate, define, or implement? If there are alternatives within reach, what can local actors do to promote them? Following is an outline of four broad theoretical premises that guided this book's approach to these questions.

1. A holistic environmental and coevolutionary perspective of history, power relations, and ecosocial change is crucial to the task of weighing prospects for the development of human settlements. At the dawn of the twenty-first century, "ecology and economy are becoming ever more interwoven—locally, regionally, nationally, and globally—into a seamless net of causes and effects" (WCED 1987, 5). Yet, even as the world's economy and the earth's ecology have become increasingly intermeshed, they remain separate in institutions and in the minds of policymakers. Current dualistic thinking divides society from nature and urban space from ecological space. It obscures the fact that humans are components of ecosystems. The task ahead, as Norgaard (1994) suggests, is to conceptually frame societies and environments as mutually interactive, coevolving systems.

2. Efforts to promote sustainable human settlements development must go well beyond the realm of management alone. The legal-institutional

terrain surrounding the production of human settlements in Mexico City embodies serious inadequacies in terms of social justice and equity.

3. The mobilization of civil society is a vital and innovative force. Of course, such mobilization is not necessarily concerned with advancing social justice and equity. Like the state, civil society is a conflict ridden and contested domain. The key point here is that historical opportunities present themselves and the mobilization of civil society can be a wellspring of empowerment. People can and do join forces to try and build healthy communities and a culture of sustainability. Empowerment helps generate hope, praxis, and social learning necessary to realize the human settlements objective enshrined in *Agenda 21:* "To achieve adequate shelter for rapidly growing populations and for the currently deprived urban and rural poor through an enabling approach to shelter development and improvement that is environmentally sound" (Robinson 1993, sec. 7.8).

4. Appropriate technology and the productive plant found within low-income settlements are increasingly important. They constitute the basis for promoting socially necessary, and ecologically sustainable, barrio-based approaches to development. In this context, the urban poor's access to land, housing, and urban services is intimately related to the necessity to secure livelihood opportunities.

The four premises correspond to the four spheres shown in figure 12.1, namely: (1) the environment, (2) the legal-institutional terrain, (3) civil society and culture, and (4) the economy and technology. The key challenge associated with each sphere is outlined in table 12.1. Political ecology lies at the nexus of the four spheres. A principle task of this book has been to document ways in which grassroots movements in Mexico City and beyond have struggled to define and implement strategies at this political ecology nexus. Political ecology movements have challenged dominant sociopolitical and economic arrangements as well as dominant values, beliefs, attitudes, and expectations (symbolic and cognitive systems). Put another way, political ecology movements—like the one in Mexico City that rallied people to transform their irregular settlements into colonias ecológicas productivas—have tried to foster a holistic ecosocial praxis. The objective in these cases has been to empower ordinary people seeking alternatives to the grip of uneven development, poverty, social injustice, and environmental degradation.

What are the prospects for grassroots movements to bring about more equitable and sustainable uses of land and other resources? The findings

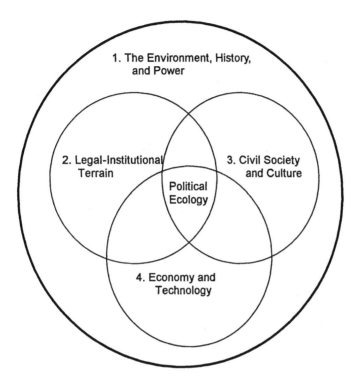

The numbers designate the four key challenges and theoretical premises of political ecology: (1) *coevolution and holism*, (2) *social justice and equity*, (3) *empowerment and community-building*, and (4) *sustainable production and reproduction*.

Figure 12.1
Political ecology: a form of praxis and empowerment for sustainable development. Source: author.

presented in this book suggest that there is room for guarded optimism. However, political struggle lies at the heart of the challenge. Until alternative movements become part of a more potent countervailing, transformative struggle, the politics of land allocation and the dynamics of irregular human settlement in Mexico City will continue making it practically impossible to keep urban sprawl from claiming more and more of the so-called ecological reserves.

Since the early 1980s, it has been the Mexican government's policy to strictly prevent any urban settlement from taking place in the greenbelt

Table 12.1
The political ecology of human settlements

Sphere of Concern	Key Challenge/Theoretical Premise
1. Environment, history, and power	*Coevolution and holism:* Human settlements are social and biogeophysical places; there is a need for insight into how these dimensions coevolve in the process of urbanization and history-making. Development theory will be of little help until it gets beyond dualisms such as structure-agency, determinism-possibilism, society-nature, and economy-ecology. To better understand how the environment and development interrelate, we need to develop a holistic worldview. Holism joins the fruits of a sociological and a biogeophysical imagination. This challenge calls for new insights and theory-building about the interplay of nature (including entropy), history, and power. In short, it calls for an understanding of coevolutionary dynamics and for the advancement—in theory and practice—of a political ecology of human settlements.
2. Legal and institutional terrain	*Social justice and equity:* Urban land and human settlements problems in Mexico City are generated by a crisis that goes beyond issues of land allocation and environmental management into broader issues concerning social justice and equity. There is a need for new approaches "that challenge not only economic rationality but also bureaucracies, in ways that encourage political pluralism and the participation by civil society in the management of its productive and vital processes" (Leff 1993). This approach calls for redefining the role of the state and local government.
3. Civil society and culture	*Empowerment and community-building:* Solutions to urban-ecological problems in Mexico City's human settlements largely depends on effective social organization at the local level. The mobilization of civil society at the grassroots and beyond generates experiences of social experimentation, social learning, and empowerment that are well-springs of praxis and opportunity. This does not mean that grassroots environmental activism and bottom-up initiatives can resolve all the world's urban-ecological ills. Rather, local initiatives and

	community empowerment are best understood as important strategies, among others, in the challenge to promote a culture and praxis of sustainability.
4. Economy and technology	*Sustainable production and reproduction:* To be sustainable, patterns of production, reproduction, and consumption must be brought into concert with the capacity of ecosystems to perform life-giving functions over the long run—that is, to generate the raw material input and to absorb the waste outputs of the human economy. Export led industrialization and the internationalization of the world economy makes this a complicated challenge. The formal sector in places like Mexico City has been unable to create enough livelihood opportunities for the urban masses. At the same time, a large portion of city dwellers live in "irregular" human settlements and have little recourse other than to rely on collective self-provisioning. Hence, planning for ecological sustainability must be made to take into account the fact that human settlements—in addition to providing living space for reproduction and consumption—are also important places for production and resource (re)generation (through both formal and informal means).

Source: Pezzoli (1997).

conservation area. Since the late 1980s, the politics of containment has taken on greater force. Yet, even with the government's intensified efforts—including the placement of miles of barbed-wire fence and stone walls at the urban-ecological perimeter; the widespread posting of billboards warning against trespassing and illegal land transactions; greater government control over land use and construction permits for all projects in rural areas; and repeat eradication of the newest illegal, incipient settlements—the urban sprawl continues. The momentum seems overwhelming.

In view of the crisis situation, it becomes difficult to think in terms of planning for ecological sustainability in Mexico City. Indeed, the concentration of urban populations in megacities around the world raises grave concerns about the future of such metroregions and the hinterlands supporting them. It may be that the increasingly urgent need to protect life-giving ecosystems and watersheds demands greater police action and a more ruthless enforcement of urban limits. Such an argument supports the current mainstream approach to the challenge: the politics of containment. But history tells us this is a failed approach.

Reaching an appropriate urban-ecological balance is not a purely technical issue; it is not an objective that can be met by even the most draconian enforcement of zoning laws. This is because the urban land problem in Ajusco derives from a broad developmental-ecological crisis facing Mexico City, and Mexico more generally. The roots of this crisis go back well beyond the late 1960s, when the failure of the import substitution industrialization began to register obvious damage. Indeed, the roots go back hundreds of years. The Valley of Mexico has been radically transformed, first by pre-Columbian societies and then, more pervasively, under the heavy hand of colonialism and resource-intensive industrialism. Roots to the contemporary urban-ecological problematic go back to the "war on water," begun in the 1500s, and more recently to the rise of the modern fossil fuel economy that enabled the city to circumvent, thus far, its bioregional constraints. But a time of reckoning has come. Concepts of development, modernization, and progress are historical products. Their meanings are not fixed. The persistence of poverty and environmental degradation on the world stage begs for new meaning and new approaches to the human development challenge.

1. Coevolution and Holism: Toward a Reconceptualization of the Urban Prospect

It is important to understand how societies and environments coevolve (Norgaard 1994). Humans are not only a part of the environment, they create it as well. Humanity is a powerful biotic factor. As Williams (1993, 26) argues: "it is possible to take the view that God may have created the world, but humans have made modern natural ecosystems." This view has roots in historical-cultural geography going back to the 1920s. It assumes that "the massive turnover of biomass needed in order to feed, clothe, shelter, and warm humans through so many millennia must be conducive to fundamental ecosystemic change"(29). The (re)production of humans and ecosystems has become an integrally interwoven and dynamic set of processes at local, regional, and global levels. This premise eschews the concept of environmental determinism. Human-nature relations are interactive; it is misleading to say that one determines the other. Of course, both humans and ecosystems do have certain limits. To ignore these limits in a theory of history making would be to embrace the equally misleading concept of cultural possiblism. What appears to be most useful—as illustrated in the excellent work by Enrique Leff, Michael Douglas, Adrian Atkinson, and other political ecologists—is an ecological and historical materialist approach that allows for dialectical analysis.

In this light, the terms of the question—is it possible to protect vital ecological zones from the pressures of urban expansion in the Valley of Mexico?—should be critiqued. The question assumes the opposition of urban and ecological space. This assumption eclipses the notion that urban and ecological space can be integrated. Mesoamerican agrarian cities of the valley—which largely sustained themselves on renewable sources of environmental low entropy—offer lessons in this respect. But current dualistic thinking divides society from nature and urban space from ecological space. It obscures the fact that humans are components of ecosystems. These observations do not imply that *all land* should be open to development as long as it is some sort of progressive green development. Clearly, there are sound reasons (ethical, social, economic, ecological) to protect a significant amount of land in as pristine a condition as possible. The point to stress here is the need for integrated and holistic approaches to urban-ecological resources. Any a priori notion of "the city" as some sort of antiecosystem is bound to be misleading. "Ecology" in the mainstream administrative view becomes a remnant under siege at the urban periphery. In this view, there is a failure to see the "ecology" within the social and economic (re)production taking place in low-income human settlements. Only the most obvious indicators of environmental damage in and around human settlements (e.g., pollution, erosion, deforestation) become part of the conceptual-visual field for policymaking.

In contrast, inquiry from a coevolutionary perspective asks how human settlements can become integral components of viable ecosystems. More specifically, when based on principles of political ecology, inquiry along coevolutionary lines asks how can we promote the technology, together with social learning and social change, necessary to bring patterns of production and consumption into concert with the capacity of ecosystems to perform life-giving functions over the long run. This approach goes beyond the technical aspects of how to regenerate the raw material input and to absorb the waste outputs of the human economy. There is an ethicosocial dimension. The long-range coevolutionary challenge is to locate strategies that foster not only *intra*generational equity but *inter*generational equity as well.

The intellectual part of this challenge calls for researchers and activists to integrate their sociological imagination with a biogeophysical imagination. As C. W. Mills (1959) astutely argued nearly four decades ago, we must develop and use our sociological imagination so that we may aim to grasp what is going on in the world and to understand what is happening in our own particular lives and the lives of others' as intersections of biography and history within society. Integrating a biogeophysical imagination with the sociological one deepens one's perspective by taking into account the interlocking ecospatial scales of life. That is, it enables a better understanding of the linkages and interactions that bind (or tear) life at various *levels:* including the milieu of an individual household; the social and political field of neighborhoods and metropolitan areas; and the spheres of local, regional, and global economy—all of which depend on stocks, flows, and sinks provided by ecosystems.

2. The Legal-Institutional Terrain: Urban Land Management and Demand for Social Justice and Equity

To explain problems associated with land use management and the development of human settlements, this book analyzed three key dimensions of the legal and institutional terrain in Mexico City. The first dimension has to do with magnitude. Over the past four decades, most of Mexico City's housing has been created through informal self-help practices outside the formal sanctions of the law. The only way for most low-income settlers to get access to land for their housing has been through illegal and semilegal means. Given the illegal or quasi-legal status of such housing, the settlements they make up have been called irregular. A similar situation exists in nearly all fast-growing cities of Asia, Africa, and Latin America.

The second dimension concerns politics and power. Despite their illegal or quasi-legal status, laws and institutions have played a key role in the development of irregular settlements. The illegality of human settlements has played into processes of political mediation and social control. Political mediation involving the urban poor's access to land is a tension-laden process; it occurs at society's frontier in relationships of power. This has become more evident as concerns about Mexico City's environmental conditions have risen.

The third dimension concerns ecology as politics. In view of the serious ecological problems facing Mexico City, the state has articulated a new *politics of containment*: legal and institutional means aimed at controlling the horizontal expansion of irregular settlements on the basis of ecological arguments. But the new environmental laws and institutions that are a part of the politics of containment have more to do with social control and political mediation than with safeguarding the environment. In many parts of Mexico City, irregular settlements continue to encroach into areas that have been declared greenbelts and ecological reserves.

Land use management problems are best understood as rooted in processes of political mediation and social control. Certain interests are served by tolerating, and sometimes by facilitating, the expansion of irregular settlements. The illegal alienation of land—by opening supplies otherwise not accessible to the majority of low-income people—has helped the government maintain political legitimacy and social stability. Over the past several decades these arrangements have been fairly stable. Millions of families have secured housing and at least the bare essentials of life. However, over the past ten years, it has become clear that certain economic, political, and biogeophysical problems are converging in a way that is creating serious stress. Such stress is clearly visible in the politics of containment enacted through the *desalojo machine* and *industry of destruction* at the city's urban-ecological frontier. Mexico City's reactive, politically mediated approach to human settlements development may have worked in the past. But the costs associated with this approach are mounting.

Besides documenting new constraints in the legal and institutional terrain, this book also brought opportunities into view. The interactive politics of environmental regulation and popular resistance around ecological and land use issues has given rise to fruitful social experimentation. State-society relations at this frontier in relationships of power involve strategies and countervailing strategies. As Jessop (1985, 20) describes it the state system can be analyzed as an ensemble of institutions whose structures and modus operandi are more open to some types of political strategy than to others. In this light, one of the lessons to come out of this book is that popular groups have to distinguish between strategies with some prospects of success and those that stand little chance. Yet this is a very tricky business. As the case of the CEP movement illustrates, winning can alter the terms of struggle. The CEP movement's close link to the land question may well have been its undoing. When ecological arguments are so closely linked to the struggle for land, the land question takes precedence. Once the threat of forced eviction passes, the ecological initiatives lose currency. Grassroots groups have to define long-haul strategies to ensure that their ecological initiatives do not fall prey to political expediency.

Community-based environmental action like that advocated by the CEP activists requires continuity and cohesion in terms of social organization. In the early stages of its development, social organization was highly cohesive throughout Los Belvederes. Given the harsh conditions and lack of services at the beginning, settlers had many reasons to cooperate and work together. For instance, they had to set up and conduct classes for their children, clear streets, lay down a provisional infrastructure for electricity and water, and generally speaking, look out for one another. Households engaged in joint decision-making on a daily basis; out of necessity they collectively produced their own livelihoods (Díaz Barriga 1991). In this context each household could be described as a proactive unit linked to other households through productive relationships in a larger household economy (Friedmann 1988, 41). However, as Los Belvederes became established over the years with more and more services secured from the state, there was less urgency for the settlers to act collectively. The waning of collective action has to be understood within the national political context, where the provision of urban services by the state in Mexico "has been one of the most powerful and subtle mechanisms of social and institutional control of everyday life" (Manuel Castells, cited in Ramírez Saiz, 1990, 235).

One way to describe this widely documented outcome is what Max Weber referred to as the expropriation of the means of administration, which results in communities loosing their capacity for collective organization and control over local development decisions. But collective organization is essential for creating jobs in the informal or cooperative survival economy. At the same time it engenders *social power*—a dynamic force essential for motivating the social experimentation and practice necessary to create sustainable livelihoods.

Conversely, a decrease in social power undermines the opportunity for low-income groups to generate wealth. This point was acknowledged by activists in Los Belvederes who wanted to create sustainable livelihoods through a productive ecology. They urged the state to expropriate the land and to establish a collective tenure arrangement, as opposed to regularization on an individual basis. The reason for this was, above all, political. As Castells has observed for similar situations, the settlers feared that standard regularization would individualize the problem—"dividing the land would create a specific relationship between each squatter and the state" (1983, 198). Activists were concerned that their movement could be fragmented and loose its internal solidarity, and that is precisely what happened. Although the state adamantly refused to consider establishing collective tenure as an alternative route to legalization, some members of the movement continued to press the case. Meanwhile, others began negotiating contracts to (re)purchase their individual lots from a cabal of developers backed by the government. This drove a wedge into the movement and seriously undermined its collective strength.

Through resistance, popular movements can help redefine concepts of development, economic values, technological efficiency, and scientific rationality; they can even help create a new economics. However, as the case of the CEP illustrates, the struggle for social transformation is constrained by the general level of organization and consciousness in the broader process of political conflict. In Mexico City, opposition movements have shaken PRI apparatus to its core, but the power to disrupt social and political relationships is not symmetrical with the power to establish new relationships and new social meaning.

People have over time forged certain expectations about politics, politicians, the institutions of society, and how these all function. Popular groups with democratic agendas must oppose not only the state's coercive apparatus and judicial system, but also the fragmenting and depoliticizing ideologies of consumerism and possessive individualism, as well as more traditional codes of behavior including *clientelism, paternalism,* and *caciquismo*, which are deeply embedded in civil society. A concerted effort is required "to create the conditions for an effectively working democracy that consists of strong representative institutions and the collective empowerment of ordinary citizens in their own communities and, beyond these boundaries, in mass based social movements and political parties" (Friedmann 1986, 22). This argument—while recognizing that radical autonomy from the state is neither strategically viable nor politically realistic—does not then imply an advocacy of political "participation" as conceptualized in the modernization paradigm. Rather it calls for a kind of "institutionalism" that rejects the passive conception of participation in favor of an active projection of grassroots change into the legal and institutional relationships of the political system itself (Foweraker 1989).[2]

3. Civil Society and Culture: Strategies for Empowerment and Community Building

This book sought to uncover the potential for innovative social change embodied in the heightening activism of civil society. The book argues that popular groups are the social agents most likely to lead the transformation of Mexican society. As Enrique Leff, Julia Lillo, and Ana Irene Carabias (1990) and others have pointed out, popular struggles in Mexico are an innovative force for social change—without social movements, no challenge from civil society will emerge able to shake the institutions of the state through which urban-ecological problems are reproduced.

The proliferation of popular movements and NGOs throughout Mexico has exerted a significant impact on political practice (Foweraker and Craig 1988a; Craig 1990; Fox and Hernández 1992; Miraftab 1996). The experimentation of popular groups with (1) different forms of association (e.g., regional coalitions, multisector, multiclass alliances, transnational networks), and with (2) different strategies to meet basic needs (e.g., productive ecology) helps create opportunities to attenuate urban-ecological problems. There are no easy top-down administrative, technical, or scientific fixes. Solutions, if they are to be found, will have to emerge from specific struggles for social change and from processes of social learning.

On its own, the force of moral persuasion directed at the controlling layers of society is bound to be inadequate. The key must be an interplay of bottom-up and top-down initiatives (Ghai 1994). The starting point for grassroots struggle may be around local issues. But as Friedmann (1992) points out, local action is constrained by the dynamics of globalization, structures of uneven development, and hostile class alliances:

> Unless these are changed as well, alternative development can never be more than a holding action to keep the poor from even greater misery and to deter the further devastation of nature. If an alternative development looks to the mobilization of civil society at the grassroots or, as Latin Americans like to say, in communities "at the base," it must also, as a second and concurrent step, seek to transform social into political power and to engage the struggle for emancipation on a larger—national and international—terrain. (viii)

The projection of grassroots struggle into the legal and institutional terrain is nascent in Mexico. As a Mexican scholar of urban social movements points out, popular groups have learned "to use the government's legal and technical language to propose viable alternatives corresponding to their demands" (Ramírez Saiz 1990b, 234). How significant are such initiatives? So far, they have only occurred at the level of neighborhoods or locally targeted plans (as in the case of Ajusco). Nonetheless, such initiatives do represent a significant advance in the forms of the urban movements' struggle (Ramírez Saiz 1989a, 1990a, 1990b). Along these lines, Leff, Carabias, and Batis underscore the significance in Mexico of diverse resource management initiatives in urban and rural communities. They argue that "[a]side from continuing to be at the margin of national development projects, these experiences are important as seedbed and laboratory in the generation of options for an environmental strategy of development" (1990, 9).

Measuring the impact of urban movements does not have to conform to the rigors of some type of "win or lose" logic. Marris (1982) provides insight into the ways incremental actions can lead toward social transformation. In terms of strategy, he states: "To counteract the repressive resilience of the established order, the hardest and most crucial task is to establish a sense of the relationships we want to create so clear, so persuasive, and so mundane that people's behavior will begin to converge upon these expectations in all the transactions of everyday life. Only then are the opportunities created by the spontaneous or contrived disruption of existing relationships exploitable" (Marris, 1982, 118). This is a crucial observation. Given the predisposition throughout society to keep faith with its legitimating principles, the power to both disrupt and to establish relationships is not symmetrical. Thus, "the outcome of any social transformation depends upon how all the diffuse and centrifugal impulses of a disintegrating order are made to *converge in a new order* predictable enough to reproduce itself" (my emphasis, Marris 1982, 121). Along such lines, Fox and Hernández (1992) note that NGOs in Mexico have made important advances but that their influence has been limited. NGOs have not yet been able to form multisectoral networks that effectively articulate social movements, grassroots economic development support work, and alternative public policy at the local and national levels (185).

To be effective in deliberate social transformation, actors must have the capacity to concert the actions of others and to overcome the resistance of vested interests. This capacity derives, in part, from the actor's position with respect to the relations of production (Callinicos 1987, 129). Said capacity also derives from an actor's relative access to what Friedmann calls the eight "bases" of social power: "life space for social reproduction, free time over and above that needed for household survival, relevant knowledge and skills, social organization, tools and instruments of production (including good health), information relevant for the exercise of knowledge and skills, social networks that relate households to each other, and financial resources" (1987, 396). To view the capacity of popular groups in this way implies that the groups' effectiveness requires "a continuing and permanent struggle for the equalization of access to the bases of social power" (396). In short, it requires collective self-empowerment.

What would this mean in practical terms where land use and the production of human settlements are concerned? Here Friedmann provides a guiding perspective: planning at the local level would aim at transforming the barrio into "a combined housing/production/living unit involving new relations of power" (1986, 27). In the case of Ajusco, one could argue that this configuration was the aim of the CEP movement. At its zenith, the CEP campaign had hundreds of families rallying under its ecological banners. At the same time, the movement had networked across the urban-rural divide and had articulated multiclass alliances (including connections with independent researcher organizations, university students, newspaper columnists, international agencies, and campesinos both locally and from the northern part of the country). However, once the land tenure program began in 1985, these connections eroded.

Although popular groups will have to continue to press the state for resources, the experience in Ajusco suggests that popular groups will also have to develop clever strategies to elicit support from external agents, other classes, and other social groups. This is what activists from the colonia Bosques del Pedregal did to build momentum for their CEP campaign. They cultivated contacts with the press to maximize positive coverage of their project, they successfully submitted grant proposals to foreign institutions and philanthropic organizations, and they joined forces with respectable research institutes. But it was not enough. They failed to generate a lasting base of support for their specific initiatives.

Improving the prospects for grassroots environmental action thus calls for establishing effective multisector networks that can articulate and scale up local initiatives (Miraftab 1996). Networks are crucial; they serve three major functions: (1) giving voice to NGOs vis-à-vis governments and the public, (2) gathering information on the NGO community and disseminating it, and (3) providing a forum for members to discuss common issues and problems (Gordon Drabek 1987). Networking can help forge stronger links between NGOs microlevel experience with specific projects and the macrolevel public policies that affect the development process. All of this would help NGOs to avoid repetition of their own and others' past failures and to promote innovation and replicability in project design and implementation.

4. Economy and Technology: Toward Sustainable Production, Reproduction, and Livelihood

To promote the principles of productive ecology, planning (including environmental planning) must take into account the importance of the barrio economy. Human settlements in Mexico City are not simply dormitories; they are also places of production and reproduction wherein social experimentation with ecodesign could reap benefits economically and ecologically. Such a view represents the beginnings of a new paradigmatic approach to human settlement research.

The colonia ecológica productiva movement in Mexico City's greenbelt zone proposed an integrated approach to urban development based on production that is "socially necessary, ecologically sound, and economically viable." This—the CEP activists argued—is in sharp contrast to the government's approach to planning that treats housing, economic, and ecological problems as separate issues. The CEP activists of Ajusco recognized that it is not possible to solve these problems separately. Faced with chronic scarcities of income and resources, they saw no alternatives outside of promoting self-reliant forms of urbanization. Their approach called for the development of appropriate technology to reduce the cost of generating employment and to recover capital investment through the rational use and reuse of natural resources. Efforts along these lines are taking place throughout the Third World. Reilly (1991, 10) observes that "as debt, structural adjustment and austerity shrink government services, organizations have less to gain from a 'diminished' state, setting aside protest and demands to concentrate on production of goods and services themselves."

From this perspective, the importance of urban services takes on a new light: "access to urban services now forms part of an ensemble of specific demands that interrelate diverse facets of social, economic and urban life. To acquire services is an urban demand, this is certain, but it is intimately related with the necessity for employment, conditions of transportation, education, health and environmental conditions" (Hiernaux, 1991, 283). So far, though, there have not been any significant attempts by the state to link the development of urban services with strategies for creating sustainable livelihoods. Pressure from grassroots environmental action will have to make this happen.

Over the past decade, the Mexican government's emphasis on economic and monetarist orthodoxy has favored a "rolling back" of state intervention in favor of promoting "free market" forces. The state seeks ways to privatize urban services now (Hiernaux 1991). But such a policy is difficult to implement for a number of reasons: costs are rising, the population has less and less income, and the government is gripped by a fiscal crisis. Privatization in the face of such problems creates tension. The heightening activism of civil society is the norm, and the state continues to use service delivery as a carrot-and-stick policy of social control. The viability of this arrangement in the near and medium term is increasingly tenuous.

In conclusion, getting to the root of irregular settlements and related ecological problems in Mexico City ultimately requires fundamental changes in the institutions that control land allocation and prices and that control the priority setting of national and municipal development goals. Moving in this direction calls for (1) fortifying the barrio economy as a means of creating sustainable livelihoods and (2) strengthening the theory and practice of proactive ecologically integrated planning. Meeting these challenges requires social learning and experimentation. It also requires creative and holistic ways of thinking about development, urbanization, economics, and ecology that can be translated into viable strategies and methods.

Mexico is entering a period in which the need for public works and government assistance for housing the urban poor has never been greater. Yet, notwithstanding the well-targeted investments of Pronasol, economic restructuring appears to be aggravating, rather than diminishing, the chronic scarcity of income and resources among Mexico's poor majority. The ecological implications of this situation are cause for concern. Environmental degradation is likely to increase as Mexico City expands (and implodes) without adequate facilities for sewage treatment, waste disposal, or policies to protect ecosystems (e.g., watersheds, wetlands, forests).

This reality demands a new approach to the production of the city in a bioregional context. Cities of the 1990s will be increasingly pressed to develop energy efficiency and to regenerate (sustain) their resource base, including local ecosystems. Rising demands for land, shelter, jobs, food, water, social justice, and equity must be reconciled with the necessity to sustain the ecological foundations of life and industry, including air basins, watersheds, soil, and biodiversity. Bottom-up pressure must continue to be brought against the state so that planning for ecological sustainability can take into account that human settlements are increasingly places for production and income/resource generation as well as places for consumption and reproduction. This challenge calls for an integrated approach to sustainable development in urban and rural areas simultaneously not just in the Valley of Mexico but regionally and around the world.

Notes

Chapter 1

1. *Environment and Urbanization* (1993).

2. *Habitat News* (1994).

3. "Peso Plummets Once More: Crisis Deepens" (1994); "Economic Nightmare May Create Political Instability" (1994); "Mexico's President Declares *Economic Emergency,* Acknowledges Real Earnings Will Drop for Most" (1995); "Zedillo Outlines His Peso-Rescue Plan, but Mexican Currency's Slide Continues" (1995).

4. According to Markiewicz (1980), the term *ejido* first became a part of Mexican agrarian reform vernacular in 1911 in a proclamation of Emiliano Zapata in which he demanded a return of the villages' ejidos.

5. Gabriel Quadri de la Torre, cited in *UnoMásUno,* February 2, 1984.

6. The interviews I am referring to include, among others, one with Dr. Hugo García Pérez, director of the Urban Development Program (Mexico City, November 4, 1987); one with Rodolfo Veloz Bañuelos, the director of territorial regularization (Dirección General de Recursos Territoriales; DGRT) (Mexico City, September 25, 1987); and one with Jesus Reyes Gamboa, the director of urban restructuring and ecological protection (DGRUPE) (Mexico City, January 22, 1993).

7. This extensive coverage was possible because of the archival resources of Servicios Informativos Procesados (SIPRO), a periodical library located at Alvaro Obregon 213, 2° Piso (Second Floor), Colonia Roma, in the Federal District. Established in the early 1980s, the SIPRO organization has clipped, classified, and made available to the general public articles from nine daily newspapers and from two magazines with national circulation.

8. The other is Xochimilco, which is also located in the southern part of the Federal District.

9. See Leff (1981, 1990a, 1994, 1995); Schteingart and Luciano d' Andrea (1991); Ibarra, Puente, and Saavedra (1987); Ibarra, Puente, and Schteingart (1984); Puente and Legorreta (1988); Coulomb and Duhau (1989); Leff, Carabias, and Batis (1990); and Hiernaux (1991).

10. Foucault's (1983, 220) use of the term *conduct* plays on its double meaning in that "*to conduct* is at the same time *to lead* others (according to mechanisms of coercion which are, to varying degrees, strict) and a way of behaving within a more or less open field of possibilities."

Chapter 2

1. Care must be taken in the use of the term *marginality*. The "myth of marginality" has long been discredited (Pearlman 1976). To be marginalized in the sense referred to here does not mean complete isolation from circuits of capital accumulation. It may well be that marginalized people contribute labor power and economic value to production or distribution that ultimately bolster the formal exchange economy. However, it would be through informal sector activities that are usually poorly remunerated and have little protection from risk and uncertainty.

2. Cited in Neil Harvey (1994). Harvey's monograph provides an excellent overview of rural modernization and poverty in Mexico. Weighing the prospects for widespread social unrest and violence, Saxe-Fernández (1994a) gives a foreboding account of the Chiapas insurrection as "the tip of the iceberg."

3. According to 1990 census data, 19.3 percent of all workers (who reported income) earned less than the minimum salary; 36.7 percent earned between one and two times the minimum salary. Nearly two-thirds (63.2 percent) reported earnings of no greater than two times the minimum salary (INEGI 1991).

4. "Zedillo Outlines Rescue Plan, Calls for Deep Sacrifices" (1995).

5. Nora Hamiliton (1994).

6. En Extrema Pobreza (1994).

7. Porras Robles (1992c).

8. Namely, Tijuana, Mexicali, Chihuahua, Ciudad Juárez, San Luis Potosí, Torreón, Aguascalientes, León, Querétaro, Culiacán, Tampico, Puebla, Toluca, Mérida, and Acapulco (INEGI 1992a, 44).

9. See Camposortega Cruz (1992); Luque González, Rodolfo Corona Cuapio, and Reina Corona Cuapio (1992); Legoretta (1992); and Coulomb (1992).

10. Speech given before the "Meeting about the Environment and Contamination" (Encuentro Sobre Contaminación y Medio Ambiente), July 4, 1994, México, D. F.

11. *Environment Watch: Latin America* (1994), 4, no. 1 (January): 1–2.

12. *Environment Watch: Latin America* (1994), 4, no. 1 (January): 1–2. Bartone et al. (1994, 104) point to the poor results of the program and suggest one of the lessons the experience offers: "One lesson from the tepid results of the program was that regulations alone can backfire—as did banning each car on a specific day of the week in Mexico City. Rather than stay off the streets, many drivers bought a second car. The added cars pushed up the costs of administration and, with more cars in the city, many of them older, the regulation ended up increasing pollution, not reducing it."

13. Fisher(1992).

14. For details about these problems and recommendations for their resolution see National Research Council (1995).

15. Land subsistence due to groundwater depletion is not unique to Mexico City. For instance, it has caused a 1.5 meter rift that cuts through the city of Zelaya, Mexico. Sinking land is also a problem in Bangkok (Bartone et al. 1994, 30).

16. Many cities face rising cost curves for water use. A recent analysis of World Bank–financed water projects indicates that the unit cost of water would more than double, and in some cases more than triple, under a new water development project (World Bank and the International Bank for Reconstruction and Development 1994, 101).

17. More than US$1 billion per year are spent subsidizing water and sanitation services in the Federal District (roughly $125 per person). The marginal price for supplying water to the Federal District is estimated at $1.00 per cubic meter of water, yet only $.10 per cubic meter is collected from consumers. The cost of water in Mexico City is one of the highest in the nation, yet cost recovery is comparatively low. Monterrey, for example, collects $.37 per cubic meter. In the United States, Phoenix collects $.25 for the same. Urban water districts in southern California purchase imported water at $.33 per cubic meter (1994 prices) and charge residential customers from $.22 to $.46 per cubic meter (National Research Council 1995).

18. The Kuala Lumpur metropolitan area has 2.5 million inhabitants, of which up to one-third live in squatter settlements.

19. Brookfield, Samad Hadi, and Mahmud's (1991) study of Kuala Lumpur is a very interesting one. It began as part of an "Ecoville" project at the University of Malaya. The researchers brought together ecological and economic factors in the study of Kuala Lumpur's urban growth and the conurbation of the Klang Valley (Yip and Low 1984).

Chapter 3

1. See Harris (1992).

2. On the subject of urban poverty—characteristics, causes, and consequences— a special issue of *Environment and Urbanization* (1995) contains a useful collection of papers.

3. The UN Preparatory Committee for the UN Conference on Human Settlements (Habitat II) defines the term as follows: "*Enablement means:* ensuring that opportunities and mechanisms for people to participate in decision-making exist at all levels; assisting individuals and groups to form effective partnerships; instilling a strong sense of public service among government leaders and managers; guaranteeing accountability, openness and transparency in decision-making processes; enhancing government's guidance responsibilities; strengthening government's responsibility to ameliorate, or improve, conditions; building leadership and man-

agement capacity; increasing the responsiveness of key groups to community needs; removing non-essential barriers to action; educating people to use information effectively; generating, disseminating and communicating accurate and timely information; and ensuring consistency and coordination through adoption of government policies and programmes that embody the spirit of enablement" (Habitat II, 1995).

4. The enabling concept was advanced for the first time by the United Nations Centre for Human Settlements in their 1986 report, titled *Global Strategy for Shelter to the Year 2000* (Habitat 1986). The Global Strategy report was subsequently adopted by the Commission for Human Settlements in 1988 and soon after that by the UN General Assembly. In an informative collection of articles on this subject, Harris (1992) notes that the World Bank moved toward institutionalizing the new urban sector orthodoxy in 1991 with the publication of *Urban Policy and Economic Development: An Agenda for the 1990s*. The United Nations followed this initiative with a complementary strategy paper, *Cities, People and Poverty: Urban Development Cooperation for the 1990s*.

5. The report failed to mention the specific time period covered by the pilot study.

6. Zedillo and Carabias, cited in *Environment Watch: Latin America.* (January 1995, 2).

7. There are now several excellent collections of current research articles in this area. For instance, see the special issue of *Environment and Urbanization* titled "Sustainable Cities: Meeting Needs" (1992) and of the *Third World Planning Review* titled "The Contribution of Cities to Sustainability" (1994). Also see Brown and Jacobson (1987); Stern, White, and Whitney (1992); Leff (1990a); and Bartone et al. (1994).

8. To list just a few, Redclift (1987); Mitlin (1992); Carley and Christie (1993); Redclift and Sage (1994); Leff (1994, 1995); Segura and Boyce (1994); CEPAL-NU (1991, 1993); Segura (1992); Pearce, Barbier, and Markandya (1989, 1990); Hardoy, Mitlin and Satterthwaite (1992); M. O'Connor (1994); Leff and Carabias (1992); Slocombe (1993); and Azuela de la Cueva (1993).

9. In a jointly published informational pamphlet (date not available), the Urban Management Programme (UMP) is described as "a ten-year (1987–1996) technical assistance programme designed to strengthen the contribution that cities and towns in developing countries make toward human development, including economic growth, environmental quality, and the reduction of poverty." The UMP works through regional offices and networks in developing countries. The Habitat is the UMP's executing agency; the World Bank is its associated agency; and the UNDP provides core funding and overall monitoring. Bilateral donors, multinational agencies, and NGOs provide other types of support.

10. The Jones and Ward article prompted a debate between the authors themselves and scholar-professionals at the World Bank. See Cohen and Leitman's (1994) rebuttal and the Jones and Ward's (1995) follow-up.

11. A number of multilateral programs support the kind of urban-environmental planning and management called for in *Agenda 21*. Three of the most notable

include the Urban Management Program (UMP), the Metropolitan Environmental Improvement Program (MEIP), and the Sustainable Cities Program (SCP). Other international initiatives in support of such efforts include the Healthy Cities Programme of WHO, the International Union of Local Authorities (IULA), the International Council for Local Environmental Initiatives (ICLEI), and the World Federation of Twin Cities, among others.

12. Habitat II (1995).

13. Habitat II, (1995).

14. Cited in *Environment Bulletin* (1994). The Bulletin is a quarterly newsletter from the Environment Department of the World Bank.

15. Cited in *Environment Bulletin* (1994).

16. Rabinovitch and Leitmann (1993, vii) point out that Curitiba was selected as a showcase to host the World Urban Forum, a meeting of mayors, city officials, and urban experts held in preparation for the 1992 UN Conference on the Environment and Development (UNCED).

17. It began with the publication of the Urban Development Law of the Federal District (DF) in the *Diario Oficial* (January 7, 1976), which decreed the necessity for an integral and comprehensive plan of the DF, including mandatory subplans for each of its sixteen wards. The process includes, most notably, the 1980 General Plan for Urban Development of the DF, the 1982 Regulatory Rule for Zoning in the DF, the 1983 Ecoplan and Ecological Program of the DF, and PRUPE (Program for the Urban Restructuring and Ecological Protection of the DF).

18. The bureaucracy includes the formation in 1983 of the General Directorate of Urban Restructuring and Ecological Protection, a new top-level office within the DF Administration (DDF) mandated to deal with the capitol's urban planning and environmental problems. It also includes the Coordinating Commission for Rural Development (COCODER)—the home base for a new force on the environmental scene, namely *ecoguardias* (ecology guards).

19. *Desalojo* is a conjugated form of the Spanish word *desalojar,* meaning to evict or dislodge.

20. Also see Meza Rodríguez (1994) and Vargas (1984).

21. I find support for this position in a range of excellent work by Mexicanist scholars, including, among others, Mumme (1992); Azuela de la Cueva (1989a, 1989b); Azuela de la Cueva and Duhau (1987); Ward (1990); Schteingart and Luciano d' Andrea (1991); Coulomb and Duhau (1989); Ramírez Saiz (1986a); Ibarra, Puente, and Saavedra (1987); and Leff (1994).

Chapter 4

1. As Ehrlich and Roughgarden (1987) describe it, *biogeochemical cycles* connect the physical world of elements with the biotic. While the earth is an open system with respect to energy, it is a closed system with respect to materials. These materials—the earth's elements—are in a constant state of flux. The elements continually

change position, phase, and combinations with one another. "The cycling of elements makes nutrients continuously available to organisms and thus makes life possible. This cycling also prevents the accumulation of elements in forms, quantities, and locations that are deleterious to organisms. Nutrients flow along paths that bind together the living and physical portions of the biosphere; these paths are called *biogeochemical cycles*" (Ehrlich and Roughgarden 1987, 553).

2. For the most part, premodern human-induced environmental changes were localized or regional in spatial extent—but, as Turner and Meyer (1993, 41) point out, there are exceptions: "There is some evidence that early hunter-gatherers eradicated the megafauna of the Western Hemisphere (Martin and Klein 1984), and burning and other flora-altering activities of paleolithic societies affected the continental ranges of biota (Sauer 1956, 1961). It has become apparent that early societies substantially altered the floristic composition of many biomes through subtle manipulations of land cover" (41).

3. For global perspective, it is useful to consider what ecologists call "net terrestrial primary production" (NTPP), the earth's total annual amount (presently more than 100 billion tons) of organic matter produced in the course of photosynthesis. Ecologists at Stanford University estimate that more than 40 billion tons of this NTPP (40 percent of the total) are somehow altered by or under the direct control of human beings (Weiner 1991, 164).

4. This does not mean that biomass-based substance economies are entirely a thing of the past. Indeed, studies conducted by researchers in India show that the majority of India's people still survive within a biomass-based subsistence economy—that is, on products obtained from plants and animals (Agarwal and Narain 1993, 242). In the poorest countries of sub-Saharan Africa and South Asia, the majority of households still depend on organic materials (timber, bamboo, and other biomass) for constructing their dwelling units (Wells 1995) and on fuelwood for their chief source of energy (Tarver 1994).

5. Of course, serious health problems can result from the misuse of sewage as a resource. In the northern part of the Valley of Mexico, untreated sewage and industrial effluent contaminates fields of alfalfa, beans, cucumbers, and chiles. Upon harvest, these crops are trucked to Mexico City—resulting in a "circle of poison" (Barry 1992, 270–271). Barry cites reports indicating that vegetables in Mexico City contain lead levels two times higher than that allowed in the United States and that at least 25 percent of fresh vegetables sold in the city are contaminated with heavy metals (271). The government's National Water Commission (CNA) considered the problem serious enough to recently prohibit the use of residual water for irrigation in seventy-five districts around the nation, especially where vegetables that are normally eaten raw are grown (*Mexico Business Monthly* 1991, 10). Approximately 220,000 hectares are affected by the ban. The CNA stated that the ban is intended to protect the general population from gastrointestinal diseases, including cholera (p. 10).

6. Estimates on these rates vary. For instance, the Subministry of Water Resources for the Valley of Mexico reported a rate of fourteen cubic meters per second,

which, according to their calculations for 1978, replaced 46 percent of all the water pumped from the valley's underground supplies (COCODER 1979, 2).

7. For literature that documents the escalation of environmental change in the global aggregate, beginning with the "Columbian Encounter," see Turner and Meyer (1993), Crosby (1972, 1986); Sale (1991), Parsons (1972), Denevan (1976), Whitmore (1991), Turner and Butzer (1992), and Turner et al. (1990).

8. Garavaglia (1992) provides a concise description of the makeup and function of a *chinampa:* "The first step towards starting a *chinampa* in the lake environment of the valley was to identify the shallow parts of the lake, using sticks as sounds. This shallow area, the *cimiento* or foundation, was marked out carefully with stakes of equal length. Layers of soil and turf were then placed one on top of the other until they were level with the surface of the water. The turf was extracted from the *ciénagas* (swamps) formed by the accumulation of aquatic plants . . . which grow in compact masses. Pieces of this mass are cut out with spades or *coas* (primitive hoes) and transported by canoe to the demarcated areas. The soil is extracted from old *chinampas* (which have risen too high above the level of the lake and have become almost uncultivable). By placing several layers of soil and vegetation on top of one another, the *chinampa* can be raised until it is some 20 to 25 centimeters above the level of the water. Now it is time to plant the willow cuttings . . . that serve to bind the soil. The willows rapidly send out shoots and the *chinampa* is ready for cultivation. Within three or four years the organic matter has decomposed and the *chinampa* is fully mature" (572).

9. The two major environmental planning measures (other than pollution abatement and control measures) presently being implemented in Mexico are (1) the conservation of the greenbelt of Ajusco, and (2) the restoration of *chinampas* in Xochimilco (Comisión para la Prevención y Control de la Contaminación Ambiental en la Zona Metropolitana del Valle de Mexico (1993). Harvard University gave a special environmental award to those doing the restoration of *chinampas* (Bermeo 1996).

10. In a study conducted by the Directorate of Ecological Planning for the Federal District, it was noted that on an annual basis, the food and drink industry in the metropolitan area of Mexico City consumes 175,193,000 liters of fuel oil, 141,341,000 cubic meters of gas, and 6,604,000 liters of diesel. The same study noted that of all the industrial plants in the valley, the city's two thermoelectric generating plants emit the most contamination. On an annual basis, the two thermoelectric plants consume 404,304,000 liters of fuel oil and 1,423,272,000 cubic meters of gas (Ballinas 1991).

11. The government estimates that within the Federal District, 67 percent of all solid waste comes from domestic end use, 9 percent from public spaces, and 24 percent from industrial and commercial sources (González Salazar 1990, 73).

12. Because these figures are aggregated, they can be misleading. For instance, per capita water consumption in Mexico City ranges from 50 to 450 liters per day. Nine percent of consumers account for 75 percent of freshwater consumption, while at least 2 million people suffer chronic shortages (Hardoy, Mitlin, and Satterthwaite 1992, 103).

13. Taking such evidence into account, Liverman (1992) argues that any shift to warmer, drier conditions in Mexico could bring nutritional and economic disaster. She notes that agriculture is already stressed by low and variable rainfall. This predicament is especially significant considering that "More than one-third of Mexico's growing population works in agriculture, a sector whose prosperity is critical to the nation's debt-burdened economy. Although only one-fifth of Mexico's cropland is irrigated, this area accounts for half of the value of the country's agricultural production, including many export crops. Irrigation districts often rely on small reservoirs and wells that deplete rapidly in dry years. The remaining rainfed cropland supports subsistence farmers and provides much of the domestic food supply. Frequent droughts already reduce harvests and increase hunger and poverty in Mexico" (351).

14. According to Ibarra, Puente, Saavedra, and Schteingart (1987, 101), 54 percent of the Valley of Mexico was forested around the time of the Aztecs. By the late 1980s, less than 15 percent of the valley was still forested.

Chapter 5

1. See "Mexican Legislators Prepare Environmental Reforms" and "Mexico City Continues Work on Environmental Law." *Environmental Watch: Latin America* (February 1995).

2. For early discussions about this involvement, see Lajous (1992), Urrutia (1992a, 1992b), and Venegas (1992).

3. See Cornelius, Gentleman, and Smith (1989).

4. Rental housing has historically lacked such patrimonial advantages; consequently, it has received relatively little attention during the past several decades of rapid urbanization (Walton 1991, 629–630).

5. Ríos Zertuche (1993).

6. See Ramírez Saiz (1989a, 1990a), Alonzo (1990), Coulomb and Duhau (1989), Schteingart and d' Andrea (1991), and Connolly (1993).

7. Until 1982, the functioning of the housing market for those with the highest incomes has been optimal in terms of supply and demand. Since 1982, times have been more difficult. Overall, the construction industry has floundered along with the rest of the economy (CIDAC 1991; Schteingart 1984; *Business Monitor International* 1992).

8. These figures refer to the total housing stock, urban and rural.

9. Gilbert and Varley (1991) point out: "Most rural people in Mexico have always housed themselves through self-help construction. In contrast, throughout the nineteenth century and up to the 1940s, Mexico's cities were dominated by tenants. Rental housing was not only the typical form of working-class accommodation but was also home for most of the middle class. Only the elite owned their own homes, usually in blocks close to the town's main square. There seems to be little

real variation between cities. Reports of conditions in most Mexican cities are agreed that renting was dominant until the middle of this century."

10. FONHAPO has acted as an external agent to facilitate self-help by providing credit to popular groups willing to sponsor it; approximately 110,000 progressive housing solutions and 75,000 serviced sites were reported to have been provided by this agency nationwide between 1983 and 1988 (Gibert and Varley 1991, 48).

11. The El Molino project is one of the few outstanding exceptions (see Suárez Pareyón 1987).

12. On an optimistic note, Quadri de la Torre (1992) argues that the impact of recent agrarian land reforms will be very favorable to the environment in both the urban and rural milieus. He says the ejido system was gangrenous, corrupt, inefficient, and ecologically devastating. He applauds the changes. Quadri de la Torre argues that in the urban context the changes will enable municipalities and the Federal District to finally create land reserves because these entities will now be able to purchase the land directly from the ejidatarios. At the same time, other individuals can buy ejido land to set it aside for conservation.

13. He reported a deficit of 22 percent in potable water infrastructure and of 50 percent in drainage infrastructure. More specifically, he said that there was an unmet need for 170.4 kilometers of underground pipes *(red secondaria)* for potable water distribution. The existing length at the time was 603.6 kilometers. He reported an unmet need for 374.1 kilometers of underground pipes for the drainage system. The existing length was 753.7 kilometers (Ríos Zertuche 1993).

14. Specifically, Ajusco Medio and Ajusco Bajo have 80 percent of Tlalpan's deficit in drainage and 44 percent in potable water (Ríos Zertuche (1993)

15. Gilbert and Varley (1991, 29) point out that in 1950 only three Mexican cities with more that 100,000 inhabitants had a majority of owner-occupied homes. By 1980, the transformation was almost complete. Practically every large Mexican city had a predominance of owners, many cities with two-thirds of their households in that category.

16. The text defines land in broad terms: "Land is normally defined as a physical entity in terms of its topography and spatial nature; a broader integrative view also includes natural resources: the soils, minerals, water and biota that the land comprises. These components are organized in ecosystems which provide a variety of services essential to the maintenance of the integrity of life-support systems and the productive capacity of the environment" (*Agenda 21,* sec. 10.1, cited in Robinson 1993, 152).

17. Section 10.1, cited in Robinson (1993, 152).

18. Cited in *La Jornada,* August 10, 1992, "La Capital" sec., p. 29.

19. In the same study, CENVI notes that 46 percent of the population in Mexico City rent their habitation; of these, roughly three-quarters use 56 percent of their household income for rent. CENVI also notes that more than 20 percent of all the public housing units in Mexico City are occupied by renters who are illegally subletting the space.

Chapter 6

1. For a thorough technical review of Mexican environmental law, see Carmona Lara (1991). For a review of environmental law and policy from a political perspective, see Mumme (1991, 1992); Mumme, Bath and Assetto (1988); and Nuccio (1991).

2. Specifically, they note that "staff at the Instituto de Investigaciones de Derecho at the Universidad Autónoma de México (UNAM), engineers at the Instituto Politécnico Nacional, demographers at the Colegio de México, officials within the Ministry of Health, the Ministry of Industry and Commerce, and the Instituto Mexicano de Ingenieros Químicos all played a part in promoting public awareness and demanding greater government attention to these problems (Mumme, Bath, and Assetto 1988, 12).

3. SEDUE undertook a nationwide series of popular forums on ecological problems to inform the public of the government's emphasis on environmental improvement and to solicit public input on environmental concerns relating to various regions and municipalities. These perspectives were to be integrated into short-, medium-, and long-term environmental plans by SEDUE (SEDUE, cited in Mumme, Bath, and Assetto 1988, 18).

4. This type of contradiction is not uncommon. As Ward (1989) points out, a rivalry among institutions headed by different *camarillas* often leads to this type of contradiction. This point serves to illustrate that although the politics of containment have a degree of integrity, they by no means comprise a well-orchestrated holistic set of policies.

5. The first stage of presentations began in September 1984. By October 1984, nineteen information booths had been set up. From November 6 through 23, 1984, public hearings were realized in thirty-one meetings conducted by the Planning Committee for the Urban Development of the Federal District. After this activity, analysis and discussion of PRUPE continued at the Urban Environment and Ecology Commission of COPLADE-DF. There were a total of 105 reunions with officials and social groups, including governors of adjoining states, representatives of campesinos and labor, institutions for research and higher education, international organizations, banking and finance institutions, professional associations, neighborhood representatives, councils from industry and commerce groups, civil associations, and ecological groups. According to official estimates, the total number of citizens who participated in the process of consultation exceeded 32,000 (DDF 1985, 12). The last stage of the process (begun in March 1985) entailed a synthesis of a set of conclusions recommended by a select group that participated in the process (including citizens and functionaries of the SPP and DDF).

6. When de la Madrid presented his National Development Plan on May 30, 1983, he asserted that this would be the only "plan" during his six-year term. From then on, the label plan has been strictly reserved for the National Development Plan. All the rest, including the master plans and subplans for the Federal District, would be termed programs.

7. Based on a series of interviews and document research, I found this to be a significant element in the perspectives of officials working in the Subdirectorate of Ajusco, the Tlalpan Ward, and the Planning Office of the Department of the Federal District.

8. Wilk (1991a, 9) notes that this growth containment strategy was complemented by two important programs: a detailed zoning code *(zonificación secundaria)* for the entire rural territory and a land reserve program (Programa de Reserva Territorial) that established specific strategies for land acquisition and urban densification programs in the urban fringes.

9. Specifically, Wilk (1991a, 18) notes that the division manages an annual budget of approximately 1,200 million pesos (approx. $US480,000) (1988), 80 percent of which go to operations and 20 percent for research and program design. An additional 3,000 million pesos (approx. $US1,200,000) were added to the 1988 budget specifically for the implementation of the growth containment strategies.

10. The first version of this program, approved in 1980 and revised in 1982, was originally titled the General Plan for the Urban Development of the Federal District. See note 6.

11. See "Busca DDF Modificar Ley de Desarrollo Urbano: Desaparecerían los ZEDEC" (1995).

12. See *La Jornada* (January 12, 1987; February 13, 1987), and Wilk (1991a).

13. These areas included eight biosphere reserves, fourteen special biosphere reserves, forty-four national parks, one protected area for wild flora and fauna, and a natural monument covering an area of 5.7 million hectares, which is almost twice the surface area of Belgium or approximately that of the state of West Virginia. (See Dirección General de Communicación Social 1991a).

14. Sources of funding for the environmental program in Mexico City in 1991 totaled: US$2.52 billion (42 percent financed by foreign loans, 58 percent by national resources). Mexico's federal budget for 1991 included an additional US$360 million for the protection of the environment in Mexico City, which are independent of the funds included in the Mexico City program (Dirección General de Communicación Social 1991a, 126).

15. For example, the parastatal enterprise Telmex—Mexico's telephone company.

16. DDF (1989b).

17. See Delegación del Tlalpan (1992).

18. DDF (1989b).

19. See "Political Changes Leave Mexican Environmental Agencies Reeling" (1993).

20. The restructuring that created SEDESOL was short lived. On December 28, 1994, the institutional terrain changed again. The environmental responsibilities of SEDESOL were shifted to a new Secretariat of Environment, Natural Resources, and Fisheries (Secretaría de Medio Ambiente, Recursos Naturales y Pesca

SMARNP). The SMARNP has three undersecretariats: (1) Natural Resources (including forestry and other resources subject to federal jurisdiction), (2) Regional Sustainable Development Planning, and (3) Fisheries.

21. See Crónica Legislativa (1993a).

22. During the 1992 Earth Summit, Mexico entered into a debt-for-nature swap with the Inter-American Development Bank. As reported by the Press Office of the president of Mexico (June 1992), the amount resulting from this operation (US$100 million) was applied in the Reforestation Program for Mexico City and the Metropolitan Area and in support of the environmental measures being implemented to solve the capital's ecological problems.

23. The price for each of the mansions was listed at US$340,000 (*UnoMásUno*, July 18, 1992).

24. "Ocuparán Zona de los Dinamos (1994).

Chapter 7

1. The interviews I am referring to include, among others, one with Dr. Hugo García Pérez, director of the Urban Development Program (November 4, 1987, Mexico City), one with Rodolfo Veloz Bañuelos, the director of territorial regularization (DGRT) (September 25, 1987, Mexico City), and most recently, one with the director of urban restructuring and ecological protection (CGRUPE) (January 22, 1993, Mexico City).

2. See "Surgen Fraccionadores Clandestinos con la Autorización de Poseedores de Tierra" (1991); Sosa (1991); Saad Sotomayor (1991); González Durán, (1991); Figuereo Osnaya (1992); "Dos Mil Cuatro Cientos Familias Viven en la Reserva Ecológica en G.A. Madero" (1992); Porras Robles (1992); and Mejía, Javier (1992).

3. See Soberón (1990, 1992).

4. In *carta abierta* (open letter) directed to government officials and to the general public, which was published in the newspaper *UnoMásUno* on January 18, 1982, a coalition of popular groups active in Ajusco made known their protest. Referring to the mass eradications of several irregular settlements and to the *delegado* of Tlalpan's statement that such eradications were necessary to nip in the bud clandestine subdivisions and fraudulent land sales, the popular groups protested against the action as a discretionary enforcement of the law. They declared that while the government "speaks of applying the law to the poor, we have never seen it applied to the rich. We have seen only the violence through which the poor have been eradicated and the cruelty in the destruction of their houses beneath the pretext that the victims had illegally acquired the land. Never have we seen the rich (politicians, businessmen and functionaries), who have monopolized great extensions of land for their exclusive subdivisions and grand estates, become the target of these laws. . . . In view of this . . . we demand that they stop the construction of luxury subdivisions, private clubs and hotels, the real culprits doing away with our forested zone and our water . . . and that [the government] regularizes our property."

5. Reino Aventura (the "Disneyland" of Mexico) is a privately run amusement park created in 1981 for the middle and upper class. Thousands of trees were destroyed to clear the site (45 hectares) and lay down acres of asphalt for parking lots. The amusement park received a full complement of infrastructure, including, water, electricity, paved roads, and so on, at a time when the Department of the Federal District (DDF) was declaring that it did not have the resources to grant even the most rudimentary infrastructure to colonias populares in the zone. A consortium of industrialists (led by the son of then regente, Carlos Hank González) financed the deal and agreed to grant 15 percent of all entrance tickets to the government (see García Pérez 1985, 95; *El Día*, January 3, 1982; *UnoMásUno*, May 21 and 23, 1981 and November 23, 1981). Based on my own observations, I know that the DDF has used these tickets as tokens of reward to influence popular groups in the area. Jardinas en la Montaña is a controversial and exclusive upper-class residential development on 700,000 square meters developed by the real estate company Somex. Other upper-class developments in the vicinity of Ajusco, including Perisur, the ritzy shopping mall, have been favored by "privileged construction"—such as the Periférico (freeway) and the Picacho-Ajusco Highway (García Pérez 1985, 50).

6. Catalina Rodríguez Lazcano and Fernando Rodríguez (1984) provide a good historical account of pueblos in the area during prehispanic times.

7. This was not unique to the zone of Ajusco. According to Cockburn (1986, 63) land on the outskirts of Mexico City held in the form of *pequeñas propiedades* has generally been of better quality than ejido land.

8. Land tenure regularization has not followed the manner spelled out under the Agrarian Law in the creation of urban ejido zones in over twenty years (Varley 1985, 93). Since the late 1960s this legal device has fallen into disuse (Azuela de la Cueva 1987, 535).

9. Prior to this concession nearly all of Ajusco was designated as a national park according to a decree issued by President Cárdenas. The concession dramatically scaled down the size of the park, from 5,000 hectares to 930 hectares.

10. One can go back even further to find evidence that the overexploitation of timber in Ajusco led to socioeconomic and ecological problems. In a letter written in 1921 by ejidatarios to government officials in the Federal District, the ejidatarios denounced the abusive exploitation of woodland by owners of the hacienda San Nicolás de Eslava, stating that it was causing erosion on their agricultural fields, which in turn undermined their productivity (Schteingart 1987, 459). These observations were later reflected in President Cárdenas's decree to designate most of the area as a national park.

Chapter 8

1. For an in-depth account of the origin and development of this ejido see Rivera (1987). Also see Lugo Medina and Bejarano González (1981) and Durand (1983).

2. Such tension is evident in a letter, dated April 13, 1983, written by the Ejidal Commission of SNT and sent to the president of Mexico. The letter called upon the president to resolve boundary disputes, to stop "invasions" and urban encroachment upon their land, and to ensure that agrarian laws are respected.

3. In this study the term *popular* refers to a heterogeneous mass of urban people that includes unemployed and semiemployed workers, recent rural migrants, workers in illegal and other precarious occupations, artisans, small business people, and women as well as men, whether employed or not (Friedmann 1989). In this respect, *popular* describes certain groups and movements with identities that are independent of any particular social class in the Marxist sense. Popular movements here are not defined in terms of members' fundamental relation to the means of production.

4. Convenio de Regularización, Secretario de Gobierno, DGRT, August 1986.

5. I obtained this information on the Teresa family and the Belvedere Association from several interviews and documents, including a series of interviews during October 1987, with Carlos Gutiérrez Topete—a university professor who had purchased land from the Teresa family—and an interview on October 6, 1987, with Irma Elizondra, the president of the Belvedere Association. Irma Elizondro was in charge of technical affairs. She set land prices and financial terms during the first stages of land tenure regularization and had a close working relationship with DDF authorities. I also referred to copies of correspondence, which consisted of letters and *amparos,* between the DDF and the Belvedere Association.

6. See the *amparo* (court injunction) filed on September 11, 1977, by the Belvedere Association against actions of the DDF, delegado of Tlalpan, and director of the DDF Regularization Office.

7. The term *amparo* means shelter or protection. According to the Political Constitution of Mexico (art. 107, sec. 2) the amparo is a means of constitutional control, exercised by a jurisdictional entity, with the objective of protecting an actor in the cases stipulated in Article 103 of the Constitution. Article 103 specifies that the juridical power of the federation is to guide the guarantee of public rights and that the people governed have the faculty to place claims and demands before a jurisdictional federal public service in order to obtain protection against violations enacted by authorities. Thus the Constitution guarantees the right of an action, the action of amparo. A *juicio de amparo* is a series of acts regulated by juridical norms. It involves the controversy and arbitration of a conflict—between "the governed" and an authority—produced by the violation of a subjective public right, committed by the latter in the offense of the former. A plaintiff presents the conflict in the form of a complaint before a jurisdictional federal authority with the aim that this federal authority judge the existence of said constitutional violation and in the affirmative case, give back to the governed the enjoyment of the guarantee and obligate the authority to respect it (Arilla Bas 1986).

8. Manuel García Beceril, the director of Injunctions and Administrative Lawsuits (Amparos y Contencioso Administrativo) en Tlalpan, told me that Rena is an unmitigated fraud. At the time I did this interview (1988), he noted that Rena

doesn't have legal title to the land, it doesn't have permission to subdivide, and it doesn't have permits for construction. García Beceril viewed his work as the defense of the *delegación.*

9. The *subdelegado jurídico y de gobierno de Tlalpan,* Juan Hernández Ayala, is paraphrased as arguing that "Just when you finish implementing a measure to contain irregular settlement, the people discover new mechanisms to *burlar* (outsmart) the authorities. Many groups have juridicial support such as amparos that don't specify size, the particular people protected, or the exact location of the site (*UnoMásUno,* March 19, 1984).

10. Indeed, this fact has sparked debate about the structure of the federal *Amparo* Law. In 1987, the president of the Supreme Court of Mexico, argued that the extraordinary increase in amparos filed between 1981 and 1986 "clearly illustrates abuse of the *Juicio de amparo,* and that such abuse is nullifying its objective." On the other hand, the president of the Mexican Academy of the Right of Amparo and other lawyers have argued that the increase is not necessarily abuse: it is due to "the increased juridical consciousness of the population" (cited in *UnoMásUno,* February 7, 1987).

Chapter 9

1. Certain activists urged on laying down a street pattern that inclined the steep grade in a zigzag fashion, but the more conventional method of extending the existing layout prevailed. As a result the streets in Bosques are so steep that vehicular access into the colonia is limited.

2. The lack of water was one of the greatest problems. It had to be carried into the settlement over long distances.

3. Francisco de la Cruz saw the movement in Bosques as an opportunity to gain support and alliances for his own movement in Campamento 2 de Octubre. Because of this activity, he is now in prison.

4. On one such march to Los Pinos (August 27, 1979) colonos were received by a special presidential commission. Out of this event the council managed to acquire a document signed by President López Portillo, in which the president called upon the DDF, SRA, and CORETT to attend to the council's demands for land tenure regularization (*El Día,* May 12, 1986). Yet, the document turned out to be of little use once the zone was (re)designated for ecological conservation in the early 1980s.

5. For a discussion of how the Junta de Vecinos functioned see Jiménez (1989) Also see Jiménez (1988).

6. See Miguel Díaz-Barriga (1990) for a full account of the struggle *colonos* faced while living along the railroad tracks.

7. There is an interesting history behind each of the elementary schools in Los Belvederes. In Dos de Octubre, *colonos* managed to get the SEP to secure an *amparo* to protect the school land. The *amparo* provided the juridical basis for authoriza-

tion of the school despite the *colonia*'s irregular status. In Bosques, grassroots activists cunningly played one state agency against another to secure their school.

8. Quoted from the grassroots document "Plan de Desarrollo para la Colonia Belvedere," authored by the Association of Settlers of Ajusco, Casa del Pueblo.

9. Arroyo Acosta et al. (1987).

10. Juicio de *Amparo,* August 1983, No. P. 177/83, *Dictamen Pericial,* p. 2.

11. The Subsecretariat of Ecology's consulted branch offices included the General Directorate of Ecological Ordering and Environmental Impact (Dirección General de Ordenamiento Ecológico e Impacto Ambiental), the Directorate of Flora and Wildlife (Dirección de Flora y Fauna Silvestres), and the General Directorate of Parks, Reserves, and Ecologically Protected Areas (Dirección General de Parques, Reservas y Areas Ecológicas Protegidas).

12. Of the 9,200 hectares targeted for expropriation, 3,000 are in the possession of the ejido San Nicolás Totolalpán. The Ejido Commission of SNT argued against the expropriation, stating the indemnification would be too small and that the inhabitants needed the land for the security of their livelihood (*UnoMásUno,* February 19, 1984).

Chapter 10

1. The role of political parties was very limited. To some extent there was a connection to the Party of the United Socialists of Mexico (Partido Socialista Unificado de México, PSUM) vis-à-vis affiliation with the National Coordinator of the Popular Urban Movement (Coordinadora Nacional del Movimiento Urbano Popular, CONAMUP). But for the most part the popular groups had little to do with electoral politics or political campaigns.

2. In July 1983, the construction of Jardines de la Montaña, an upper-class residential development, was begun not far from Los Belvederes. Jardines was built in a wooded area that had been part of a park. Earlier in 1983 the developers of Jardines got permission from the Subsecretariat of Forestry and Wildlife (Subsecretaría Forestal y de la Fauna) to clear 135 trees if they paid a fine. But in reality, closer to 10,000 small trees were chopped down on the 700,000-square-meter site (*El Día,* July 4, 1983). The state justified granting permission to the developers on the grounds that the occupants could afford to pay for their own infrastructure and that the development had its own waste treatment facility.

3. Quote taken from a bulletin printed October 25, 1983, by activists of the cooperative Bosques del Pedregal en Lucha.

4. For example, the march and demonstration that concluded in front of the offices of the Center of Social Communication (Centro de Comunicaciones Sociales; CENCOS) on June 7, 1984 (*El Día,* June 8, 1984).

5. For example, the front arranged a press conference *(rueda de prensa)* in front of the Superior Court of Justice (Tribunal Superior de Justicia) of the DDF on October 16, 1984 (*La Jornada,* October 16, 1984).

6. One example is the grievance letter addressed to President de la Madrid, and "to public opinion," printed in *UnoMásUno* and signed by the front on August 14, 1984.

7. See *El Día* (August 16, 1984), *La Jornada* (September 23, 1984), and *La Jornada* (September 24, 1984).

8. Certified with the Mexican government under invention number 6758/101432/ 189208. Secofin marca registrada, expediente no. 238477, folio 103169.

9. These observations are substantiated, according to Mena (1987, 552), in studies realized by investigators in the UAM-Iztapalapa under the direction of Dr. David Muñoz (1984–1985) and by National Laboratories for Industrial Development (Laboratorios Nacionales de Fomento Industrial; LANFI), Departamento de Biotecnología, under the direction of Dr. Gustavo Viniegra Gonzalez (1987).

10. For example, see "Intentan crear la primera Colonia Ecológica de Mexico en el Ajusco" (1984), "Surge la Primera Colonia Ecológia del País" (1984), "La Hermosa Gente: Bosques del Pedregal" *Ovaciones*, (1984), "Piden que el DDF Tome en Cuenta los Asentamientos en el Ajusco" (1984), and "La Ecología como Arma Política" (1984).

11. GTA (1992a, 1992b)

12. An alternative to the SIRDO technology is called the Ecoreactor, an anaerobic digestor. There are forty-eight Ecoreactors in place around the country (López 1995). For a comparative perspective on initiatives centered on reuse and recycling in the urban milieu, see Pacheco (1992), Pearlman (1990), Rabinovitch (1992), Smit and Nasr (1992), and Brown (1992).

13. Subdelegación del Ajusco (1985c).

14. The fact that the residents of Mexico City do not elect their own mayor is one of the great contradictions that opposition groups have mobilized to change (see Ward 1989). On July 31, 1996, Mexico's Chamber of Deputies unanimously approved a series of constitutional reforms that impact federal electoral processes and the governing structure of the Federal District. In 1997, the residents of the Federal District directly elected an executive.

15. When this satellite office was first created it was a subward, an administrative organ with more clout than a subdirectorate. In 1987, the delegado of Tlalpan, David Galindo, cut 350 employees from the ward's payroll while trimming down the bureaucracy. Citing the budget crisis, Galindo announced that it was necessary to dissolve five of the seven subwards that had existed up until that time and that the subwards of Ajusco and Coapa had to be reduced in status to subdirectorates (*La Jornada*, May 17, 1987).

16. Subdelegación del Ajusco (1985, 15).

17. Interview with the director of the Community Development Program, Subdirectorate of Ajusco, September 23, 1987, Tlalpan, Federal District.

18. This is not the individual's real name. I use a pseudonym here to protect the innocent.

19. Evidence of intensiveness are statements such as "his origin and political formation grows out of relations with CONAMUP, the Popular School of Tacuba (Prearatoria Popular de Tacuba), Association of Unemployed Medical Professionals (the Asociación de Médicos Desempleados), the Popular Worker and Campesino Union (Union Obrera y Campesina Popular), and with the National Coordination of Unions (Coordinación Nacional de Sindicatos.)"

20. Of course, this was publically denied by the government (*El Día,* August 26, 1987). Popular groups pressured the state about Martínez's captivity, and in October 1987 gobernación agreed to review his and other cases (*La Jornada,* October 3, 1987).

21. This information is based on my interviews with a developer who was a member of the Belvedere A.C. (October 28, 1987, Tlalpan, Federal District) and on my interview with Irma Elizondra, the president of the Belvedere A.C. who was in charge of technical affairs and determining payment schedules (October 6, 1987, Tlalpan, Federal District). According to these sources, Teresa sold 80 percent of their land to several buyers in the zone during the 1960s (they kept 20 percent for themselves). These buyers, in turn, sold the land in lots of 1,000 square meters to those people who formed the Belvedere Association. When the regularization process got under way, these middle- and upper-class members of the Belvedere Association could not legally prove they owned the land. Faced with this problem, members of the Belvedere Association joined with several other landlords in the zone and formed a real estate coalition (*El Día,* December 22, 1986). The objective was to coordinate claims and to minimize the possibility of disputes. The real estate coalition agreed to allow the Belvedere Association to collect all land payments in the name of Teresa. Teresa's claim to the land was recognized by the state, even though Teresa had already sold most of it off. The arrangement was that Teresa would keep 35 percent of all incoming land payments, while the other landlords would divide the remaining 65 percent among themselves.

22. Outlining these demands, the coalition submitted a petition to authorities at the Tlalpan Ward and at DGRT (see *La Jornada,* August 27, 1986).

23. Interview with Ernesto Bravo, August 1, 1986, Bosques del Pedregal.

24. The program might have been implemented more quickly were it not for administrative turnover: two changes of delegados, a change of the government secretary (Secretario del gobierno), and a change of the director of DART (*La Jornada,* October 7, 1986).

25. See *El Día* (August 16, 1986), *UnoMásUno* (August 16, 1986), *La Jornada* (September 10, 1986), *La Jornada* (October 25, 1986), *Excelsior* (January 3, 1987) *La Jornada* (January 3, 1987), *UnoMásUno* (March 4, 1987).

Chapter 11

1. Zedillo (1994).

2. He named the Sierra de Guadalupe, parts of Texcoco, Sierra de Santa Catarina, the Sierra de Chinahuatzin, Ajusco, and the Desierto de los Leones.

3. For a sampling of newspaper coverage from 1991 to 1996 see "Surgen Fraccionadores Clandestinos con la Autorización de Posseedores de Tierra" (1991), Saad Sotomayor (1991), "Dos Mil Cuatro Cientos Familias Viven en la Reserva Ecológica" (1992), "Habrá Invasión a Zonas Verdes" (1994), "Parque Nacional en Peligro" (1995), "La Zona Rural y de Conservación Está Sujeta a Fuertes Presiones por el Crecimiento de la Mancha Urbana" (1995), and "Toman Vecinos delegación" (1996).

4. "El Desalojo Es la Unica Vía" (1994)

5. "Invaden Familias el Ajusco Medio" (1994).

6. "Habrá Invasión a Zonas Verdes" (1994).

7. Notable instances are written up in the following articles, all printed in the "Ciudad y Metropoli" section of the Mexico City–based newspaper *Reforma*: Ramos (1995), "Denunciará el PVEM Ecocidio en la Ciudad"; Rodríguez (1995), "Destruyen Zona Verde"; Páramo and Medina (1995), "Preservarán la Ecología de la Sierra de Guadalupe"; Carrasco, Vicenteño, and Joyner (1995), "Desalojan a 350"; López (1995), "Promoverán en la Ladera Programa de Ecoturismo"; Hidalgo (1995), "Seguirán Desalojos en Zonas de Reserva"; and Hidalgo and Carrasco (1995) "Desalojan la Ladera."

8. "Environmental Protection in Mexico City" (1994).

9. A select list of networks that link city-based groups focused on environmental planning and management include (1) the Habitat International Coalition, an NGO network active in more than sixty countries; (2) the Megacities Project; (3) Asia Pacific 2000, a UN-sponsored network for NGOs active in Asian cities; (4) MEDCITIES, a network among Mediterranean cities; (5) CIUDAD and IIED-America Latina, both active in Latin America; and (6) ENDA-Tiers Monde, and Environmental Liaison Center International (ELCI)—both active in Africa and elsewhere. See Bartone et al. (1994) for more detail on these networks.

10. Since 1990, a number of new multilateral programs have emerged globally with the express purpose of supporting urban environmental planning and community-based development initiatives. They include, among others, (1) the Urban Management Programme (UMP), (2) the Metropolitan Environmental Improvement Programme (MEIP), a UNDP-funded effort executed by the World Bank; (3) the Sustainable Cities Programme (SCP) launched by Habitat as the global counterpart of the Asian MEIP and an operational arm of the UMP. See Bartone et al. (1994) for more detail on these programs.

11. As already noted, the 1985 earthquakes in Mexico City mark a turning point in the heightening mobilization of civil society (see Ramírez Saiz 1989b).

12. For some of the best work, see Leff (1981, 1990a, 1995); Schteingart and Luciano d' Andrea (1991); Ibarra, Puente, Saavedra, and Schteingart (1987); Ibarra, Puente and Schteingart (1984); Puente and Legorreta (1988); Coulumb and Duhau (1989).

Chapter 12

1. *Habitat News* (1994).

2. In the context of developed countries, see Cole (1992) for an interesting and thorough review of environmental poverty law as an empowering type of legal advocacy.

References

"A Proyectos Ecológicos, lo que Obtenga de la Concesión de Reino Aventura: Camacho." (1992). *UnoMásUno*, June 4.

Agarwal, Anil and Sunita Narain (1993). "Towards Green Villages." In *Global Ecology: A New Arena of Political Conflict*, edited by Wolfgang Sachs (pp. 242–256). Atlantic Highlands, New Jersey: Zed.

Aguayo Quezada, Sergio and Bruce Michael Bagley (eds.) (1990). *En Busca de la Seguridad Perdida: Aproximaciones a la Seguridad Nacional Mexicana*. México, D.F.: Siglo Veintiuno Editores.

Aguilar, Adrian Guillermo (1988). "La Política Urbana y el Plan Director de la Ciudad de México: ¿Proceso Operativo o Fachada Política?" Unpublished manuscript. Instituto de Geografía, Universidad Nacional Autónoma de México, México, D.F.

Aguilar, A. M. (1985). "Crisis and Development Strategies in Latin America." *Development and Peace*. 6:17–32.

Aguirre, A. (1986). "Tasa de Crecimiento Poblacional de 1% en el Año 2000: Una Meta Inalcanzable." *Estudios Demográficos y Urbanos*. 1 (3): 443–474.

Aguilar, Guillermo Adrián and Guillermo Olvera L. (1991). "El Control de la Expansión Urbana en la Ciudad de México: Conjeturas de un Falso Planteameinto." *Estudios Demográficos y Urbanos*. 16 (1) 89–115.

Alonzo, Jorge (1980). *Lucha Urbana y Acumulación de Capital*. México, D.F.: Centro de Investigaciones Superiores del INAH.

Alonzo, Jorge (1990). "Aproximaciones a los Movimientos Sociales." In *Crisis, Conflicto y Sobrevivencia: La Sociedad Urbana en México*, edited by Guillermo de la Peña, Juan Manuel Durán, Augustín Escobar, and Javier García de Alba (pp. 421–438). Guadalajara: Universidad de Guadalajara and CIESAS.

Amaral de Sampaio, Maria Ruth (1994). "Community Organization, Housing Improvements and Income Generation: A Case Study of *Favelas* in São Paulo. *Habitat International*. 18 (4): 81–97.

Angel, S., R., S. Tanphiphat Archer, and E. Wegelin (eds.) (1983). *Land for Housing the Poor*. Singapore: Select Books.

Archer, Ray W. (1994). "Urban Land Consolidation for Metropolitan Jakarta Expansion, 1990–2010." *Habitat International*. 18 (4): 37–52.

Arendt, Hannah (1958). *The Human Condition*. Chicago: University of Chicago Press.

Arilla-Bas, Fernando (1986). *El Juicio de Amparo*. México, D.F.: Editorial Kratos.

Arroyo Acosta, Eligio, J. M. Correa Garcia, E. L. Gonzalez Villegas, L. Mosta Díaz, and A. Mora Alva (1987). *Belvedere La Tierra es de Quien la Habita*. Masters' thesis, México D.F., Arquitectura Autogobierno.

Atkinson, Adrian (1991). *Principles of Political Ecology*. London: Belhaven.

Atkinson, Adrian (1992). "The Urban Bioregion as 'Sustainable Development' Paradigm." *Third World Planning Review*. 14 (4): 327–354.

Atkinson, Adrian (1994). "The Contribution of Cities to Sustainability." *Third World Planning Review*. 16 (2): 97–101.

Atkinson, Adrian and Chamniern Paul Vorratnchaiphan (1994). "Urban Environmental Management in a Changing Development Context: The Case of Thailand." *Third World Planning Review*. 16 (2): 147–169.

Audefroy, Joël (1994). "Eviction Trends Worldwide—And the Role of Local Authorities in Implementing the Right to Housing." *Environment and Urbanization*. 6 (1): 8–24.

Austin (1994). "The Austin Memorandum on the Reform of Article 27, and its impact on the Urbanization of the Ejido in México." *Bulletin of Latin American Research*. 14 (3): 327–335.

Azuela de la Cueva, Antonio and Emilio Duhau (1987). "De la Economía Política de la Urbanización a la Sociología de las Políticas Urbanas." *Sociológica*. 2 (4): 41–69.

Azuela de la Cueva, Antonio (1993). *Desarrollo Sustentable: Hacia una Política Ambiental*. México, D.F.: UNAM, Coordinación de Humanidades.

Azuela de la Cueva, Antonio (1982). "El Desarrollo Urbano y la Tesis de la Función Social de la Propiedad." *Derecho y Sociedad Mexicana*. 3 (5): 65–86.

Azuela de la Cueva, Antonio (1987). "Low-Income Settlements and the Law in Mexico City." *International Journal of Urban and Regional Research*. 11 (4): 523–542.

Azuela de la Cueva, Antonio (1989a). "El Significado Jurídico de la Planeación Urbana en México." In *Una Década de Planeación Urbano-Regional en México, 1978–1988*, edited by Gustavo Garza (pp. 55–78). México, D.F.: Colegio de México.

Azuela de la Cueva, Antonio (1989b). *La Ciudad, La Propiedad Privada y el Derecho*. México, D.F.: Colegio de México and Programas Educativos, S.A.

Azuela de la Cueva, Antonio (1989c). "Una Torre de Babel para el Ajusco: Territorio, Urbanización y Medio Ambiente en el Discurso Jurídico Mexicano." In *Servicios Urbanos, Gestión Local y Medio Ambiente*, edited by Martha Schteingart and Luciano d' Andrea (pp. 205–229). México, D.F.: Colegio de México and CE.R.FE.

Bachrach, P. and M. S. Baaratz (1962). "The Two Faces of Power." *American Political Science Review.* 56: 107–122.

Bachrach, P. and M. S. Baaratz (1963). "Decisions and Non-Decisions: An Analytical Framework." *American Political Science Review.* 57.

Bachrach, P. and M. S. Baaratz (1970). *Power and Poverty.* New York: Oxford University Press.

Ballinas, Victor (1991). "La Ciudad y el Medio Ambiente." *La Jornada.* May 26.

Ballinas, Victor (1995). "La atomización de la tenencia de la tierra." *La Jornada.* April 3.

Banamex-Accival, Grupo Financiero (1996). *Review of the Economic Situation of Mexico.* 72 (845) April.

Bandyopadhyay, J. and V. Shiva (1988). "Political Economy of Ecology Movements." *Economic and Political Weekly,* June 11, pp. 1223–1232.

Barkin, David (1990). *Distorted Development: Mexico in the World Economy.* San Francisco: Westview.

Baross, P. (1983). "The Articulation of Land Supply for Popular Settlements in Third World Cities." In *Land for Housing the Poor,* edited by S. Angel, R. Archer, S. Tanphiphat, and E. Weglin (pp. 180–220). Singapore: Select.

Baross, P. and J. Van Der Linden (1990). *The Transformation of Land Supply Systems in Third World Cities.* London: Avebury.

Barry, Tom (1995). *Zapata's Revenge: Free Trade and the Farm Crisis in Mexico.* Boston: South End.

Barry, Tom (ed.) (1992). *Mexico: A Country Guide.* Albuquerque: Inter-Hemispheric Education Resource Center.

Bartelmus, Peter (1994). *Environment, Growth, and Development: The Concepts and Strategies of Sustainability.* New York Routledge.

Bartone, Carl (1994). "Urban Sanitation, Sewage and Wastewater Management: Responding to Growing Household and Community Demand." Unpublished paper prepared for the Second Annual World Bank Conference on Environmentally Sustainable Development, "The Human Face of the Urban Environment," September 19–21.

Bartone, Carl, Janis Bernstein, Josef Leitmann, and Hochen Eigen (1994). *Toward Environmental Strategies for Cities: Policy Considerations for Urban Environmental Management in Developing Countries.* Urban Management Programme Policy Paper no. 18. Washington, D.C.: World Bank.

Bartone, Carl and Emilio Rodríguez (1993). "Watershed Protection in the São Paulo Metropolitan Region: A Case Study of an Issue-Specific Urban Environmental Management Strategy." In *Infrastructure Notes.* Urban no. UE-9. Washington, D.C.: World Bank.

Batley, Richard (1994). "Privatization and Cities: Can Government Manage It?" *The Urban Age.* 2 (4): 3.

Belvedere, Casa del Pueblo (1985). "De lucha utopia a la realidad." Unpublished bulletin distributed by the Association of Ajusco Settlers (Asociación de Colonos del Ajusco, Mexico City).

Benítez Badillo, Griselda (1986). *Arboles y Flores del Ajusco.* México, D.F.: Instituto de Ecología and Museo de Historia Natural de la Ciudad de México.

Benítez Badillo, Griselda (1987). "Efectos del Fuego en la Vegetación Herbácea de un Bosque de Pinus Hartwegii de la Sierra del Ajusco." In *Aportes a la Ecología Urbana de la Ciudad de México,* edited by Eduardo H. Rapoport and Ismael R. López-Moreno (pp. 111–152). México, D.F.: Noriega Editores and Editorial Limusa.

Benítez, Griselda, Alicia Chacalo, and Isabelle Barois (1987). "Evaluación Comparativa de la Pérdida de la Cubierta Vegetal y Cambios en el Uso del Suelo en el Sur de la Ciudad de México." In *Aportes a la Ecología Urbana de la Ciudad de México,* edited by Eduardo H. Rapoport and Ismael R. López-Moreno (pp. 193–228). México, D.F.: Noriega Editores and Editorial Limusa.

Bennett, Vivienne (1993). *The Politics of Water: Urban Protest, Gender and Power in Monterrey, Mexico.* Pittsburgh: Pittsburgh University Press.

Berdan, Frances, F. (1982). *The Aztecs of Central Mexico: An Imperial Society.* San Diego: Harcourt Brace Jovanovich College Publishers.

Bermeo, Ariadna (1995a). "Faltan Estudios para Iniciar Carretera—INE." *Reforma,* March 4, sec. B, "Ciudad y Metrópoli," p. 2.

Bermeo, Ariadna (1995b). "Preservarán Zona Forestal el DDF." *Reforma,* March 16, Sec. B, "Ciudad y Metrópoli," p. 4.

Bermeo, Ariadna (1995c). "Proponen un Plan Rector de Ecología: Estudiarán en Foro Aire, Suelo, Agua y Preservación del Ambiente del Valle de México." *Reforma.* February 16, sec. B, "Cindad y Metrópoli," p. 2.

Bermeo, Ariadna (1996). "Xochimilco, Rescate Constante: Premiado por la Universidad de Harvard." *Reforma.* April 28, sec. B, "Ciudad y Metrópoli," p. 10.

Best, Gustavo (1990). "Un Desarrollo Energético Alternativo y la Gestión del Medio Ambiente." In *Medio Ambiente y Desarrollo en México.* Vol. 2, edited by Enrique Leff (pp. 453–488). México, D.F.: Centro de Investigaciones Interdisciplinarias en Humanidades, Universidad Nacional Autónoma de México, and Grupo Editorial.

Bhalla, A. S. (ed.) (1992). *Environment, Employment and Development.* A Study of the ILO World Employment Programme. Geneva: International Labour Office.

Blaikie, P. and H. Brookfield (1987). "Defining and Debating the Problem." In *Land Degradation and Society,* edited by P. Blaikie and H. Brookfield (pp. 1–26). London: Methuen.

Blair Smith, Elliot (1993). "Power Plays." *San Deigo Union Tribune,* June 20, sec. 1, p. 1.

Bloom, David E. and Adi Brender (1993). "Labor and the Emerging World Economy." *Population Bulletin.* 48 (2).

Bordon, Alejandra and Arturo Páramo (1996). "Anuncian Programa de Prevención para el DF: Afinan Combate al Cólera," *Reforma*. April 18, sec. B, "Ciudad y Metrópoli," p. 1.

Borman, F. H. (1976) "An Inseparable Linkage: Conservation of Natural Ecosystems and the Conservation of Fossil Energy." *BioScience*. 26:754–760

Botkin, D. B. (1990). *Discordant Harmonies*. New York: Oxford University Press.

Boutros-Ghali, Boutros (1995). "Democracy: A Newly Recognized Imperative." *Global Governance: A Review of Multilateralism and International Organizations*. 1 (1): 3–1.

Boyden, Steven, Sheelagh Millar, Ken Newcombe, and Beverley O'Neill (1981). *The Ecology of a City and Its People: The Case of Hong Kong*. Canberra: Australian National University Press.

Brecher, Jeremy and Tim Costello (1994). *Global Village or Global Pillage: Economic Reconstruction from the Bottom Up*. Boston: South End.

Brennan, Ellen M. (1993). "Urban Land and Housing Issues Facing the Third World." In *Third World Cities: Problems, Policies and Prospects*, edited by John D. Kasarda and Allan M. Parnell (pp. 74–91). Newbury Park: Sage.

Brenton, Tony (1994) *The Greening of Machiavelli: The Evolution of International Politics*. London: Earthscan Publications.

Bromley, Ray (1993). "Small-Enterprise Promotion as an Urban Development Strategy." In *Third World Cities: Problems, Policies and Prospects*, edited by John D. Kasarda and Allan M. Parnell (pp. 120–133). Newbury Park: Sage.

Brookfield, Harold, Abdul Samad Hadi, and Zaharah Mahmud (1991). *The City in the Village: The In-situ Urbanization of Villages, Villagers and Their Land Around Kuala Lumpur, Malaysia*. New York: Oxford University Press.

Brown, L. B. and J. L. Jacobson (1987). *The Future of Urbanization: Facing the Ecological and Economic Constraints*. World Watch Paper no. 77. Washington, D.C.: World Watch Institute.

Brown, Lester R, Holly Brough, Alan Durning, Christopher Flavin, Hilary French, Jodi Jacobson, Nicolas Lenssen, Marcia Lowe, Sandra Postel, Michael Renner, John Ryan, Linda Starke and John Young (1992). *State of the World*. New York: W. W. Norton.

Browne, Enrique (1992). "War on Waste and Other Urban Ideals for Latin America." In *Rethinking the Latin American City*, edited by Richard M. Morse and Jorge E. Hordoy (pp. 110–130). Washington, D.C.: Woodrow Wilson Center Press.

Brundtland, Gro Harlem (1990). "Global Change and Our Common Future." In *One Earth, One Future: Our Changing Global Environment*, edited by Cheryl Simon Silver and Ruth S. DeFries for the National Academy of Sciences. Washington, D.C.: National Academy Press.

Bryant, Raymond B. (1992). "Political Ecology: An emerging Research Agenda in Third-World Studies." *Political Geography*. 11, no. 1 (January): 12–36.

Burgess, Rod (1982). "Self-Help Housing Advocacy: A Curious Form of Radicalism: A Critique of the Work of John F.C. Turner." In *Self-Help Housing: A Critique,* edited by Peter Ward (pp. 56–99). New York: Mansell.

Burgess, Rod (1985). "Problems in the Classification of Low-Income Neighborhoods in Latin America." *Third World Planning Review.* 7 (4).

Burnstein, Janis D. (1994). "Land Use Considerations in Urban Environmental Management." Urban Management Programme Paper no. 12. Washington, D.C.: World Bank.

"Busca DDF Modificar Ley de Desarrollo Urbano: Desaparecerían los ZEDEC" (1995). *Reforma,* February 23, special edition.

Bush, V. P. (1987). "Proyecciones de la Población de la Zona Metropolitana de la Ciudad de México." In *Atlas de la Ciudad de México,* edited by Gustavo Garza (pp. 410–414). México, D.F.: DDF and Colegio de México.

Business Monitor International (1992). *Mexico 1992: Annual Report on Government, Economy, Industry and Business, with Forecasts to 1993.* London: Business Monitor International.

Bystydzienski, Jill M. (ed.) (1992). *Women Transforming Politics: Worldwide Strategies for Empowerment.* Bloomington: Indiana University Press.

Cadman, David and Geoffrey Payne (eds.) (1990). *The Living City: Towards a Sustainable Future.* New York: Routledge.

Calderón, Julio A. (1986). "Luchas por la Tierra en la Ciudad de México (1980–1984): El Caso de las Zonas Ejidales y Comunales." Master's thesis in sociology. Facultad Latinoamericana de Ciencias Sociales, México, D.F.

Callinicos, A. (1987). *Making History: Agency, Structure and Change in Social Theory.* Cambridge, Mass: Basil Blackwell.

Calthorpe, Peter (1993). *The Next American Metropolis: Ecology Community, and the American Dream.* New York: Princeton Architectural Press.

Camposortega Cruz, Sergio (1992). "Evolución y Tendencias Demográficas de la ZMCM." In *La Zona Metropolitana de La Ciudad de México: Problemática Actual y Perspectivas Demográficas y Urbanas,* edited by CONAPO (pp. 3–16). México, D.F.: Dirección General de Estudios de Población.

Cano, David (1992a). "Presentarán Hoy en la ARDF el Anteproyecto del Nuevo Reglamento de Desarrollo Urbano y Uso del Suelo." *UnoMásUno,* July 15, "Valle de México, sec.," p. 11.

Cano, David (1992b). "Toda Decisión sobre el Reglamento del Uso del Suelo Deberá Ser Aprobada por la ARDF." *UnoMásUno,* July 17, "Valle de México" sec., p. 11.

Carabias Lillo, Julia (1995) "Orientaciones Generales de Política de Medio Ambiente, Recursos Natruales y Pesca." *SEMARNAP* (México, D.F.), February 15.

Carabias Lillo, Julia and Herrera, A. (1986). "La Ciudad y Su Ambiente." *Cuadernos Políticos.* 45:56–69.

Carley, Michael and Ian Christie (1993). *Managing Sustainable Development.* Minneapolis: University of Minnesota Press.

Carlos, Manuel L. (1981). *State Policies, State Penetration and Ecology: A Comparative Analysis of Uneven Development and Underdevelopment in Mexico's Micro Agrarian Regions.* Working Paper in U.S.-Mexican Studies, no. 19. San Diego: Center for U.S.-Mexican Studies, University of California.

Carmona Lara, María del Carmen (1991). *Derecho Ecológico.* México, D.F.: Universidad Nacional Autónoma de México and Instituto de Investigaciones Jurídicas.

Carr, Barry and Ricardo Anzaldúa Montoya (eds.) (1986). *The Mexican Left, the Popular Movements, and the Politics of Austerity.* Monograph Series no. 18. San Diego: Center for U.S.-Mexican Studies.

Carrasco, Jorge, David Vicenteño, and Alfredo Joyner (1995). "Desalojan a 350: Deshacen Campamento en Operación Hormiga." *Reforma,* March 8, sec. B, "Ciudad y Metrópoli."

Castells, Manuel (1977). *The Urban Question: A Marxist Approach.* Cambridge: MIT Press.

Castells, Manuel (1983). *The City and the Grassroots.* Los Angeles: University of California Press.

Castells, Manuel (1989). *The Informational City: Information Technology, Economic Restructuring and the Urban Regional Process.* Cambridge, Mass.: Basil Blackwell.

Castells, Manuel (1993). "The Informational Economy and the New International Division of Labor." In *The New Global Economy in the Informational Age: Reflections on Our Changing World,,* edited by Martin Carnoy, Manuel Castells, Stephen S. Cohen, and Fernando Henrique Cardoso (pp. 15–43). University Park, Pa.: Pennsylvania State University Press.

Castillejos, Margarita (1988). "Efectos de la Contaminación Ambiental en la Salud de Niños Escolares en Tres Zonas del Area Metropolitana de la Ciudad de México." In *Medio Ambiente y Calidad de Vida,* edited by Sergio Puente and Jorge Legorreta (pp. 301–333). México, D.F.: DDF and Plaza & Janés.

Castillejos, Margarita (1991). "La Contaminación Ambiental en México y Sus Efectos en la Salud Humana." In *Servicios Urbanos, Gestión Local y Medio Ambiente,* edited by Martha Schteingart and Luciano d' Andrea (pp. 187–204). México, D.F.: Colegio de México and CE.R.FE.

Castillo Berthier, Héctor (1991). "Desechos, Residuos, Desperdicios: Sociedad y Suciedad." In *Servicios Urbanos, Gestión Local y Medio Ambiente,* edited by Martha Schteingart and Luciano d' Andrea (pp.137–147). México, D.F.: Colegio de México and CE.R.FE.

Castillo Berthier, Héctor, Margarita Camarena, and Alicia Ziccardi (1987). "Basura: Procesos de Trabajo e Impactos en el Medio Ambiente Urbano." *Estudios Demográficos y Urbanos.* 2 (3): 513–543.

Castro, J. F. (1991). "El Cholera: Una Plaga Rediviva." *Gaceta Médica de México* 127 (5): 395–398.

Centro de Estudios Políticos (ed.) (1986). "El Movimiento Urbano Popular." *Estudios Políticos.* 4/5 (1–4): 128.

CEPAL-NU (Comisión Económica para América Latina y el Caribe, Naciones Unidas) (1991). *El Desarrollo Sustentable: Transformación Productiva, Equidad y Medio Ambiente.* New York: United Nations.

CEPAL-NU (1993). *Procedimientos de Gestión para un Desarrollo Sustentable: Aplicables a Municipios, Microregiones y Cuencas.* Santiago, Chile: CEPAL-NU.

Chambers, Robert (1995). "Poverty and Livelihoods: Whose Reality Counts?" *Environment and Urbanization.* 7 (1): 173–204.

Cheru, Fantu (1995). "Blessing or Curse? The Effects of SAPs on the Urban Poor in Africa." *Countdown to Istanbul.* 1:5–7.

CHF (Cooperative Housing Foundation) (1990). "Environmentally Sound Shelter in Informal Urban Settlements." *Ekistics.* 57 (342): 122–126.

CHF (1992). *Partnership for a Livable Environment.* Washington, D.C.: CHF.

Childers, Erskine and Brian Urquhart (1994). *Renewing the United Nations System.* Uppsala, Sweden: Dag Hammarsk Jöld Foundation.

Chiras, Daniel D. (1991). *Environmental Science: Action for a Sustainable Future,* 3rd edition. Redwood City, California: The Benjamin Cummings Publishing Co., Inc.

Choucri, Nazli (ed.) (1993). *Global Accord: Environmental Challenges and International Responses.* Cambridge, Mass.: MIT Press.

CIDAC (Centro de Investigación para el Desarrollo, A.C.) (1991). *Vivienda y Estabilidad Política: Alternativas para el Futuro.* Mexico City: Editorial Diana.

Cleeland, Nancy (1991). "Mexico City Gives Common Folk a Breather with Oxygen Kiosks." *San Diego Tribune,* March 9, sec. A, p. 3.

Cockcroft, James D. (1983). *Mexico: Class Formation, Capital Accumulation, and the State.* New York: Monthly Review Press.

Cockburn, J. A. C. (1986). *Luchas por la Tierra en la Ciudad de México: 1980–1984.* Masters' thesis. Facultad Latinoamericana de Ciencias Sociales (FLACSO), México, D.F.

COCODER (1979). *Plan Ajusco Síntesis General.* México D.F.: DDF

Cohen, Michael A. and Josef L. Leitmann (1994). "Will the World Bank's Real 'New Urban Policy' Please Stand Up?" Comment on "The World Bank's 'New' Urban Management Programme: Paradigm Shift or Policy Continuity?" by Gareth A. Jones and Peter M. Ward. *Habitat International.* 18 (4): 117–126.

Cohen, Shaul E. (1994). "Greenbelts in London and Jerusalem." *Geographical Review.* 84 (1): 74–89.

Colby, Michael E. (1989). *The Evolution of Paradigms of Environmental Management in Development.* Monograph prepared for the World Bank, Strategic Planning and Review Department, WPS 313. Washington, D.C.: PPR Dissemination Center.

Cole, Luke W. (1992). "Empowerment as the Key to Environmental Protection: The Need for Environmental Poverty Law." *Ecology Law Quarterly.* 19 (4): 619–684.

Colosio, Luis Donaldo (1992). "Environment and Development in Mexico." *Mexico on the Record.* 1(7): 5–6.

Comisión Coordinadora para el Desarrollo Agropecuario del Distrito Federal (1979). *Plan Ajusco: Síntesis General.* México, D.F.: México.

Comisión Coordinadora para el Desarrollo Agropecuario del Distrito Federal. (1981) *Ajusco, Mirador de México.* México, D.F.: México.

Comisión Coordinadora para el Desarrollo Agropecuario del Distrito Federal COCODA (1982a). *Documentos sobre la Probemática Agraria del Distrito Federal.* México, D.F.: México.

Comisión Coordinadora para el Desarrollo Agropecuario del Distrito Federal (1982b). *Memoria, 1978–1982.* México, D.F.: México.

Comisión Metropolitana para la Prevención y Control de la Contaminación Ambiental en el Valle de México (1992). *¿Qué Estamos Haciendo para Combatir la Contaminación del Aire en el Valle de México?* México, D.F.: Pinacoteca 2000.

Comisión Metropolitana para la Prevención y Control de la Contaminación Ambiental en el Valle de México (1993). *Ciudad de México: Respuestas a un Reto Murdial.* Mexico, D.F.: Pinacoteca 2000.

Comisión para la Prevención y Control de la Contaminación Ambiental en la Zona Metropolitana del Valle de México (1993). *Ciudad de México: Respuestas a un Reto Mundial.* Mexico City, D.F.: DDF.

"Comunicado Rebelde: Las Acciones Militares a la Violencia Institucionalizada: EPR" (1996). *Excelsior,* August 30.

CONAPO (Consejo Nacional de Población) (ed.). (1992). *La Zona Metropolitana de la Ciudad de México: Problemática Actual y Perspectivas Demográficas y Urbanas.* México, D.F.: Dirección General de Estudios de Población.

Connolly, Pricilla (1982). "Uncontrolled Settlements and Self-Build: What Kind of Solution? The Mexico City Case." *Self-help Housing: a Critique,* edited by P. Ward (pp. 141–174). London: Mansell.

Connolly, Pricilla (1989). "Programa Nacional de Desarrollo Urbano y Vivienda, 1984: ¿Desconcentración Planificada o Descentralización de Carencias?" In *Una Década de Planeación Urbano-Regional en México, 1978–1988,* edited by Gustavo Garza (pp. 103–120). México, D.F.: Colegio de México.

Connolly, Pricilla (1990). "Housing and the State in Mexico." *Housing Policy in Developing Countries,* edited by G. Shidlo (pp. 5–32). London: Routledge.

Connolly, Pricilla (1993). " 'The Go-Between': CENVI, A Habitat NGO in Mexico City." *Environment and Urbanization.* 5 (1): 68–90.

Contreras, Antonio (1993). "Compromiso Renovado con los que Menos Tienen." *Gaceta de Solidaridad.* 3 (67): 22.

Contreras Domínguez, Wilfrido (1988). "El Agua ¿Recurso Renovable?" In *Medio Ambiente y Calidad de Vida,* edited by Sergio Puente and Jorge Legorreta (pp. 251–262). México, D.F.: DDF and Plaza & Janés.

Contreras Salcedo, Jamie (1996). "Arranca Hoy la Avi; Vivienda Digna y Seguridad en la Tenencia a las Familias." *Excelsior.* May 16.

Cook, Maria Lorena, Kevin J. Middlebrook, Juan Molinar Horcasitas (eds.) (1994). *The Politics of Economic Restructuring: State-Society Relations and Regime Change in Mexico.* San Diego: Center for U.S.-Mexican Studies, University of California.

COPEVI (Centro Operacional de Poblamiento y Vivienda). (1977). *Investigación sobre la Vivienda.* México: COPEVI.

COPLADE (1987). Collection of papers on "Regularización de la Tenencia de la Tierra Dentro de la Zona del Ajusco." México, D.F.: México.

COPLAMAR (Coordinación General del Plan Nacional de Zonas Deprimidas y Grupos Marginados) (1982). *Vivienda: Necesidades Esenciales en México (Situación Actual y Perspectivas al Año 2000).* México, D.F.: Siglo Veintiuno editores.

Cordera Campos, Rolando and Enrique González Tiburcio (1991). "Crisis in Transition in the Mexican Economy." In *Social Responses to Mexico's Economic Crisis of the 1980s,* edited by Mercedes González de la Rocha and Augstín Escobar Latapí (pp. 19–56). U.S.-Mexico Contemporary Perspectives Series, 1., San Diego: Center for U.S.-Mexican Studies.

Cornelius, Wayne A. (1975). *Politics and the Migrant Poor in Mexico City.* Stanford: Stanford University Press.

Cornelius, Wayne A. (1976). "The Impact of Cityward Migration on Urban Land and Housing Markets." In *The City in Comparative Perspective,* edited by John Walton and L. H. Masotti (pp. 240–266). New York: Halsted.

Cornelius, Wayne A. (1992). "The Politics and Economics of Reforming the Ejido Sector in Mexico: An Overview and Research Agenda." *Lasa Forum.* 23 (3): 3–8.

Cornelius, Wayne A. (1994). Editorial *Foreign Policy Magazine.* 34: (January): 9.

Cornelius, W. and A. Craig, (1988). *Politics in Mexico: An Introduction and Overview.* Reprint series 1. 2d. ed. San Diego: center for US-Mexican Studies, University of California.

Cornelius, W., J. Gentleman, and P. H. Smith, (eds.) (1989). *Mexico's Alternative Political Futures.* Monograph Series no. 30. San Diego: Center for U.S.-Mexican Studies.

Coulomb, René (1991). "La Participación de los Servicios Urbanos: ¿Estrategias de Sobrevivencia o Prácticas Autogestionarias?" In *Servicios Urbanos, Gestión Local y Medio Ambiente,* edited by Martha Schteingart and Luciano d' Andrea (pp. 265–280). México, D.F.: Colegio de México and CE.R.FE.

Coulomb, René (1992). "El Acceso a la Vivienda en la Ciudad de México." In *La Zona Metropolitana de la Ciudad de México: Problemática Actual y Perspectivas*

Demográficas y Urbanas, edited by Consejo Nacional de Población (CONAPO) (pp. 157–178). México, D.F.: Dirección General de Estudios de Población.

Coulomb, René and Emilio Duhau (eds.) (1989). *Políticas Urbanas y Urbanización de la Política.* México, D.F.: División de Ciencias Sociales y Humanidades, and Universidad Autónoma Metropolitana, and Eón editores.

Craig, Ann L. (1990). "Institutional Context and Popular Strategies." In *Popular Movements and Political Change in Mexico,* edited by Joe Foweraker and Ann L. Craig (pp. 271–284). Boulder, Colo.: Lynne Rienner.

Cremoux, Raúl (ed.) (1991). *Mexico City in Figures: Pórtico de la Ciudad de México.* México, D.F.: Grupo Gráfico Romo.

Crónica Legislativa (1993a). "Nueva Formas de Gobierno para la Capital de la República." *Crónica Legislativa.* 2, no. 11 (October–November): 3–11.

Crónica Legislativa (1993b). "Nueva Ley de Asentamientos Humanos." *Crónica Legislativa.* 2, no. 10 (August–September): 47–50.

Cronon, William (1991). *Nature's Metropolis: Chicago and the Great West.* New York: W. W. Norton.

Crosby Jr., Alfred W. (1972). *The Columbian Exchange: Biological and Cultural Consequences of 1492.* Westport, Conn.: Greenwood.

Crosby, Jr., Alfred W. (1986). *Ecological Imperialism: The Biological Expansion of Europe.* Westport, Conn.: Greenwood.

Cuevas, Karina (1995). "Harán Plan Conurbado de Vivienda." *Reforma.* February 22, sec. B, "Ciudad y Metrópoli," p. 4.

Cullen, M. and S. Wollery (eds.) (1982). *World Congress on Land Policy: 1980.* Lexington, Mass.: Lexington.

Cymet, David (1992). *From Ejido to Metropolis, Another Path: An Evaluation on Ejido Property Rights and Informal Land Development in Mexico City.* American University Studies Series no. 21 in Regional Studies. Vol. 6. San Francisco: Peter Lang.

Daly, Herman E. (1991). *Steady-state Economics,* 2nd ed. Washington, D.C.: Island Press.

Daly, Herman E. and John B. Cobb Jr. (1994). *For the Common Good: Redirecting the Economy toward Community, the Environment, and a Sustainable Future.* 2nd edition, updated and expanded, Boston: Beacon.

Darling, Juanita (1991). "Cynics See Political Motivation for Mexico's Move on Pollution." *Los Angeles Times* (San Diego), April 15, sec. A. p. 6.

Dasmann, Raymond F. (1985). "Achieving the Sustainable Use of Species and Ecosystems." *Landscape Planning.* 12:211–219.

Davalos, Renato (1996). "Condiciones Inadecuadas en Una de Cada 4 Viviendas, Admite un Plan de Gobierno." *Excelsior.* May 17.

Davis, Diane E. (1994). *Urban Leviathan: Mexico City in the Twentieth Century.* Philadelphia: Temple University Press.

DDF (1982) *Plan Parcial de Desarrollo Urbano, Delegación Tlalpan.* México, D.F.: DDF.

DDF (1983). *Programa de Desarrollo de la Zona Metropolitana.* México, D.F.: DDF.

DDF (1985). *Programa de Reordinación Urbana y Protección Ecológica* (PRUPE). México, D.F.: DDF.

DDF (1986). *Programa de Integración Social del Territorio: Belvederes, Tlalpan.* México, D.F.: DDF.

DDF (1987). *Programa General de Desarrollo Urbano del Distrito Federal (1987–1988).* México, D.F.: DDF.

DDF (1991). AGUA 2000: Estrategia para la Cuidad de México, D.F.: DDF.

DDF (1989a). "Decreto por el que se Establece como Zona Prioritaria de Preservación del Equilibrio Ecológico y se Declara Zona Sujeta a Conservación Ecológica, como Area Natural Protegida, la Superficie de 727-61-42 Hectáreas." *Diario Official,* June 28.

DDF (1989b). "Programa de Manejo de la Zona Sujeta a Conservación Ecológica." *Gaceta Oficial del Departamento del Distrito Federal.* 2, no. 26 (December 25): 5–10.

de la Peña, Guillermo, Juan Manuel Durán, Agustín Escobar, and Javier García de Alba (eds.) (1990). *Crisis, Conflicto y Sobrevivencia: La Sociedad Urbana en México.* Guadalajara: Universidad de Guadalajara and CIESAS.

de Rojas, José Luis (1986). *México Tenochtitlán: Economía y Sociedad en el Siglo XVI.* México, D.F.: Fondo de Cultura Económica.

Delegación del Tlalpan (1992). *Parque Ecológico de la Ciudad de Mexico.* México, D.F.: DDF.

Delegación de Tlalpan (1993). "Boletín Informativo," January.

Delgado, Javier (1991). "Valle de México: El Crecimiento por Conurbaciones." *Revista Interamericana de Planificación.* 24 (94): 226–253.

Denevan, W. M. (1976) *The Native Population of the Americas in 1492.* Madison, Wisc.: The University of Wisconsin Press.

Devas, Nick and Carole Rakodi (1993). *Managing Fast Growing Cities: New Approaches to Urban Planning and Management in the Developing World.* New York: Longman Scientific and Technical and John Wiley.

Dewalt, Billie R., Martha W. Rees, and Arthur D. Murphy (1994). *The End of the Agrarian Reform in Mexico.* Transformation of Rural Mexico Series no. 3. La Jolla, Calif.: Center for U.S.-Mexican Studies, University of California San Diego.

Di Pace, Maria, Sergio Federovisky, Jorge E. Hardoy, Jorge H. Morello, and Alfredo Stein (1992). "Latin America." In *Sustainable Cities: Urbanization and the Environment in International Perspective,* edited by Richard Stern, Rodney White, and Joseph Whitney (pp. 205–227). Boulder, Colo.: Westview.

Díaz Barriaga, Miguel (1991). "Urban Politics in the Valley of Mexico: A Case Study of Urban Movements in the Ajusco Region of Mexico City, 1970–1987." Ph.D. dissertation, Stanford University.

Díaz Barriaga, Miguel (1993). "Ejido Urbanization in Mexico City: A Case study of the Ajsuco Region." Manuscript, Swathmore College, Dept. of Anthropology.

Díaz-Betancourt, Martha, Ismael López-Moreno, and Eduardo H. Rapoport (1987). "Vegetación y Ambiente Urbano en la Ciudad de México." In *Aportes a la Ecología Urbana de la Ciudad de México,* edited by Eduardo H. Rapoport and Ismael R. López-Moreno (pp. 13–71). México, D.F.: Noriega Editores and Editorial Limusa.

Dirección General de Comunicación Social (1991a). *Mexican Agenda:* México, D.F.: Dirección de Publicaciones.

Dirección General de Comunicación Social (1991b). *Tercer Informe.* México, D.F.: Dirección de Publicaciones.

Dirección General de Comunicación Social (1992). *Cuarto Informe.* México, D.F.: Dirección de Publicaciones.

Dirección General de Reordenación Urbana y Protección Ecológica (1987). *Programa General de Desarrollo Urbano del Distrito Federal: 1987–1988.* México, D.F.: DDF.

Doebele, William A. (1985). "Land Policy and Shelter" Paper prepared for the U.N. International Year of Shelter of the Homeless conference.

Doebele, William A. (1987). "The Evolution of Concepts of Urban Land Tenure in Developing Countries." *Habitat International* 11 (1): 7–22.

Dogan, M. and J. D. Kasarda (eds.) (1988). *A World of Giant Cities.* Newbury Park, Calif.: Sage.

"Dos Mil Cuatro Cientos Familias Viven en la Reserva Ecológica en G.A. Madero" (1992). *UnoMásUno,* August 6.

Douglass, Mike (1987). "The Future of Cities on the Pacific Rim." *Comparative Urban and Community Research.* 2:67.

Douglass, Mike (1989a). "The Environmental Sustainability of Development: Coordination, Incentive and Political Will in Land Use Planning for the Jakarta Metropolis." *Third World Planning Review.* 11 (2): 211–237.

Douglass, Mike (1989b). "The Future of Cities on the Pacific Rim." *Pacific Rim Cities in the World Economy,* edited by M. P. Smith (pp. 9–67). Comparative Urban and Community Research vol. 2. New Brunswick, New Jersey: Transaction.

Douglass, Mike (1992). "The Political Economy of Urban Poverty and Environmental Management in Asia: Access, Empowerment and Community Based Alternatives." *Environment and Urbanization.* 4 (2): 9–32.

Douglass, Mike and Yok-shiv F. Lee (1996). "City and Community: Toward Environmental Sustainability." Chap. 6 in *World Resources 1996–97,* edited by World Resource Institute (pp. 125–148). Washington, D.C.: World Resource Institute.

Douglass, Mike and Malia Zoghlin (1994). "Sustaining Cities at the Grassroots: Livelihood, Environment and Social Networks in Suan Phlu, Bangkok." *Third World Planning Review.* 16 (2): 171–200.

Drakakis-Smith, David (1995). "Third World Cities: Sustainable Urban Development." *Urban Studies,* 32: nos. 4–5 (May): 659–677.

Dreyfus, Hubert L. and Paul Rabinow (1983). *Michel Foucault: Beyond Structuralism and Hermeneutics.* Chicago: University of Chicago Press.

Duhau, Emilio (1985). "Reordenamiento Urbano y Desconcentración Territorial: Comenbtarios al Programa de Desarrollo de la Zona Metropolitana y la Región Central" *Revista A.* 6 (15): 97–106.

Duhau, Emilio (1991a). "Gestión de los Servicios Urbanos en México: Alternativas y Tendencias." In *Servicios Urbanos, Gestión Local y Medio Ambiente,* edited by Martha Schteingart and Luciano d' Andrea (pp. 83–107). México, D.F.: Colegio de México and CE.R.FE.

Duhau, Emilio (1991b). "Tierras Ejidales y Políticas de Suelo en la Ciudad de México." *Medio Ambiente y Urbanización.* 9 (34): 43–58.

Dunkerley, H. with the assistance of C. Whitehead (ed.) (1983). *Urban Land Policy: Issues and Opportunities.* A World Bank publication. New York: Oxford University Press.

Duran, Gonzales Pablo (1991). "Devastan la Zona Ecológica en San Lorenzo Acopilco, en Cuajimalpa." *Excelsior.* May 30.

Durand, J. (1983). *La Ciudad Invade al Ejido.* México, D.F.: Ediciones de la Casa Chata.

Earth Negotiations Bulletin (1996). 11, no. 37 (June): 1–2.

Eckersley, Robyn (1992). *Environmentalism and Political Theory: Toward an Ecocentric Approach.* Albany: State University of New York Press.

Eckstein, Susan (1990). "Poor People versus the State and Capital: Autonomy of a Successful Community Mobilization for Housing in Mexico City." *International Journal of Urban and Regional Research.* 14:274–296.

"Economic Nightmare May Create Political Instability" (1994). *San Diego Union-Tribune,* December 28, sec. A, p. 1.

Edel, Matthew (1992). "Latin American Urban Studies: Beyond Dichotomy." In *Rethinking the Latin American City,* edited by Richard M. Morse and Jorge E. Hordoy (pp. 66–82). Washington, D.C.: Woodrow Wilson Center Press.

Ehrlich, Paul and Anne H. Ehrlich (1991). *Healing the Planet: Strategies for Resolving the Environmental Crisis.* Menlo Park, Calif.: Addison-Wesley.

Ehrlich, Paul R. and Jonathan Roughgarden (1987). *The Science of Ecology.* New York: MacMillan.

Eisenstadt, S. N. (ed.) (1968). *Max Weber: On Charisma and Institution Building.* Chicago: University of Chicago Press.

"El Desalojo Es la Unica Vía: Preocupa a Cocoder Invasiones en Zonas Ecológicas" (1994). *Reforma* (México, D.F.), September 11, sec. B, "Ciudad y Metrópoli, p. 7.

"El D.F. Ya Llegó a Sus Límites: Ríos Zertuche." (1992). *UnoMásUno,* June 23, p. 26.

"En Extrema Pobreza; Más de la Mitad de los Habitantes del Valle de México, sin Calidad de Vida." (1994). *Reforma*, September 10, "Ciudad y Metrópoli" sec., p. 7B.

Engle, Robert and Joan G. Engle (eds.) (1990). *Ethics of Environment and Development: Global Challenge, International Response.* Tucson: University of Arizona Press

Environment and Urbanization (1992). Special issue, "Sustainable Cities: Meeting Needs, Reducing Resource Use and Recycling, Re-Use and Reclamation." 4, no. 2 (October): 3–8.

Environment and Urbanization (1993). Book notes. 5, no. 2 (October): 186.

Environment and Urbanization (1995). 7, no. 1 (April): 57–76.

Environment Watch: Latin America (1994–1995). News and analysis for business and policy professionals from Cutter Information Corp. January–February.

Environmental Committee of the IDB (Inter-American Development Bank) (1993). *1992 Annual Report on the Environment and Natural Resources.* Washington, D.C.: IDB.

"Environmental Protection in Mexico City: The IDB, the Government and the Residents of Sierra Santa Catarina" (1994). *The Other Side of Mexico* (Mexico City), no. 34 (January–February): 9–10.

EPA (U.S. Environmental Protection Agency) (1992). "Mexico's Environmental Laws and Enforcement." *Mexico Trade and Law Reporter*, March, 9–12.

"Es Urgente Erradicar el Fecalismo en la Zona" (1987). *Crónica: Voz del Ajusco.* A news publication of the Tlalpan District, México, D.F. 1 (5): 4.

Escobar Latapí, Agustín and Mercedes González de la Rocha (1995). "Crisis, Restructuring and Urban Poverty in Mexico." *Environment and Urbanization.* 7 (1): 57–76.

Esquinca, Vianey and Ernesto Núñez (1995). "Advierten Derroche de Agua." *Reforma* (México, D.F.), March 8, sec. B.

Ezcurra, Exequiel and Marisa Mazari-Hiriart (1996). "Are Megacities Viable? A Cautionary Tale from Mexico City." *Environment.* 38 (1).

Ezcurra, Exequiel and Carlos Montaña (1990). "Los Recursos Naturales Renovables en el Norte Arido de México." In Vol. 1 of *Medio Ambiente y Desarrollo en México,* edited by Enrique Leff (pp. 297–327). México, D.F.: Centro de Investigaciones Interdisciplinarias en Humanidades, Universidad Nacional Autónoma de México, and Grupo Editorial.

Faber, Daniel (1992). "The Ecological Crisis of Latin America: A Theoretical Introduction." *Latin American Perspectives.* 19 (1): 3–16.

Falkenmark, M. and R. A. Suprato (1992). "Population-landscape Interactions in Development: A Water Perspective to Environmental Sustainability." *Ambio.* 21(10): 31–36.

"Fecalismo al Aire Libre, Problema de Salud Pública en México: SSA." (1991). *La Jornada,* April 22, "La Capital sec.," p. 15.

Figuereo Osnaya, Luis (1992). "Venta de Residencias en Zonas Ecológicas de Cuajimalpa." *UnoMásUno,* July 18.

Findley, Sally E. (1993). "The Third World City: Development Policy and Issues." In *Third World Cities: Problems, Policies and Prospects,* edited by John D. Kasarda and Allan M. Parnell (pp. 1–31). Newbury Park: Sage.

Finger, Matthias (1993). "Politics of the UNCED Process." In *Global Ecology: A New Arena of Political Conflict,* edited by Wolfgang Sachs (pp. 36–48). New Jersey: Zed.

Firman, Tommy and Ida Ayu Indira Dharmapatni (1994). "The Challenge to Sustainable Development in Jakarta Metropolitan Region." *Habitat International.* 18 (3): 79–94.

Fisher, Jonathan (1992). "A Conversation with Mexico's President." *International Wildlife.* 22 (5): 48–51.

Foucault, M. (1983). "The Subject and Power." In *Michel Foucault: Beyond Structuralism and Hermeneutics,* second ed. edited by Hubert L. Dreyfus and Paul Rabinow. England: Harvester.

Foweraker, Joe (1982). "Accumulation and Authoritarianism on the Pioneer Frontier of Brazil." *The Journal of Peasant Studies.* 10 (1): 95–117.

Foweraker, Joe (1989). "Popular Movements and the Transformation of the System." In *Mexico's Alternative Political Futures,* edited by Wayne A. Cornelius, Judith Gentleman, and Peter H. Smith (pp. 109–130). Monograph Series no. 30. San Diego: Center for U.S.-Mexican Studies.

Foweraker, Joe (1990a). "Popular Movements and Political Change in Mexico." In *Popular Movements and Political Change in Mexico,* edited by Joe Foweraker and Ann L. Craig (pp. 3–20). Boulder, Colo.: Lynne Rienner.

Foweraker, Joe (1990b). "Popular Organization and Institutional Change." In *Popular Movements and Political Change in Mexico,* edited by Joe Foweraker and Ann L. Craig (pp. 43–58). Boulder, Colo.: Lynne Rienner.

Foweraker, Joe and Ann A. Craig (eds.) (1988a). *Popular Movements and Political Change in Mexico.* Boulder, Colo.: Lynne Rienner.

Foweraker, Joe and Ann A. Craig (eds.) (1988b). *Popular Movements and the Transformation of the Mexican Political System: Proposal for a Research Workshop.* La Jolla, Calif: Center for U.S.-Mexican Studies, University of California, San Diego.

Fox, Jonathan and Gustavo Gordillo (1989). "Between State and Market: The Campesinos' Quest for Autonomy." In *Mexico's Alternative Political Futures,* edited by Wayne A. Cornelius, Judith Gentleman, and Peter H. Smith (pp. 131–172). Monograph Series no. 30. San Diego: Center for U.S.-Mexican Studies.

Fox, Jonathan and Luis Hernández (1992). "Mexico's Difficult Democracy: Grassroots Movements, NGOs, and Local Government." *Alternatives.* 17 (2): 165–208.

Frías, Carlos (1989). "Experiencias de Gestión del Habitat por Organizaciones Populares Urbanas." *Cuadernos CIDAP* (Lima, Peru). 4: 1–39.

Friedmann, John (1986). "The World City Hypothesis." *Development and Change.* 17:69–83.

Friedmann, John (1987). *Planning in the Public Domain: From Knowledge to Action.* Princeton, N.J.: Princeton University Press.

Friedmann, John (1988). *Life Space and Economic Space: Essays in Third World Planning.* New Brunswick, N.J.: Transaction.

Friedmann, John (1989). "The Latin American Barrio Movement as a Social Movement: Contribution to a Debate." *International Journal of Urban and Regional Research* 13 (3): 501–510.

Friedmann, John (1992). *Empowerment: The Politics of Alternative Development.* Cambridge, Mass.: Blackwell.

Friedmann, John and Haripriya Rangan (eds.) (1993) *In Defense of Livelihood: Comparative Studies on Environmental Action.* West Hartford, Conn.: UNRISD and Kumarian.

Friedmann, John and Mauricio Salguero (1988). "The Political Economy of Survival and Collective Self-Empowerment in Latin America: A Framework and Agenda for Research." In *Comparative Urban and Community Studies.* 4 (2): 62–87.

Friedmann, John and Clyde Weaver (1979). *Territory and Function: The Evolution of Regional Planning.* London: Edward Arnold.

Friedmann, Marci, Jeffery Nield, and Timothy Weed (1994). "Community Involvement in the Protection of a Threatened Latin American Ecosystem: The Antisanna Ecological Reserve." *Journal of Environment and Development.* 3 (1): 153–161.

Friedrich, Otto (1984). "Mexico City: The Population Curse." *Time Magazine,* August 6, pp. 24–35.

Fuglesang, Andreas, and Dale Chandler, and Daya Akuretiyagama (1993). *Participation and Empowerment through Organizational Development: What Can We Learn from the Grameen Bank?* West Hartford, Conn: Kumarian.

Furedy, Christine (1992). "Garbage: Exploring Non-Conventional Options in Asian Cities." *Environment and Urbanization.* 4 (2): 42–60.

Gamboa de Buen, Jorge (1992). Speech delivered to Third Commission of Land Use of the Second Assembly of Representatives of the Federal District, January 14, México, D.F.

Gamboa de Buen, Jorge and José A. Revah Locouture (1991). "Servicios Urbanos y Medio Ambiente: El Caso de la Ciudad de México." In *Servicios Urbanos, Gestión Local y Medio Ambiente,* edited by Martha Schteingart and Luciano d' Andrea (pp. 375–388). México, D.F.: Colegio de México and CE.R.FE.

Garavaglia, Juan Carlos (1992). "Human Beings and the Environment in America: On 'Determinism' and 'Possibilism.' " *International Social Science Journal.* 134 (November): 569–577.

García Hurtado, Alvaro García D'Acuña, and Eduardo García D'Acuña (1981). "Las Variables Ambientales en la Planificación de Desarrollo." In *Estilos de Desarrollo y Medio Ambiente en la América Latina,* edited by Osvaldo Sunkel and Nicolo Gligo (pp. 433–470). México, D.F.: Fondo de Cultura Económica.

García Lascuraín, María (1988). "Calidad de Vida en la Periferia de la Zona Metropolitana de la Ciudad de México." In *Medio Ambiente y Calidad de Vida,* edited by Sergio Puente and Jorge Legorreta. (pp. 109–136). México, D.F.: DDF and Plaza & Janés.

García Pérez, Agustín (1985). *La Delegación: Política Urbana y Participación Popular: El Caso de la Delegación de Tlalpan (1970–1982).* Bachelor's thesis in Sociology. Universidad Nacional Autónoma de México, México D.F.

Garinca, Raúl López (1987). "Ajusco Medio." *Cronica, Voz del Ajusco.* A publication of the Tlalpan Ward, Mexico City. August.

Garza, Gustavo (1986). "Planeación Urbana en México en Período de Crisis (1983–1984)." *Estudios Demográficos y Urbanos.* 1 (1): 73–96.

Garza, Gustavo (1989a). "Introducción: Imagen de la Planeación Territorial en México." In *Una Década de Planeación Urbano-Regional en México, 1978–1988,* edited by Gustavo Garza (9–24). México, D.F.: Colegio de México.

Garza, Gustavo (1989b). "La Política de Parques y Ciudades Industriales en México: Etapa de Expansión, 1971–1987." In *Una Década de Planeación Urbano-Regional en México, 1978–1988,* edited by Gustavo Garza (pp. 177–210). México, D.F.: Colegio de México.

Garza, Gustavo (ed.) (1987). *Atlas de la Ciudad de México.* México, D.F.: DDF and Colegio de México.

Garza, Gustavo (ed.) (1989c). *Una Década de Planeación Urbano-Regional en México, 1978–1988.* México, D.F.: Colegio de México.

Garza, G. and Schteingart, M. (1978). "Mexico City: The Emerging Megalopolis." In vol. 6 of *Latin American Urban Research,* edited by Wayne Cornelius and Robert Kemper, (pp. 51–86). Beverley Hills: Sage.

Garza, Gustavo and Martha Schteingart (1986). "Mexico City: Industrial Dynamics and the Formation of Space in a Semi-Peripheral Metropolis." Manuscript. El Colegio de México, Mexico, D.F.

Gaventa, J. (1980). *Power and Powerlessness: Quiescence and Rebellion in an Appalachian Valley.* Urbana: University of Illinois Press.

Gaye, Malick (1992). "The Self-Help Production of Housing and the Living Environment in Dakar, Senegal." *Environment and Urbanization.* 4 (2): 101–108.

Geisse, Guillermo (1981). "Políticas Realistas de Tierra Urbana en America Latina." *Revista Eure,* 23 (December): 30–37.

Geisse, Guillermo (1983). "Latin America." *World Congress on Land Policy, 1980,* edited by M. Cullen and S. Woolery. Lexington, Mass.: Lexington.

Geisse, Guillermo (1992). "Urban Alternatives for Dealing with Crisis." In *Rethinking the Latin American City,* edited by Richard M. Morse and Jorge E. Hordoy (pp. 208-222). Washington, D.C.: Woodrow Wilson Center Press.

Geisse, Guillermo and Francisco Sabatini (1981). "Renta de la Tierra: Heterogenei-dad Urbana y Medio Ambiente" *Estilos de Desarrollo y Medio Ambiente en la América Latina,* edited by O. Sunkel and N. Gligo. México, D.F.: Fondo de Cultura Económica.

Georgescu-Roegen, Nicolas (1971). *The Entropy Law and the Economic Process.* Cambridge, Mass.: Harvard University Press.

Ghai, D. (1993). Foreword in *In Defense of Livelihood: Comparative Studies on Environmental Action,* edited by John Friedmann and Haripriya Rangan (pp. vii–viii). West Hartford, Conn.: UNRISD and Kumarian.

Ghai, D. (1994). "Environment, Livelihood and Empowerment." *Development and Change.* 25 (1): 1–11.

Giddens, Anthony (1982). *Profiles and Critiques in Social Theory.* London: Mac-millan.

Giddens, Anthony (1983). *A Contemporary Critique of Historical Materialism.* Vol. 1, *Power, Property and the State.* Berkeley and Los Angeles: University of California Press.

Giddens, Anthony (1986). *Central Problems in Social Theory: Action, Structure and Contradiction in Social Analysis.* London: Macmillan.

Gil Elizondo, Juan R. (1987). "Planeación del Desarrollo Urbano de la Ciudad de México." In *Atlas de la Ciudad de México,* edited by Gustavo Garza (pp. 395–400). México, D.F.: DDF and Colegio de México.

Gilbert, Alan G. (1989). *Housing and Land in Urban Mexico.* Monograph Series no. 31. San Diego: Center for U.S.-Mexican Studies.

Gilbert, Alan G. (1991). "Self-Help Housing during Recession: The Mexican Expe-rience." In *Social Responses to Mexico's Economic Crisis of the 1980s,* edited by Mercedes González de la Rocha and Agustín Escobar Latapí (pp. 221–241). U.S.-Mexico Contemporary Perspective Series no. 1. San Diego: Center for U.S.-Mexican Studies.

Gilbert, Alan G. (1992). "Third World Cities: Housing, Infrastructure and Servic-ing." *Urban Studies.* 29, (3/4): 435–460.

Gilbert, Alan G. (1993). "Third World Cities: The Changing National Settlement System." *Urban Studies.* 30 (4/5): 721–740.

Gilbert, Alan G. (1994). *The Latin American City.* London: Latin American Bureau.

Gilbert, Alan G. and Ann Varley (1991). *Landlord and Tenant: Housing the Poor in Urban Mexico.* London: Routledge.

Gilbert, Alan G. and Peter Ward (1985). *Housing the State and the Poor: Policy and Practice in Three Latin American Cities.* Cambridge: Cambridge University Press.

Goldrich, Daniel and David V. Carruthers (1992). "Sustainable Development in Mexico? The International Politics of Crisis or Opportunity." *Latin American Perspectives,* 19 (1): 97–122.

Goliber, Thomas J. (1994). Foreword in *Urbanization in Africa: A Handbook,* edited by James D. Tarver (pp. xv–xviii). London: Greenwood.

González, Amador (1996). "Inflación." *La Jornada,* May 10.

Gonzalez, F. B. (1983). "La Irregularidad de la Tenencia de la Tierra en las Colonias Populares: 1976–1982." *Revista Mexicana de Sociologia.* 3 (July–September): 797–829.

González-Romero, Alberto, Patricia Balina-Tessaro, and Sergio Alvarez-Cárdenas (1987). "Estudio sobre una Comunidad de Roedores en una Zona Agrícola del Valle de México." In *Aportes a la Ecología Urbana de la Ciudad de México,* edited by Eduardo H. Rapoport and Ismael R. López-Moreno (pp. 153–191). México, D.F.: Noriega Editores and Editorial Limusa.

Gónzales Durán, Pablo. "Devastan la Zona Ecológica en San Lorenzo Acopilco, en Cuajimalpa." *Excelsior.* May 30.

González de la Rocha, Mercedes and Agustín Escobar Latapí (eds.) (1991). *Social Responses to Mexico's Economic Crisis of the 1980s.* U.S.-Mexico Contemporary Perspectives Series no. 1. San Diego: Center for U.S.-Mexican Studies.

González Salazar, Gloria (1990). *El Distrito Federal: Algunas Problemas y Su Planeación.* México, D.F.: Instituto de Investigaciones Económicas.

Gordon, C. (ed.) (1980). *Power/Knowledge: Selected Interviews and Other Writings by Michel Foucault.* New York: Pantheon.

Gordon Drabek, Anne (ed.) (1987). "Development Alternatives: The Challenge for NGOs." *World Development,* supplement. 15 (Autumn). iix–xv.

Gorz, André (1980). *Ecology as Politics.* Translated by Patsy Vigderman and Jonathan Cloud. Boston: South End Press.

Gotica Villegas, Beatriz, Lucia Pulido Vega, and Jose Avila Bravo (1985). *Ajusco: Conservación Ecológica y Desarrollo Urbano.* Masters' thesis, México D.F., Universidad Autónoma Nacional de México (UNAM).

Gottlieb, Robert (1993). *Forcing the Spring: The Transformation of the American Environmental Movement.* Covelo, Calif.: Island.

Graisbord, Boris (1989). "Zona Metropolitana de la Ciudad de México: Fragmentación Política y Planeación del Valle Cuautitlán-Texcoco." In *Una Década de Planeación Urbano-Regional en México, 1978–1988,* edited by Gustavo Garza (pp. 287–302). México, D.F.: Colegio de México.

Greenpeace México (1995). "Nuevo ley ambiental: reisgos y retos." *Este País.* June, pp. 58–60.

Grena Kliass, Rosa (1990). "Planning and Conservation: Green Areas and the Environmental Quality of the City of São Paulo." *Third World Planning Review.* 12 (4): 351–360.

Grubb, Michael, Matthias Koch, Abby Munson, Francis Sulliban, and Koy Thomson (1993). *The Earth Summit Agreements: A Guide and Assessment.* A Publication of the Energy and Environmental Programme, the Royal Institute of International Affairs. London: Earthscan.

Grupo Tecnología Alternativa (1984). "Colonia Ecológica Productiva." unpulbished community design document. Bosques del Pedregal, Mexico City, D.F.

Grupo de Tecnología Alternativa (1990). "Proyecto Bosques, Ajusco." unpublished community design document. Bosques del Pedregal, Mexico City, D.F.

GTA (Grupo de Tecnología Alternativa) (1992a). "Reporte Final del Programa de Investigación IDRC/GTA/SIRDO." Report prepared for the International Center for Development Research, Project no. 3-P-88-0104. México, D.F.: GTA.

GTA (Grupo de Tecnología Alternativa) (1992b). *The SIRDO Integral System for Recycling Organic Waste from México*. México, D.F.: GTA, S. C.

Gudynas, Eduardo (1990). "The Search for an Ethic of Sustainable Development in Latin America." In *Ethics of Environment and Development: Global Challenge, International Response*, edited by Ronald Engle and Joan G. Engle (pp. 139–149). Tucson: University of Arizona Press.

Guerrero, Chiprés (1996). "El Ejército Disfraza de Civiles las Bajas Recibidas, Aseguran." *La Jornada*, August 25.

Guevara Sada, Sergio and Patricia Moreno Casaola (1987). "Areas Verdes de la Zona Metropolitana de la Ciudad de México." In *Atlas de la Ciudad de México*, edited by Gustavo Garza (pp. 231–236). México, D.F.: DDF and Colegio de México.

Guillermo Aguilar, Adrián and Guillermo Olvera L. (1991). "El Control de la Expansión Urbana en la Ciudad de México: Conjeturas de un Falso Planteameinto." *Estudios Demográficos y Urbanos*. 16 (1): 89–115.

Guillermoprieto, Alma (1990). "Letter from Mexico City." *The New Yorker*, September 17, pp. 93–104.

Habitat (1976). *Report of HABITAT: United Nations Conference on Human Settlements*, May 31–June 11, Vancouver. New York: United Nations.

Habitat (1983). *Land for Housing the Poor: Report of the United Nations Seminar of Experts on Land for Housing the Poor*. Thallberg and Stockholm: March.

Habitat (1986). *Global Strategy for Shelter to the Year 2000*. New York: Oxford University Press.

Habitat (United Nations Centre for Human Settlements) (1987). *Global Report on Human Settlements*. New York: Oxford University Press.

Habitat (1990). *People, Settlements, Environment and Development*. Nairobi, Kenya: Habitat.

Habitat (1996a) *Global Report on Human Settlements*. New York: United Nations.

Habitat (1996b) *Habitat Agenda*. New York: United Nations.

Habitat (1996c). *The Global Plan of Action, United Nations Conference on Human Settlements, Istanbul, June 3–14, 1996*. Draft (revision number 3). New York: United Nations.

Habitat II (1995). Draft statement of principles and commitments and global plan of action, Report of the Secretary General, January 11, 1995.

Habitat News (1994). 16, no. 1 (April): 10.

"Habrá Invasión a Zonas Verdes: Falta Plan Habitacional—Cocoder." (1994) *Reforma,* September 27, sec. B, "Ciudad y Metrópoli," p. 7.

Hahn, Ekhart and Udo E. Simmons (1991). "Ecological Urban Restructuring." *Ekistics.* 58 (348): 199–209.

Halffter, Gonzalo (1991). *El Concepto de Reserva de la Biosfera.* Proceedings of the seminar on biological diversity in Mexico. no. 1. México, D.F.: Universidad Nacional Autónoma de México and World Wildlife Federation.

Hall, Peter (1984). *The World Cities.* London: Butler & Tanner.

Halla, Francos (1994). "A Coordinating and Participatory Approach to Managing Cities: the Case of the Sustainable Dar es Salaam Project in Tanzania." *Habitat International.* 18 (3): 19–31.

Hamilton, Nora (1994). Presentation at the Center for U.S. Mexican Studies, November 30, La Jolla, California.

Hardoy, E. Jorge (1992a). "Introduction." In *Rethinking the Latin American City,* edited by Richard M. Morse and Jorge E. Hordoy (pp. xi–xvii). Washington, D.C.: Woodrow Wilson Center Press.

Hardoy, E. Jorge (1992b). "Theory and Practice of Urban Planning in Europe, 1850–1930: Its Transfer to Latin America." In *Rethinking the Latin American City,* edited by Richard M. Morse and Jorge E. Hordoy (pp. 20–49). Washington, D.C.: Woodrow Wilson Center Press.

Hardoy, Jorge E., Diana Mitlin, and David Satterthwaite (1992). *Environmental Problems in Third World Cities.* London: Earthscan.

Hardoy, Jorge E. and David Satterthwaite (1981). *Shelter, Need, and Response: Housing, Land and Settlement Policies in Seventeen Third World Nations.* New York: John Wiley.

Hardoy, Jorge E. and David Satterthwaite (1990a). "The Future City." In *The Poor Die Young,* edited by J. E. Hardoy, S. Cairncross, and D. Satterthwaite. London: Earthscan.

Hardoy, Jorge E. and David Satterthwaite (1990b). "Urban Change in the Third World: Are Recent Trends a Useful Pointer to the Urban Future?" In *The Living City: Towards a Sustainable Future,* edited by David Cadman and Geoffrey Payne (75–109). New York: Routledge.

Harris, Nigel (ed.) (1992). *Cities in the 1990s: The Challenge for Developing Countries.* Highlights from a workshop in which representatives of international aid agencies and governments in developing countries considered the new urban policy statements of the World Bank, the United Nations Development Programme, and the United Nations Centre for Human Settlements, November 1991. New York: St. Martin's Press.

Harvey, David. (1982). *The Limits to Capital.* Chicago: University of Chicago Press.

Harvey, Neil (1994). *Rebellion in Chiapas: Rural Reforms, Campesino Radicalism, and the Limits to Salinismo.* Transformation of Rural Mexico Series no. 5. La Jolla, Calif.: Center for U.S.-Mexican Studies, University of California, San Diego.

Held, David, James Anderson, Bram Gieben, Stuart Hall, Laurence Harris, Paul Lewis, Noel Parker, and Ben Turok (eds.) (1983). *States and Societies.* New York: New York University Press.

Heredia, Carlos A. and Mary E. Purcell (1995). "La Polarización de la Sociedad Mexicana." *Este País.* 54:2–26.

Heridia, Carlos (1994). "Democracy and Justice: Essential for True Social Development." *The Other Side of Mexico.* 34 (January–February): 5.

Hernández, Aída and Alicia Castillo (1992). "Semana Ecológica del Ajusco." *Oikos.* A publication of the Centro de Ecología, Universidad Nacional Autónome de México, México D.F. 15 (May–June): 4.

Hernández, Anabel (1996). "Se acerca DF a ser Estado." *Reforma.* April 18, sec. B, "Ciudad y Metrópoli," p. 3.

Hernández, Luis (1994). "Revenge of the Mayans: The Roots of Mexico's Inevitable Indian Uprising." *The Other Side of Mexico.* 34 (January–February): 7.

Herrera, Amílcar O. (1981). "Desarrollo, Medio Ambiente y Generación de Tecnologías Apropiadas." In *Estilos de Desarrollo y Medio Ambiente en la América Latina,* edited by Osvaldo Sunkel and Nicolo Gligo (pp. 558–589). México, D.F.: Fondo de Cultura Económica.

Herrera Legarreta, Ana (1990). "Contaminación en Aire, Agua y Suelo en la Ciudad de México." In vol. 1 of *Medio Ambiente y Desarrollo en México,* edited by Enrique Leff (pp. 547–580). México, D.F.: Centro de Investigaciones Interdisciplinarias en Humanidades, UNAM, and Grupo Editorial.

Herrera-Revilla, I., R. Medina-Bañuelos, J. Carrillo-Rivera, and E. Vázquez-Sánchez (1994). "Diagnóstico del Estado Presente de las Aguas Subterráneas de la Ciudad de México y Determinación de Sus Condiciones Futuras." Contracto no. 3-33-1-6684. México, D.F.: DGCOH, DDF, Instituto de Geofísica, and Universidad Nacional Autónoma de México.

Hesselberg, Jan (1995). "The Need for a Bold Shelter Strategy." *Countdown to Istanbul.* 1:5–7.

Hidalgo, Jorge A. (1995). "Seguirán Desalojos en Zonas de Reserva." *Reforma,* sec. B, "Ciudad y Metrópoli."

Hidalgo, Jorge Arturo (1996). "Advierten del Regreso de Invasión a Predios." *Reforma.* April 29, sec. B, "Ciudad y Metrópoli," p. 2.

Hidalgo, Jorge A., Jorge Carrasco (1995). "Desalojan la Ladera: Se Enfrentan Policías con Habitantes del Area Ecológica de Contreras." *Reforma,* sec. B, "Ciudad y Metrópoli."

Hiernaux, Daniel (1989). "La Planeación de la Ciudad de México: Logros y Contradicciones." In *Una Década de Planeación Urbano-Regional en México,*

1978–1988, edited by Gustavo Garza, (pp. 235–254). México, D.F.: Colegio de México.

Hiernaux, Daniel (1991). "Servicios Urbanos, Grupos Populares y Medio Ambiente en Chalco, México." In *Servicios Urbanos, Gestión Local y Medio Ambiente*, edited by Martha Schteingart and Luciano d' Andrea (pp. 281–304). México D.F.: Colegio de México and CE.R.FE.

Hildyard, Nicholas (1993). "Foxes in Charge of the Chickens." In *Global Ecology: A New Arena of Political Conflict*, edited by Wolfgang Sachs (pp. 22–35). New Jersey: Zed.

Hirschman, Albert (1984). *Getting Ahead Collectively: Grassroots Experiences in Latin America.* Cambridge: Cambridge University Press.

Hoenderdos, W. H. and H. C. M. Verbeek (1989). "The Low-Income Housing Market in Mexico: Three Cities Compared." In *Housing and Land in Urban Mexico,* edited by Allan Gilbert (pp. 51–64). Monograph Series no. 31. San Diego: Center for U.S.-Mexican Studies.

Holmberg, Johan (ed.) (1992). *Making Development Sustainable: Redefining Institutions, Policy, and Economics.* Washington, D.C.: Island Press.

Hosaka, Mitsuhiko (1993). "Sharing Local Development Experiences Transnationally: Networking in Support of Community and Local Initiatives." *Environment and Urbanization* 5 (1): 132–147.

Houghton, R. A. and J. E. Hobbie (1983). "Changes in the Carbon Content of Terrestrial Biota and Soils between 1860 and 1980: A Net Release of CO_2 to the Atmosphere." *Ecological Monographs* 53:235–262.

Howe, Deborah A. (1993). "Growth Management in Oregon." In *Growth Management: The Planning Challenge of the 1990s,* edited by Jay M. Stein (pp. 61–75). Newbury Park: Sage.

Ibarra, Valentín, Sergio Puente, and Fernando Saavedra (eds.) (1987). *La Ciudad y el Medio Ambiente in América Latina: Seis Estudios de Caso. (Proyecto Ecoville).* México, D.F.: Colegio de México.

Ibarra, Valentín, Sergio Puente, Fernando Saavedra, and Martha Schteingart (1987). "La Ciudad y el Medio Ambiente: El Caso de la Zona Metropolitana de la Ciudad de México." In *La Ciudad y el Medio Ambiente in América Latina,* edited by Valentín Ibarra, Sergio Puente, and Fernando Saavedra. México, D.F.: Colegio de México.

Ibarra, Valentín, Sergio Puente, and Martha Schteingart (1984). "La Ciudad y el Medio Ambiente." *Demografía y Economía.* 1 (57): 110–143.

IIED-AL (Instituto Internacional de Medio Ambiente y Desarrollo-América Latina) with CEA (Center for Advanced Studies) and GASE(Group for the Analysis of Ecological Systems) (1992). "Sustainable Development in Argentina." *Environment and Urbanization.* 4 (1): 37–52.

"Indicadores de Desarrollo Social" (1992). *El Financiero.* April 20.

INEGI (Instituto Nacional de Estadística, Geografía Informática) (1991). *XI Censo General de Población y Vivienda, 1990.* México, D.F.: INEGI.

INEGI (1992a). "México: Economic and Social Information." Section by National Soliderity Program *INEGI International Review*, special issue. 4 (1): 87–92.

INEGI (1992b). *Tlalpan: Cuaderno de Información Básica Delegacional*. 1992 ed. México, D.F.: INEGI.

"Intentan Crear la Primera Colonio Ecológica de Mexico en el Ajusco" (1984). *El Día*, April 27.

Instituto Internacional de Medio Ambiente y Desarrollo and IIED-AL (1992). "NGO Profile." *Environment and Urbanization*. 4 (2): 155–167.

Inter-American Development Bank, Environmental Management Committee. (1992). *1992 Annual Report on the Environment and Natural Resources*.

Inter-American Foundation (1991). "NGOs Face the Challenge of a New Decade," *Grassroots Development*, special issue. 15 (2).

"Invaden Familias el Ajusco Medio: Intentan Fraccionar el Centro de Capacitación Ecológica Infantil." (1994). *Reforma*, September 20, sec. B, "Ciudad y Metrópoli" p. 7.

"Invaden Zonas Ecológicas: Afectan Asentamientos Mil 778 hectáreas" (1994). *Reforma*, September 4, Sec. B, "Ciudad y Metrópoli," p. 7.

Iracheta Cenecorta, Alfonso X. (1989). "Diez Años de Planeación del Suelo en la Zona Metropolitana de la Ciudad de México." In *Una Década de Planeación Urbano-Regional en México, 1978–1988*, edited by Gustavo Garza (pp. 255–286). México, D.F.: Colegio de México.

Jahnke, Marlene (ed.) (1995). "United Nations Activities, Forty-ninth Session: First Part." *Environmental Policy and Law*. 25 (1/2): 2–14.

Jansson, AnnMari, Monica Hammer, Carl Folke, and Robert Costanza (eds.) (1994). *Investing in Natural Capital: The Ecological Economics Approach to Sustainability*. Covelo, Calif.: Island.

Jarque, Carlos M. (1992). "Reflexiones sobre el Papel de la Informática en México." *Instituto Nacional de Estadística, Geografía e Informática*. 15 (4): 3–7.

Jessop, Bob (1982). *The Capitalist State: Marxist Theories and Methods*. New York: New York University Press.

Jessop, Bob (1985). "The State and Political Strategy." Paper presented at IPSA Conference (International Political Science Association) Paris.

Jiménez, Edith (1988). "New Forms of Community Participation in Mexico City: Success or Failure?" *Bulletin of Latin American Research*. 7 (1): 66–96.

Jiménez, Edith (1989). "A New Form of Government Control over *Colonos* in Mexico City." In *Housing and Urban Land in Mexico*, edited by Alan Gilbert (pp. 157–172). Monograph Series no. 31. San Diego: Center for U.S.-Mexican Studies.

Jiménez, Gerardo and María Luisa Pérez (1996). "Toman vecinos delegación." *Reforma*. May 23, sec. B "Ciudad y Metrópoli," p. 1.

Jiménez Muñoz, Jorge; María Luisa Picard-Ami, Manuel Soria López, Guadalupe Reyes Domínguez, and Eliseo Moyao Morales (1986). *La Ciudad: De Monumento Histórico a Laberinto Social*. México, D.F.: Casa Ciudad.

Joekes, Susan. (1994). "Gender, Environment and Population." *Development and Change*. Special issue: "Development and the Environment: Sustaining People and Nature," edited by Dharam Ghai. 25 (1): 137–166.

Johnston, R. J. (1994). "World Cities in a World-System: Conference, Sterling, VA, April 1993." *International Journal of Urban and Regional Research*. 18 (1): 150–152.

Jones, Gareth A. and Peter M. Ward (1994). "The World Bank's 'New' Urban Management Programme: Paradigm Shift or Policy Continuity?" *Habitat International*. 18 (3): 33–51.

Jones, Gareth A. and Peter M. Ward (1995). "The Blind Men and the Elephant: A Critic's Reply." *Habitat International*. 19 (1): 61–72.

Joochul, Kim (1991). "Urban Redevelopment of Greenbelt-Area Villages: A Case Study of Seoul, Korea." *Bulletin of Concerned Asian Scholars*. 23 (2): 20–29.

Joyner, Alfredo et al. (1996). "Hay 44 polvorines en el DF." *Reforma*. April 23, sec. B "Ciudad y Metrópoli," p. 1.

Jusidman, Clara (1988). "Empleo y Mercados de Trabajo en el Area Metropolitana de la Ciudad de México, 1975–1988." In *Medio Ambiente y Calidad de Vida*, edited by Sergio Puente and Jorge Legorreta (pp. 225–250). México, D.F.: DDF and Plaza & Janés.

Kandell, Jonathan (1988). *La Capital: The Biography of Mexico City*. New York: Random House.

Kasarda, John D. and Allan M. Parnell (1993a). "Introduction: Third World Urban Developmemt Issues." In *Third World Cities: Problems, Policies and Prospects*, edited by John D. Kasarda and Allan M. Parnell (pp. ix–xii). Newbury Park: Sage.

Kasarda, John D. and Allan M. Parnell (eds.) (1993b). *Third World Cities: Problems, Policies and Prospects*. Newbury Park: Sage.

Kates, R. W., B. L. Turner, II, and W. C. Clark (1990). "The Great Transformation." In *The Earth as Transformed by Human Action*, edited by B. L. Turner, II, W. C. Clark, R.W. Kates, J. F. Richards, J. T. Mathews, and W. B. Meyer. New York: Cambridge University Press.

Keating, Michael (1993). *The Earth Summit's Agenda for Change*. Geneva: Centre for Our Common Future.

Kiy, Richard (1995). "Environment a Priority with Mexico." *Border Links*. 4 (1): 3.

Knight, Alan (1990). "Historical Continuities in Social Movements." In *Popular Movements and Political Change in Mexico*, edited by Joe Foweraker and Ann L. Craig (pp. 78–102). Boulder, Colo.: Lynne Rienner.

Koloditch, James (ed.) (1993). "Industry Subsector Analysis of Solid/Hazardous Waste Management Equipment and Services in Mexico." Research Report prepared by the American Embassy in the United States and the Foreign Commercial Service in México, D.F.

Kombe, Jackson W. M. (1994). "The Demise of Public Urban Land Management and the Emergence of Informal Land Markets in Tanzania: A Case of Dar-es-Salaam City." *Habitat International.* 18 (1): 23–43.

Korten, David C. (1990). *Getting to the 21st Century: Voluntary Action and the Global Agenda.* West Hartford: Kumarian.

Kumar, Ranjit and Barbara Murck (1992). *On Common Ground: Managing Human-Planet Relationships.* New York: John Wiley.

"La Ecología como Arma Política" (1984). *La Jornada,* December 2.

"La Hermosa Gente: Bosques del Pegregal" (1984). *Ovaciones,* June 29.

"La Urbanización del Ajusco." (1994) *Reforma,* sec. B "Ciudad y Metrópoli," p. 7.

"La Zona Rural y de Conservación Está Sujeta a Fuertes Presiones por el Crecimiento de la Mancha Urbana" (1995). *La Jornada,* April 3.

Lafragua, José and Manuel Orozco y Berra ([1853] 1987). *La Ciudad de México.* México, D.F.: Editorial Porrúa.

Lajous, Luz (1992). "El Futuro Político de la Ciudad de México." *Examen.* 3 (36): 25–26.

Latin American and Caribbean Commission on Development and the Environment (1990). *Our Own Agenda.* New York, N.Y.: United Nations Development Program.

Leff, Enrique (1981). "Sobre la Articulación de las Ciencias en la Relación Naturaleza-Sociedad." In *Biosociología y Articulación de las Ciencias,* edited by Enrique Leff. México, D.F.: Universidad Autónoma de México.

Leff, Enrique (1990a). "The Global Context of the Greening of Cities." In *Green Cities: Ecologically Sound Approaches to Urban Space,* edited by David Gordon (pp. 55–66). New York: Black Rose.

Leff, Enrique (1990b). "Introducción a una Visión de los Problemas Ambientales de México. In vol. 1 of *Medio Ambiente y Desarrollo en México,* edited by Enrique Leff (pp. 7–73). México, D.F.: Centro de Investigaciones Interdisciplinarias en Humanidades, Universidad Nacional Autónoma de México, and Grupo Editorial.

Leff, Enrique (ed.) (1990c). *Medio Ambiente y Desarrollo en México.* Vols. 1 and 2. México, D.F.: Centro de Investigaciones Interdisciplinarias en Humanidades, Universidad Nacional Autónoma de México, and Grupo Editorial.

Leff, Enrique (1993). "Marxism and the Environmental Question: From the Critical Theory of Production to an Environmental Rationality for Sustainable Development." *Capitalism, Nature, Socialism: A Journal of Socialist Ecology.* 4 (1): 45–66.

Leff, Enrique (1994). *Ecología y Capital: Racionalidad Ambiental, Democracia Participativa y Desarrollo Sustentable.* México, D.F.: Siglo Vientiuno Editores.

Leff, Enrique (1995). *Green Production: Toward an Environmental Rationality,* translated by Margaret Villanueva. New York: Guilford.

Leff, Enrique and Julia Carabias (1992). *Cultura y Manejo Sustentable de los Recursos Naturales.* México, D.F.: Universidad Nacional Autónoma de México.

Leff, Enrique, Julia Carabias, and Ana Irene Batis (eds.) (1990). *Recuros Naturales, Técnica y Cultura: Estudios y Experiencias para un Desarrollo Alternativo.* A Seminar Series Report of the Centro de Investigaciones Interdisciplinarias en Humanidades, Coordinación de Humanidades, Universidad Nacional Autónoma de México. México, D.F.: SEDUE and PNUMA/ORPALC.

Legorreta, Jorge (1992). "El Medio Ambiente, Asentamientos Ilegales y su Impacto en la Calidad de Vida." In *La Zona Metropolitana de La Ciudad de México: Problemática Actual y Perspectivas Demográficas y Urbanas,* edited by CONAPO (pp. 127–140). México, D.F.: Dirección General de Estudios de Población.

Leitmann, Josef (1994). "The World Bank and the Brown Agenda: Evolution of a Revolution." *Third World Planning Review.* 16 (2): 117–128.

Leitmann, Josef, Carl Bartone, and Janis Bernstien (1992). "Environmental Management and Urban Development: Issues and Options for Third World Cities." *Environment and Urbanization.* 4 (2): 131–140.

Leman, Edward and John E. Cox (1991). "Sustainable Urban Development: Strategic Considerations for Urbanizing Nations." *Ekistics.* 58 (348): 216–224.

Leonard, A. (1984). *Seeds.* New York: Graphic Impressions.

Leopold, Aldo (1966). *A Sand County Almanac.* New York: Ballantine Books.

Lewin, John (1981). *Housing Cooperatives in Developing Countries.* New York: John Wiley.

Linden, J. Van Der (1994). "Editorial: Where Do We Go from Here?" *Third World Planning Review.* 16 (3): 223–229.

Lindner, Warren (ed.) (1989). "Sustainable Development: From Theory to Practice." *Development,* special issue. 2 (3).

Linn, Johannes F. (1983). *Cities in the Developing World: Policies for their Equitable and Efficient Growth.* A World Bank publication. New York: Oxford University Press.

Liverman, Diana M. (1990a). "Seguridad y Medio Ambiente en Mexico." In *En Busca de la Seguridad Perdida,* edited by S. Aguayo and B. Bagley (pp. 223–263). México, D.F.: Siglovienteuno.

Liverman, Diana M. (1990b). "Vulnerability to Global Environmental Change." In *Understanding Global Environmental Change: The Contributions of Risk Analysis and Management.* Report on an International Workshop. Worchester Mass.: Clark University, Earth Transformed Program.

Liverman, Diana M. (1992). "Global Change in Mexico." *Earth and Mineral Sciences.* 60 (4): 71–76.

Liverman, Diana M. (1993). "Environment and Security in Mexico." In *Mexico: In Search of Security,* edited by Bruce M. Bagley and Sergio Aquayo Quezada (pp. 211–246). New Brunswick, N.J.: Transaction.

Liverman, Diana M. and Karen L. O'Brien (1991). "Global Warming and Climate Change in Mexico." *Global Environmental Change.* 1 (5): 351–364.

Logan, Kathleen (1988). "Women's Political Activity and Empowerment in Latin American Urban Movements." In *Urban Life: Readings in Urban Anthropology,* edited by George Grelch and Walter Zenner (pp. 343–358). 2d ed. Illinois: Waveland.

Logan, Kathleen (1990). "Women's Participation in Urban Protest." In *Popular Movements and Political Change in Mexico,* edited by Joe Foweraker and Ann L. Craig (pp. 150–159). Boulder, Colo.: Lynne Rienner.

López, Jorge X. (1995). "Promoverán en la Ladera Programa de Ecotruismo." *Reforma,* February 20, sec. B, "Ciudad y Metrópoli."

López, Jorge X. (1996). "Descartan Escasez de Agua." *Reforma.* March 19, sec. B "Ciudad y Metrópoli," p. 1.

López, Jorge X. and Jorge A. Hidalgo (1996) "Dará DF Brinco Poblacional." *Reforma.* May 28, sec. B "Ciudad y Metrópoli," p. 1.

López-Moreno, Ismael R. (ed.) (1991). *El Arbolado Urbano de la Zona Metropolitana de la Ciudad de México.* Xalapa, Veracruz: Programme on Man and the Biosphere (MAB, UNESCO) and Instituto de Ecología A.C.

Lowe, Marcia D. (1991). "Shaping Cities: The Environmental and Human Dimensions." Worldwatch Paper no. 105. Washington, D.C.: WorldWatch Institute.

Lowenthal, Abraham (1994). "Growth and Equity: Charting a New Course." *Hemisfile.* 5 (4): 1–2.

Lugo Medina L. G. and F. Bejarano (1981). "La Acción del Estado, el Capital y la Formación de las Colonias Populares, en la Transformación Urbana de las Tierras Ejidales en las Delegaciones de Magdalena Contreras y Tlalpan: El Caso de la Colonia Popular Miguel Hidalgo." Bachelor's thesis. Universidad Iberoamericana, México, D.F.

Lukes, Steven (1974). *Power: A Radical View.* London: Macmillan.

Luque González, Rodolfo, and Reina Corona Cuapio (1992). "El Perfil de la Migración en la ZMCM." In *La Zona Metropolitana de la Ciudad de México: Problemática Actual y Perspectivas Demográficas y Urbanas,* edited by CONAPO (pp. 21–32). México, D.F.: Dirección General de Estudios de Pablación.

Lustig, Nora (1992). "Equity and Growth in Mexico." In *Towards a New Development Strategy for Latin America: Pathways from Hirschman's Thoughts,* edited by Simón Teitel (pp 219–258). Washington, D.C.: Inter-American Development Bank.

Lusugga Kironde, J. M. (1995). "Access to Land by the Urban Poor in Tanzania: Some Findings from Dar es Salaam." *Environment and Urbanization.* 7 (1): 77–96.

Lutz, Ellen (1990). *Human Rights in Mexico: A Politics of Impunity.* Los Angeles: Americas Watch.

Lynch, Kenneth (1994). "Urban Fruit and Vegtable Supply in Dar es Salaam." *The Geographical Journal.* 160 (3): 307–318.

Mabogonje, Akin L. (1994). "Introduction: Cities and Africa's Economic Recovery." In *Urbanization in Africa: A Handbook,* edited by James D. Tarver (pp. xxi–xxxii). London: Greenwood.

MacNeill, James, John E. Cox, and Ian Johnson (1991). "Sustainable Development: The Urban Challenge." *Ekistics.* 58 (348): 195–198.

MacNeill, James, Peiter Winsemius, and Taizo Yakushiji (1991). *Beyond Interdependence: The Meshing of the World's Economy and the Earth's Ecology.* New York: Oxford University Press.

Mainwaring, Scott (1987). "Urban Popular Movements, Identity and Democratization in Brazil" *Comparative Political Studies.* 20, no. 2 (July).

Malpezzi, S. (1990). "Urban Housing and Financial Markets: Some International Comparisons." *Urban Studies* 27: 971–1022.

Marcuse, Peter (1992). "Why Conventional Self-Help Projects Won't Work." In *Beyond Self-Help Housing,* edited by Kosta Mathéy (pp. 15–21). New York: Profil Verlag.

Marris, Peter (1982). *Community Planning and Conceptions of Change.* Boston: Routledge and Kegan Paul.

Martin, P. S. and R. G. Klien (1984). *Quaternary Extinctions: A Prehistoric Revolution.* Tucson: University of Arizona Press.

Markiewicz, Dana (1980). *Ejido Organization in Mexico, 1934–1976.* UCLA Latin American Center Publications Special Studies vol. 1 Los Angeles: UCLA Latin American Center Publications.

Martínez-Alier, Juan and Lori Ann Thrupp (1992). "A Political Ecology of the South." *Latin American Perspectives.* 19 (1): 148–152.

Massolo, Alejandra and Lucila Díaz Ronner (1985a). "Consumo y Lucha Urbana en la Ciudad de México: Mujeres Protagonistas." *De la Metrópoli Mexicana.* 6 (15): 135–151.

Massolo, Alejandra and Lucila Díaz Ronner (1985b) *Doña Jovita: Una Mujer en el Movimiento Urbano Popular.* México, D.F.: Mujeres para el Diálogo.

May, Richard (ed.) (1989). *The Urbanization Revolution: Planning a New Agenda for Human Settlements.* New York: Plenum.

Mazari, M. and M. D. Mackay (1993). Potential Groundwater Contamination by Organic Compounds in the Mexico City Metropolitan Area. *Environmental Science and Technology.* 27 (5): 794–802.

McDonnell, Mark J. and Steward T. A. Pickett (eds.) (1993). *Humans as Components of Ecosystems: The Ecology of Subtle Human Effects and Populated Areas.* New York: Springer-Verlag.

McKee, David L. (1994). *Urban Environments in Emerging Economies.* Westport, Conn.: Praeger.

Meffert, Karin (1993). "Co-operative Self-Help Housing: The Case of El Molino in Mexico City." In *Beyond Self-Help Housing,* edited by Kosta Mathéy (pp. 323–340). New York: Mansell.

Mejía, Javier (1992). "La Venta Ilegal e Invasiones de Tierras, Grave Amenaza para Zonas de Reserva Ecológica del D.F." *UnoMásUno,* November 14.

Mele, P. (1986). "El Espacio Industrial entre la Ciudad y la Región," Documento de Investigación no. 3. Instituto de Ciencias, Universidad Autónoma de Puebla.

Mena Abraham, Josefina (1987). "Tecnología Alternativa, Transformación de Desechos y Desarrollo Urbano." *Estudios Demográficos y Urbanos.* 2, no. 3, September–December).

Mena Abraham, Josefina (1990). "Proyecto Bosques del Pedregal, Ajusco Medio: Programa de Investigación de la Evaluación del Proceso SIRDO." Mexico City, D.F.: Grupo de Tecnológia Alternativa (GTA) and Centro Internacional de Investigación para el Desarrollo.

Méndez, Rosa María (1995). "Explica Gobierno Programa de Vivienda: Se Agota la Reserva Territorial.—Elizondo." *Reforma,* February 3, sec. B, "Ciudad y Metrópoli," p. 5.

Mendieta y Núñez, Lucio (1985). *El Problema Agrario de México.* México, D.F.: Editorial Porrúa.

Mendoza Pichardo, Maribel (1993). "Zertuche en la Asamblea de Representantes: Propuso una Seria de Iniciativas para el Desarrollo de la Demarcación." *Noti Tlalpan* (a news publication of the Tlalpan district, México, D.F.), January 15.

Menéndez Garza, Fernando (1991). "Mexico City's Program to Reduce Air Pollution." In *Economic Development and Environmental Protection in Latin America,* edited by Joseph S. Tulchin and Andrew I. Rudman (pp. 103–107). Boulder, Colo.: Lynne Rienner.

Merino y Guevara, Hector (1991). "Provisión de Agua y Drenaje en las Ciudades Mexicanas: Un Reto Permanente." In *Servicios Urbanos, Gestión Local y Medio Ambiente,* edited by Martha Schteingart and Luciano d' Andrea (pp. 121–130). México, D.F.: Colegio de México and CE.R.FE.

"Mexican Cities Begin Privatizing Municipal Waste Collection" (1993). *Environment Watch: Latin America.* 3, no. 1 (January): 6.

México (1936). "Decreto que Declara Parque Nacional Cumbres del Ajusco." *Diario Oficial,* September 23, México, D.F.: México.

México (1982). *Plan de Desarrollo Ubano: Distrito Federal.* México, D.F.: DDF.

México (1984). Programa de Conservación del Ajusco, Delegación de Tlalpan. México, D.F.: México.

México (1986). *Dirección General de Reordinación Urbana y Protección Ecológica.* México, D.F.: DDF.

México (1996). "Programa Estatal de Protección al Ambiente 1996–1999." Toluca, México: Gobierno del Estado de México.

Mexico Business Monthly (1991). (November).

Mexico Business Monthly (1992). Vol. 1, no. 12.

"Mexico City's Air Pollution Continues, Industry Criticizes Contingency Programs" (1994). *Environment Watch: Latin America.* 4, no. 1 (January): 1–2.

"Mexico Publishes Environmental Norms Program for 1994" (1994). *Environment Watch: Latin America,* supplement. 4, no. 5 (May): 1.

"Mexico Slows Growth: More Families Opting for Fewer Children, Cutting Forecasts of Population Gain." (1994). *San Diego-Union Tribune,* August 9, sec. A, p. 1.

"Mexico's President Declares *Economic Emergency,* Acknowledges Real Earnings Will Drop for Most" (1995). *Los Angeles Times,* January 4, sec. A, p. 1.

Meza Rodríguez, Víctor (1994). "La Conservacíon de la Naturaleza en México: Historia y Porvenir." *Examen.* 5, no. 56 (July): 36–37.

Milbrath, Lester W. (1989). *Envisioning a Sustainable Society: Learning Our Way Out.* Albany: State University of New York Press.

Mills, C. Wright (1959). *The Sociological Imagination.* New York: Oxford University Press.

Miraftab, Faranak (1996) "Old Enemies, New Rivals, Future Partners? The Case of State-NGO Relations in Mexico." Working Paper no. 672. Institute of Urban and Regional Development, University of California, Berkeley.

Mitchell, Bruce (1994). "Sustainable Development at the Village Level in Bali, Indonesia." *Human Ecology.* 22 (2): 189–211.

Mitlin, Diana (1992). "Sustainable Development: A Guide to the Literature." *Environment and Urbanization.* 4 (1): 111–124.

Mitlin, Diana and Jane Bicknel (1992a). Editors' introduction, "Sustainable Cities." *Environment and Urbanization,* special issue, "Sustainable Cities: Meeting Needs, Reducing Resource Use and Recycling, Re-Use and Reclamation." 4 (2): 3–8.

Mitlin, Diana and Jane Bicknel (1992b). "Will UNCED Sustain Development?" *Environment and Urbanization,* special edition: "Sustainable Development and the Global Commons: A Third World Re-assessment." 4 (1): 3–7.

Mitlin, Diana and David Satterthwaite (1990). "Human Settlements and Sustainable Development: The Role of Human Settlements and of Human Settlements Policies in Meeting Development Goals and in Addressing the Issues of Sustainability at Global and Local Levels." Paper prepared for the Human Resources Program of the International Institute for Environment and Development, Nairobi, Habitat.

Mitlin, Diana and John Thompson (1995). "Participatory Approaches in Urban Areas: Strengthening Civil Society or Reinforcing the Status Quo?" *Environment and Urbanization.* 7 (1): 231–250.

Moctezuma, Pedro (1987). "Apuntes sobre la Política Urbana y el Movimiento Popular en México." *Sociológica.* 2 (4): 133–142.

Moguel, Julio and Enrique Velázquez (1991). "Social Organization and Ecological Struggle in a Region of Northern Mexico." Unpublished manuscript. Universidad Nacional Autónoma de México.

Moreno Mejía, Sergio (1987). "Sistema Hidráulico del Distrito Federal." In *Atlas de la Ciudad de México,* edited by Gustavo Garza (pp.183–186). México, D.F.: DDF and Colegio de México.

Morfin, G. A. and M. Sánchez (1984). "Controles Jurídicos y Psicosociales en la Producción del Espacio Urbano para Sectores Populares en Guadalajara." *Encuentro.* 1: 115–41.

Morse, Richard M. and Jorge E. Hardoy (1992). *Rethinking the Latin American City.* Baltimore: John Hopkins University Press.

MRP (Movimiento Revolucionario del Pueblo) (1983). "Elementos de Línea Política para el Movimiento Urbano Popular." Mimeograph. México, D.F.

Mumford, Lewis (1961). *The City in History: Its Origins, Its Transformations, and Its Prospects.* New York: Harcourt Brace Jovanovich.

Mumford, Lewis (1972). "The Natural History of Urbanization." In *The Ecology of Man: An Ecosystem Approach,* edited by Robert Leo Smith (pp. 140–152). New York, N.Y.: Harper and Row

Mumme, Steven P. (1991). "Clearing the Air: Environmental Reform in Mexico." *Environment.* 33 (10): 6–11.

Mumme, Steven P. (1992). "System Maintenance and Environmental Reform in Mexico: Salinas's Preemptive Strategy." *Latin American Perspectives.* 19 (1): 123–143.

Mumme, Steven P., Richard C. Bath, and Valarie J. Assetto (1988). "Political Development and Environmental Policy in Mexico." *Latin American Research Review.* 23 (1): 7–34.

Muteshi, Jacinta (1995). "Collaborative Alliances: The Environment, Women and the Africa 2000 Network." *Environment and Urbanization.* 7 (1): 205–218.

National Research Council, Academia Nacional de la Investigación Científica, A.C., and Academia Nacional de Ingeniería, A.C. (1995). *Mexico City's Water Supply Improving the Outlook for Sustainability.* Washington, D.C.: National Academy Press. Available from the National Academy Press (tel.: 1-800-624-6242).

Navarro B., Bernardo (1989). La Urbanización Popular en la Ciudad de México. México, D.F.: Instituto de Investigaciones Económicas and Universidad Nacional Autónoma de México, Editorial Nuestro Tiempo.

N'Dow, Wally (1995). "Editorial." *Countdown to Istanbul.* 1:2.

Negrete, María E., Boris Graizbord, and Crescencio Ruiz (1993). "Población, Espacio y Medio Ambiente en la Zona Metropolitana de la Ciudad de México." Colegio de México, Serie Cuadernos de Trabajo no. 2. México, D.F.: Colegio de México.

Nelson, A. (1988). "An Empirical Note on How Regional Urban Containment Policy Influences an Interaction between Greenbelt and Exurban Land Markets." *Journal of the American Planning Association.* 54:178–184.

NEXOS (1983). "La Larga Marcha de los Ecólogos Mexicanos: Entrevista con Arturo Gómez Pompa." *NEXOS.* 6 no. 6 (September): 25–29.

Nientied, P. and J. Van der Linden (1985). "Approaches to Low-Income Housing in the Third World: Some Comments." *International Journal of Urban and Regional Research.* 9 (3): 311–329.

Nocedal, Jorge (1987). "Las Comunidades de Pájaros y su Relación con la Urbanización en la Ciudad de México." In *Aportes a la Ecología Urbana de la Ciudad de México,* edited by Eduardo H. Rapoport and Ismael R. López-Moreno (pp. 73–110). México, D.F.: Noriega Editores and Editorial Limusa.

Norgaard, Richard B. (1994). *Development Betrayed: The End of Progress and a Coevolutionary Revisioning of the Future.* New York: Routledge.

Nuccio, Richard A. (1991). "The Possibilities and Limits of Environmental Protection in Mexico." In *Economic Development and Environmental Protection in Latin America,* edited by Joseph S. Tulchin and Andrew I. Rudman (pp. 109–122). Boulder, Colo.: Lynne Rienner.

Nuccio, Richard A., Angelina M. Ornelas, and Iván Restrepo (1990). "Mexico's Environment and the United States." In *In the U.S. Interest,* edited by Janet Welsh Brown (pp. 19–53). Boulder, Colo.: Westview.

Nuccio, Richard A., Angelina M. Ornelas, and Iván Restrepo (1993). "Mexico's Environment: Securing the Future." In *Mexico: In Search of Security,* edited by Bruce M. Bagley and Sergio Aquayo Quezada (pp. 247–286). New Brunswick, N.J.: Transaction.

Nussbaum, Martha (1995). "Human Capabilities, Female Human Beings." In *Women, Culture and Development: A Study of Human Capabilities,* edited by M. Nussbaum and J. Glover. Oxford: Clarendon.

Oberlander H. P. (1985). *Land. The Central Human Settlement Issue.* Human Settlements Issue no. 7. Vancouver: University of British Columbia Press.

O'Connor, James (1988). "Capitalism, Nature, Socialism: A Theoretical Introduction." *Capitalism, Nature, Socialism: A Journal of Socialist Ecology.* 1:11–38.

O'Connor, Martin (1989). "Codependency and Indeterminancy: A Critique of the Theory of Production." *Capitalism, Nature, Socialism: A Journal of Socialist Ecology.* 3:33–57.

O'Connor, Martin (ed.) (1994). *Is Capitalism Sustainable? Political Economy and the Politics of Ecology.* New York: Guilford.

"Ocuparán Zona de los Dinamos: Piden Comuneros de Magdalena Contreras Terrenos para Urbanizar" (1994). *Reforma,* September 27, sec. B, "Ciudad y Metrópoli," p. 7.

Olayo, Ricardo (1995). "El DDF, en Busca de Terrenos Subutilizados para Construir Viviendas." *La Jornada.* November 17.

Oppenheimer, Andres (1996). "Mexican Rebels Cut a Bloody Swath." *Maimi Herald,* August 30.

Organisation for Economic Co-operation and Development (1995). *Planning for Sustainable Development: Country Experiences.* Washington, D.C.: OECD Publications and Information Center.

Orozco y Berra, Manuel ([1854] 1973). *Historia de la Ciudad de México: Desde Su Fundación hasta 1854.* A collection of articles, originally published by Orozco y Berra in *Diccionario Universal de Historia y Geografía (1854).* Compiled by the Seminario de Historia Urbana del Departamento de Investigaciones Históricas del INAH. México, D.F.: Secretaría de educación Pública.

Pacheco, Cristina (1995). "Tepoztlán, otro parque nacional en peligro." *La Jornada.* July 20.

Pacheco, Margarita (1992). "Recycling in Bogotá: Developing a Culture for Urban Sustainability." *Environment and Urbanization.* 4 (2): 74–79.

Pradilla Cobos, Emilio (1995a) "Una Ciudad sin Rumbo." *La Jornada,.* July 12.

Pradilla Cobos, Emilio (1995b) "Regiones y Ciudades en el Plan Nacional de Desarrollo 1995–2000." *Coyuntura.* No. 64 (October): 26–36.

Palacios L., Juan José (1989). "La Insuficiencia de la Política Regional en México: Patrones de Asignación de la Inversión Pública Federal, 1959–1986." In *Una Década de Planeación Urbano-Regional en México, 1978–1988,* edited by Gustavo Garza (pp. 155–176). México, D.F.: Colegio de México.

Palacios L., Juan José (1990). "Economía Subterránea en América Latina: ¿Alternativa Obligada de Supervivencia O Mecanismo Ilegal de Producción?" In *Crisis, Conflicto y Sobrevivencia: La Sociedad Urbana en México,* edited by Guillermo de la Peña, Juan Manuel Durán, Agustín Escobar, and Javier García de Alba. Guadalajara: Universidad de Guadalajara and CIESAS.

Páramo, Arturo, Ariadna Bermeo, and Jorge X. López (1995). "La Muralla Verde de la Ciudad." *Reforma,* March 13, sec. B, "Ciudad y Metrópoli," p. 6.

Páramo, Arturo and Gabriela Medina (1995). "Preservarán la ecología de la Sierra de Guadalupe: Evitarán Especular con Tierras;" *Reforma,* March 2, sec. B, "Ciudad y Metrópoli."

"Parque Nacional en Peligro" (1995). *La Jornada.* July 20.

Parsons, J. J. (1972). "Spread of African Pasture Grasses to the American Tropics." *Journal of Range Mange.* 25:12–17.

Parsons, T. (1949). *The Structure of Social Action: A Study in Social Theory.* 2d ed. New York: Free Press.

Payne, G. (1984). *Low-Income Housing in the Developing World: The role of Sites and Services and Settlement Upgrading.* New York: John Wiley.

Payne, G. (1989). *Informal Housing and Land Subdivisions in Third World Cities: A Review of the Literature.* Oxford: CENDEP.

Pearce, David, Edward Barbier, and Anil Markandya (1989). *Blueprint for a Green Economy.* London: Earthscan.

Pearce, David, Edward Barbier, and Anil Markandya (1990). *Sustainable Development: Economics and Environment in the Third World.* London: Earthscan.

Pearlman, Janice (1976). *The Myth of Marginality: Urban Proverty and Politics in Rio de Janerio.* Berkeley and Los Angeles: University of California Press.

Pearlman, Janice (1986). "The Informal Sector and the Dynamics of Housing Policy Evolution." Paper presented at seminar The Role of the Informal Sector in National Housing Production, June 18–20, USAID, Rio de Janerio.

Pearlman, Janice (1990). "Mega-Cities: Innovations for Sustainable Cities of the 21st Century." Speech delivered at the International Summit on the Environment, April 18–20, Los Angeles.

Pease García, Henry (1992). "Experiments in Democratizing a City Management." In *Rethinking the Latin American City,* edited by Richard M. Morse and Jorge E. Hordoy (pp. 162–178). Washington, D.C.: Woodrow Wilson Center Press.

Peattie, Lisa (1982). "Some Second Thoughts on Sites and Services" *Habitat International.* 6:131–139.

Perló Cohen, Manuel. (1981). *Estado, Vivienda y Estructura Urbana en el Cardenismo: El Caso de la Ciudad de México.* Cuadernos de Investigación Social no. 3. México, D.F.: Instituto de Investigaciones Sociales, Universidad Nacional Autónoma de México.

"Peso Plummets Once More: Crisis Deepens." (1994). *San Diego Union-Tribune,* December 28, sec. A, p. 1.

Petrich, Blanche (1996). "Prioridad en el Popo: Aprender a vivir bajo el volcán." *La Jornada.* June 3.

Pezzoli, Janice (1992). "The Practice of Agriculture in the Aztec Empire and New Spain: A Comparison and Contrast from an Ecological Perspective." Unpublished paper. University of California, San Diego.

Pezzoli, Keith (1987). "The Urban Land Problem and Popular Sector Housing Development in Mexico City." *Environment and Behavior.* 19 (3): 371–397.

Pezzoli, Keith (1991). "Environmental Conflicts in the Urban Milieu: the Case of Mexico City." In *Environment and Development in Latin America: The Politics of Sustainability,* edited by David Goodman and Michael Redclift (pp. 205–229). New York: Manchester University Press.

Pezzoli, Keith (1993). "Sustainable Livelihoods in the Urban Milieu: A Case Study from Mexico City." In *In Defense of Livelihood: Comparative Studies on Environmental Action,* edited by John Friedmann and Haripriya Rangan (pp. 127–154). West Hartford, Conn.: UNRISD and Kumarian.

Pezzoli, Keith (1997). "Sustainable Development: A Transdisciplinary Overview of the Literature." *Journal of Environmental Planning and Management.* 40(5): 549–601.

Pickett, Steward T. A. and Mark J. McDonald (1993). "Humans as Components of Ecosystems: A Synthesis." In *Humans as Components of Ecosystems: The Ecology of Subtle Human Effects and Populated Areas,* edited by Mark J. McDonald and Steward T. A. Pickett, (pp. 310–316). New York: Springer-Verlag.

Pickett, Steward T. A., V. T. Parker, and P. L. Fiedler (1992). "The New Paradigm in Ecology: Implications for Conservation Biology above the Species Level." In *Conservation Biology,* edited by P. L. Fiedler and S. K. Jain (pp. 65–88). New York: Chapman and Hall.

"Piden Que el DDF Tome en Cuenta los Asentamientos en el Ajusco" (1984). *El Día,* August 31.

Pillsbury, Barbara (1994). "Environmental Protection in Mexico City: The IDB, the Government and the Residents of Sierra Santa Catarina." *The Other Side of Mexico.* 34 (January–February): 9.

Pinsky, Barry (1993). "Settlements Information Network (SINA)." *Environment and Urbanization* 5 (1): 155–161.

Polanyi, M. (1966). *The Tacit Dimension.* Gardena, N.Y.: Anchor.

"Political Changes Leave Mexican American Environmental Agencies Reeling" (1993). *Environment Watch: Latin America.* 3, no. 11 (December): 11–22.

Ponting, Clive (1991). *A Green History of the World: The Environment and Collapse of Great Civilizations.* New York: Penguin.

"Popocatepetl Volcano Spews Black Ash." (1994). *San Diego Union-Tribune,* December 26, sec. A, p. 1.

Porras Robles, Angel (1992a). "El D.F.: Acosado por Marchas y Plantones." *UnoMásUno,* November 14.

Porras Robles, Angel (1992b). "Participan Autoridades en la Invasión de Zonas de Reserva Ecológica en Iztapalapa." *UnoMásUno,* August 21.

Porras Robles, Angel (1992c). "Serán Derribados Más de 150 Mil Arboles en Ajusco, Topilejo y Desierto de los Leones: Están Afectados por el Gusano Barrenador." *UnoMásUno,* July 8.

Portes, Alexandro (1989). "Latin American Urbanization during the Years of the Crisis." *Latin American Research Review.* 25:7–44.

Portes, Alejandro, and John Walton (1981). *Labor, Class, and the International System.* New York: Academic Press.

Portes, Alejandro, Manuel Castells, and Lauren A. Benton (eds.) (1989). *The Informal Economy: Studies in Advanced and Less-Developed Countries.* Baltimore: Johns Hopkins University Press.

Pozas Garza, M. (1989). "Land Settlement by the Urban Poor in Monterrey." In *Housing and Land in Urban Mexico,* edited by Alan Gilbert (pp. 65–78). Monograph Series no. 31. San Diego: Center for U.S.-Mexican Studies.

Pratap Chatterjee and Matthias Finger (1994). *The Earth Brokers: Power, Politics and World Development.* New York: Routledge.

Price, Marie (1994). "Ecopolitics and Environmental Nongovernmental Organizations in Latin America." *Geographical Review.* 84 (1): 42–58.

Puente, Sergio (1988). "La Ciudad Material de Vida en la Zona Metropolitana de la Ciudad de México: Hacia un Enfoque Totalizante." In *Medio Ambiente y Calidad de Vida,* edited by Sergio Puente and Jorge Legorreta (pp. 13–107). México, D.F.: DDF and Plaza & Janés.

Puente, Sergio and Jorge Legorreta (eds.) (1988). *Medio Ambiente y Calidad de Vida.* México, D.F.: DDF and Plaza & Janés.

Puig, Carlos (1990). "Reaparece el Cículo Vicioso Protesta Social-Represión: El Director de Americas Watch Fundamenta el Informe sobre Violación de Derechos Humanos." *Proceso.* 712 (June 25): 13–19.

Pye-Smith, Charlie, Grazia Borrini Feyerabend, and Richard Sandbrook (1994). *The Wealth of Communities: Stories of Success in Local Environmental Management.* West Hartford, Conn: Kumarian.

Quadri de la Torre, Gabriel (1991a). "Ajusco: Distrito Federal." *Examen*. 3 (27): 44–45.

Quadri de la Torre, Gabriel (1991b). "Una Breve Crónica del Ecologismo en México." In *Servicios Urbanos, Gestión Local y Medio Ambiente,* edited by Martha Schteingart and Luciano d' Andrea (pp. 337–353). México, D.F.: Colegio de México and CE.R.FE.

Quadri de la Torre, Gabriel (1992). "Biodiversidad y Desarrollo." *Examen*. 4 (38): 5–9.

Quadri de la Torre, Gabriel (1993a). "Agua: Economía y Sustentabilidad en la Ciudad de México." *Examen*. 4 (2): 37–40.

Quadri de la Torre, Gabriel (1993b). "La Teoría del Ecologismo en una Sola Colonia." *Examen*. 5 (54): 37–39.

Quadri de la Torre, Gabriel (1994a). "La Apuesta Ecólogica de Zedillo." *Examen*. 6 (63): 41–42.

Quadri de la Torre, Gabriel (1994b). "Conservación Ecólogica: Condiciones y Conflictos." *Examen*. 5 (59): 38–39.

Rabinovitch, Jonas (1992). "Curitiba: Towards Sustainable Urban Development." *Environment and Urbanization*. 4 (2): 62–73.

Rabinovitch, Jonas and Josef Leitmann (1993). *Environmental Innovation and Management in Curitiba, Brazil* UNDP, Habitat, and World Bank Urban Management Program UMP Working Paper Series no. 1. Reprint, Washington, D.C.: UNDP, UNCHS, International Bank for Reconstruction and Development, and World Bank–UMP.

Rakodi, Carole (1990). "Can Third World Cities be Managed?" In *The Living City: Towards a Sustainable Future,* edited by David Cadman and Geoffrey Payne (pp. 111–125). New York: Routledge.

Ramírez Saiz, Juan Manuel (1983). *Carácter y Contradicciones de la Ley General de Asentamientos Humanos*. México, D.F.: Universidad Nacional Autónoma de México and Imprenta Aldena.

Ramírez Saiz, Juan Manuel (1986a). "Organizaciones Populares y Lucha Política." *Cuadernos Politicos*. 45 (January–March): 38–55.

Ramírez Saiz, Juan Manuel (1986b). "Reivindicaciones Urbanas y Organización Popular. El Caso de Durango." *Estudios Demográficos y Urbanos*. 3:399–422.

Ramírez Saiz, Juan Manuel (1989a). *Actores Sociales y Proyecto de Ciudad*. México, D.F.: Plaza y Valdés Editores.

Ramírez Saiz, Juan Manuel (1989b). "Los Objetivos de la Ley General de Asentamientos Humanos." In *Una Década de Planeación Urbano-Regional en México, 1978–1988,* edited by Gustavo Garza (pp. 27–54). México, D.F.: Colegio de México.

Ramírez Saiz, Juan Manuel (1990a). "Estrategias de Sobrevivencia y Movimientos Sociales." In *Crisis, Conflicto y Sobrevivencia: La Sociedad Urbana en México,* edited by Guillermo de la Peña, Juan Manuel Durán, Agustín Escobar, and Javier

García de Alba, (pp. 469–474). Guadalajara: Universidad de Guadalajara and CIESAS.

Ramírez Saiz, Juan Manuel (1990b). "Urban Struggles and their Political Consequences." In *Popular Movements and Political Change in Mexico*, edited by Joe Foweraker and Ann L. Craig (pp. 234–246). Boulder, Colo.: Lynne Rienner.

Ramírez Saiz, Juan Manuel (1992). "Planning for Sustainable Urban Development: Cities and Natural Resource Systems in Developing Countries." *Third World Planning Review*. 14 (4): 409–413.

Ramos, Claudia (1995). "Denunciará el PVEM Ecocidio en la Ciudad: Dicen que la Tala Continúa en Zonas de Conservación." *Reforma*, March 16, sec. B, "Ciudad y Metrópoli."

Rapoport, Eduardo H. and Ismael R. López-Moreno (eds.). (1987). *Aprotes a la Ecología Urbana de la Ciudad de México.* Sponsored by the Instututo de Ecología, the Museo de Historia Natural de la Ciudad de México, and the Program on Man and the Biosphere (MAB, UNESCO). México, D.F.: Noriega Editores and Editorial Limusa.

Redclift, Michael (1987). *Sustainable Development.* New York: Methuen.

Redclift, Michael and Colin Sage (1994). *Strategies for Sustainable Development, Local Agendas for the Southern Hemisphere.* New York: Wiley.

Reding, Andrew (1988). "Mexico at a Crossroads." *World Policy Journal.* 5 (Fall): 643.

Rees, William E. (1992). "Ecological Footprints and Appropriated Carrying Capacity: What Urban Economics Leaves Out." *Environment and Urbanization.* 4 (2): 121–130.

Reid, Anne (1985). "Las Chinampas de Iztapalapa: Su Technológia, Historia y Desaparición." *Revista A.* VI (15): 183–192, Universidad Autónoma Metropolitana, Unidad Azcapotzalco, División de Ciencias Sociales y Humanidades.

Reilly, Charles (1991). "When do Environmental Problems Become Issues? Whose Issues? And Who Manages Them Best?" Paper based on remarks at the UNRISD Workshop on Sustainable Development through People's Participation in Resource Management, May 1990. Geneva. Washington D.C.: Inter-American Foundation.

Reilly, Charles (ed.) (1995). *New Paths to Democratic Development in Latin America: The Rise of NGO-Municipal Collaboration.* Boulder, Colo.: Lynne Rienner.

Reyes Osorio, Sergio (1975). "Hacia una política de organización económica en el sector rural." In *Los Problemas de la Organización Campesina*, edited by Iván Restrepo Fernáez (pp 25–41). México, D.F.: Editorial Campesina.

Rhodes, Robert E. (1991). "The World's Food Supply at Risk." *National Geographic.* 179 (4): 75–105.

Richardson, Harry W. (1993). "Efficiency and Welfare in LDC Mega-Cities." In *Third World Cities: Problems, Policies and Prospects,* edited by John D. Kasarda and Allan M. Parnell (pp. 32–54). Newbury Park, California: Sage.

Richerson, Peter J. (1993). "Humans as Components of the Lake Titicaca Ecosystem: A Model System for the Study of Environmental Deterioration." In *Humans as Components of Ecosystems: The Ecology of Subtle Human Effects and Populated Areas*, edited by Mark J. McDonnell and Steward T. A. Pickett. New York: Springer-Verlag.

Ríos Zertuche, Francisco (1993). "Mensaje del Lic. Francisco Ríos Zertuche con Motivo de su Compasercia ante la ARDF, el Día 7 de Enero de 1993." Speech delivered to the Assembly of Representatives of the Federal District (ARDF), January 7, México, D.F.

Riva Palacio, Enrique (1987). "Contaminación del Ecosistema de la Ciudad de México." In *Atlas de la Ciudad de México*, edited by Gustavo Garza (pp. 229–230). México, D.F.: DDF and Colegio de México.

Rivera Lona, M. (1987). "La Transformación del Suelo Ejidal en Suelo Urbano: El Caso del Ejido de San Nicolás Totolapán." Bachelor's thesis in sociology. Universidad Nacional Autónoma de México, México, D.F.

Robinson, Dennis and David Wilk (1991). "Understanding the Patterns of Growth." *Land Lines.* (May): 1–2. Cambridge, Mass.: Lincoln Institute of Land Policy.

Robinson, Nicolas A. (ed.) (1993). *Agenda 21: Earth's Action Plan.* IUCN Environmental Policy and Law Paper no. 27. New York: Oceana.

Rocha, Alberto (1992). "Peligra la Zona con Desaparecer por el Crecimiento Urbano y la Sobreexplotación de Mantos Acuíferos." *Excelsior.* January 4.

Rodgers, Gerry (1989). "Introduction: Trends in Urban Poverty and Labor Market Access." In *Urban Poverty and the Labor Market: Access to Jobs and Incomes in Asian and Latin American Cities,* edited by Gerry Rogers (pp. 1–33). Geneva: International Labour Office.

Rodgers, Gerry (ed.) (1989b). *Urban Poverty and the Labor Market: Access to Jobs and Incomes in Asian and Latin American Cities.* Geneva: International Labour Office.

Rodríguez, Cecilia (1991). "The World's Most Polluted City." *Los Angeles Times,* April 21, sec. M, p. 6.

Rodríguez, Cynthia (1995). "Destruyen Zona Verde." *Reforma,* March 18, sec. B, "Ciudad y Metrópoli."

Rodríguez, D. (1986). "La Organización Popular ante el Reto de la Reconstrucción." *Revista Mexicana de Ciencias Políticas y Sociales.* 123 (January–March): 59–80.

Rodríguez, V. (1987). *The Politics of Decentralization in Mexico: Divergent Outcomes of Policy Implementation.* Ph.D. dissertation. University of California, Berkeley.

Rodríguez Lazcano, Catalina and Fernando Rodríguez (1984). *Tlalpan.* Colección Delegaciones Políticas México, D.F.: Departmento del Distrito Federal.

Rohter, Larry (1989). "Mexico City's Filthy Air: World's Worst, Worsens." *New York Times,* April 12, sec. A, p. 4.

Rolnik, Raquel (1995). "Urban Planning in the 1990s." *Countdown to Istanbul*. No. 2 (April): 6.

Rondinelli, Dennis A. and John D. Kasarda (1993). In *Third World Cities: Problems, Policies and Prospects*, edited by John D. Kasarda and Allan M. Parnell (pp. 92–119). Newbury Park: Sage.

Ros, Jaime (1996). Prospects for Growth and the Environment in Mexico in the 1990s. *World Development*. 24, (2): 307–324.

Rosenau, James N. (1995). "Governance in the Twenty-first Century." *Global Governance: A Review of Multilateralism and International Organizations*. 1 (1): 13–43.

Rostow, Walt W. (1956). "The Takeoff into Self-Sustained Growth." *The Economic Journal*. 66 (261): 25–48.

Rostow, Walt W. (1980). *The World Economy: History and Prospect*. Austin: University of Texas Press.

Saad Sotomayor, Patricia (1991). "Industriales Talan las Zonas Boscosas de las Reservas Ecológicas Indispensables." *Excelsior,* May 19.

Sabloff, Jeremy (1989). *The Cities of Ancient Mexico: Reconstructing a Lost World*. New York: Thames and Hudson.

Sachs, Ignacy (1982). *Ecodesarrollo: Desarrollo sin Destrucción*. México, D.F.: Colegio de México.

Sachs, Ignacy (1987). *Development and Planning*. New York: Cambridge University Press.

Sachs, Ignacy (ed.) (1993). *Global Ecology: A New Arena of Political Conflict*. New Jersey: Zed.

Safa, Patricia (1990). "La Crisis de la Ciudad, Movimientos Urbanos y Necesidades Socioculturales: El Caso de Santo Domingo de Los Reyes." In *Crisis, Conflicto y Sobrevivencia: La Sociedad Urbana en México,* edited by Guillermo de la Peña, Juan Manuel Durán, Agustín Escobar and Javier García de Alba (pp. 439–351). Guadalajara: Universidad de Guadalajara and CIESAS.

Salazar Ramírez, Hilda (1992). "El Medio Ambiente y la Participación Ciudadana: El Foro Mexicano de la Sociedad Civil para Río 92." *El Cotidiano* 47 (May): 11–15.

Sale, Kirkpatrick (1991). *The Conquest of Paradise: Christopher Columbus and the Columbian Legacy*. New York: Plume.

Sanders, William T., Jeffery R. Parson, and Robert S. Santley (1979). *The Basin of Mexico: Ecological Processes in the Evolution of a Civilization*. New York: Academic Press.

Sanderson, Steven (1986). *The Transformation of Mexican Agriculture: International Structure and the Politics of Rural Change*. Princeton, N.J.: Princeton University Press.

Sandoval, Juan Manuel (1991). "Los Nuevos Movimientos Sociales y Medio Ambiente en México." In *Servicios Urbanos, Gestión Local y Medio Ambiente,* edited by Martha Schteingart and Luciano d' Andrea (pp. 305–335). México, D.F.: Colegio de México and CE.R.FE.

Sassen, Saskia (1991). *The Global City: New York, London, Tokyo.* Princeton, N.J.: Princeton University Press.

Sassen, Saskia (1994). *Cities in a World Economy.* Thousand Oaks, Calif.: Pine Forge.

Satterthwaite, David (1991). "Urban and Industrial Environmental Policy and Management." In *Development Research: The Environmental Challenge,* edited by James T. Winpenny (pp. 119–136). London: Overseas Development Institute.

Satterthwaite, David (1996). "Secondary Cities: Latin America's Boom Towns." *Hemisfile.* 7(3): 1–2, 12.

Sauer, C. O. (1956). "The Agency of Man on Earth." In *Man's Role in Changing the Face of the Earth,* edited by W. L. Thomas, Jr. (pp. 49–69). Chicago: University of Chicago.

Sauer, C. O. (1961). "Fire and Early Man." *Paideuma: Mitteilungen zur Kulturkunde.* 7:399–407.

Sayer, Andrew (1992). *Method in Social Science: A Realist Approach.* 2nd ed. New York: Routledge.

Saxe-Fernández, John (1994). "The Chiapas Insurrection: Consequences for Mexico and the United States." *International Journal of Politics, Culture and Society.* 8 (2): 325–342.

Saxe-Fernández, John (1994b). "Globalization: Processes of Integration and Disintegration." *International Journal of Politics, Culture and Society.* 8 (2): 203–224.

Sayeg Helú, Jorge (1975). *La Creación del Distrito Federal.* México, D.F.: DDF, Secretaría de Obras y Servicios.

Schmink, M. (1984). "Administración Comunitaria del Reciclamiento de Desechos: El Sirdo." Publication of Seeds no. 8 (Seeds/ P.O. Box 3923/ Grand Central Station/ NY, NY 10163).

Schteingart, Martha (1983). "La Promoción Inmobiliario en el Area Metropolitana de la Ciudad de México (1960–1980)." *Demografía y Economía.* 17 (53): 83–105.

Schteingart, Martha (1984). "El Sector Inmobiliario y la Vivienda en la Crisis." *Comercio Exterior.* 34, no. 8 (August).

Schteingart, Martha (1986). "Movimientos Urbano-Ecológicos en la Ciudad de México: El Caso del Ajusco." *Estudios Políticos.* 5 (1): 17–23.

Schteingart, Martha (1987). "Expansión Urbana, Conflictos Sociales y Deterio Ambiental en la Ciudad de México: El Caso del Ajusco." *Estudios Demográficos y Urbanos.* 2 (3): 449–477.

Schteingart, Martha (1989a). "The Environmental Problems Associated with Urban Development in Mexico City." *Environment and Urbanization.* 1 (1).

Schteingart, Martha (1989b). *Los Productores del Espacio Habitable: Estado, Empresa y Sociedad en la Ciudad de México.* México, D.F.: Colegio de México.

Schteingart, Martha (1991). "Los Servicios Urbanos en el Contexto de la Problemática Ambiental." In *Servicios Urbanos, Gestión Local y Medio Ambiente,* edited by Martha Schteingart and Luciano d' Andrea (pp. 69–79). México, D.F.: Colegio de México and CE.R.FE.

Schteingart, Martha (1993). "The Role of the Informal Sector in Providing Urban Employment and Housing in Mexico." In *Urban Management: Policies and Innovations in Developing Countries,* edited by G. Shabbir Cheema and Sandra E. Ward. Westport, Conn.: Praeger.

Schteingart, Martha and Luciano d' Andrea (eds.) (1991). *Servicios Urbanos, Gestión Local y Medio Ambiente.* México, D.F.: Colegio de México and CE.R.FE.

Scott, David C. (1994). "Mexico: The Importance of Being Ernesto." *Hemisfile. 5* (5): 1–3.

Secretaría de Gobernación (1988). *Ley General del Equilibrio y la Protección al Ambiente.* México, D.F.: México.

SEDUE (Secretaría de Desarrollo Urbano y Ecología) (1982). *Planeación Democrática y Ecología.* Mexico, D.F.: México.

SEDUE (1984). *Programa Nacional de Desarrollo Urbano y Vivienda: 1984–1988* and *Programa de Ecología 1984–1988.* México, D.F.: México.

SEDUE (1985). *Recopilación y Sistematización de la Legislación Ambiental Mexicana.* México, D.F.: SEDUE.

SEDUE (1986). "Informe sobre el Estado del Medio Ambiente en México, D.F.: Subsecretaría de Ecología.

SEDUE (1987). "Vivienda y Ecología." *El Día,* anniversary supplement, September 7.

SEDUE (1989). "Reglamento de la Ley General del Equilibrio Ecológico y la Protección al Ambiente en Materia de Impacto Ambiental." *Gaceta Ecológica.* 1 (June): 1–95.

SEDUE (1990). *Programa Nacional de Desarrollo Urbano: 1990–1994.* México, D.F.: México.

Segura, Olman (1992). *Desarrollo Sostenible y Políticas Económicas en América Latina.* San José, Costa Rica: Editorial DEI.

Segura, Olman and James K. Boyce (1994). "Investing in Natural and Human Capital in Developing Countries." In *Investing in Natural Capital: The Ecological Economics Approach to Sustainability,* edited by AnnMari Jansson, Monica Hammer, Carl Folke, and Robert Costanza (pp. 479–489). Covelo, Calif.: Island.

Selby, H. A., A. D. Murphy, S. A. Lorenzen, I. Cabrera, A. Castaneda, and I. Ruiz Love (1990). *The Mexican Urban Household: Organizing for Self-Defense.* Austin: University of Texas Press.

Semo, Ilán and Juan Manuel Sandoval (1983a). "Cuando el Progreso Nos Alcance: Hacia una Crítica de la Ecología Política." *El Buscón.* 1, no. 5 (July/August): 49–58.

Semo, Ilán and Juan Manuel Sandoval (1983b). "Cuando el Progreso Nos Alcance: el Ecologismo de Estado." 1, no. 6 (September/October): 107–116.

Serrania Alvarez, Lázaro (1984). "Continúan las Fastuosas Construcciones en el Ajusco." *UnoMásUno,* February 17, Metrópoli sec.

Serrano Gutiérrez, Alicia (1993a). "En el Programa del Pronasol Participaron 64 Colonias en Ajusco Medio." *Noti Tlalpan* (a news publication of the Tlalpan District, México, D.F.), January 8.

Serrano Gutiérrez, Alicia (1993b). "Se Entregaron 727 Escrituras a Residentes de Tlalpan." *Noti Tlalpan* (a news publication of the Tlalpan District, México, D.F.), January 8.

Setchell, Charles A. (1995). "The Growing Environmental Crisis in the World's Megacities: The Case of Bangkok." *Third World Planning Review.* 17 (1): 1–18.

Silva, Samuel (1994). "How Many People, How Much Wealth?" *Inter-American Development Bank Report.* July, p. 9.

Silva Herzog, Jesús (1964). *El Agrarismo Mexicano y la Reforma Agraria.* México, D.F.: Fondo de Cultrura Económica.

Simmie, James (1993). *Planning at the Crossroads.* London: University College London Press.

Singh, Naresh (1995). *Empowerment for Sustainable Development: Towards Operational Strategies.* Winnipeg, Manitoba, Canada: International Institute for Sustainable Development.

Sitarz, Daniel (1993). *Agenda 21: The Earth Summit Strategy to Save Our Planet.* Boulder, Colo.: Earth.

Slater, David (ed.) (1985). *New Social Movements and the State in Latin America.* Amsterdam: CEDLA.

Slocombe, Scott D. and Thaddeus C. Trzyna (1993). *What Works? An Annotated Bibliography of Case Studies of Sustainable Development.* Published for the Commission on Environmental Strategy and Planning of IUCN, and the World Conservation Union. ICUN-CESP working paper no. 5. Sacramento, Calif.: International Center for the Environment and Public Policy.

Smit, Jac and Joe Nasr (1992). "Urban Agriculture for Sustainable Cities: Using Wastes and Idle Land and Water Bodies as Resources." *Environment and Urbanization.* 4 (2): 141–152.

Soberón, Jorge (1990). "Restauración Ecológica en el Ajusco Medio." *Oikos.* 5 (September–October): 4. (a publication of the Centro de Ecología, Universidad Nacional Autónoma de México, México, D.F.).

Soberón, Jorge (1992). "Voluntad Política 'Notable': Convertir Lomas del Seminario en Reserva Ecológica." *UnoMásUno,* August 18.

Sosa, Iván (1991). "Inexorable, el Avance de la Mancha Urbana Sobre las Areas de Conservación Ecológica." *El Financiero,* April 5.

SPP (Secretaría de Programación y Presupuesto (1983). *Plan Nacional de Desarrollo: 1983–1988.* México, D.F.: SPP.

SPP (1989). *Plan Nacional de Desarrollo: 1989–1994.* México, D.F.: SPP.

Stern, Richard, Rodney White, and Joseph Whitney (eds.) (1992). *Sustainable Cities: Urbanization and the Environment in International Perspective.* Boulder, Colo.: Westview.

Stuart, Gene S. (1988). *America's Ancient Cities.* Washington, D.C.: Special Publications Division of the National Geographic Society.

Suárez Pareyón, Alejandro (1987). *El Programa de Vivienda del Molino: Una Experiencia Autogestiva de Urbanización Popular.* México, D.F.: Centro de la Vivienda y Estudios Urbanos, A.C. (CENVI).

Subdelegación del Ajusco (1985a). "Plan de Trabajo." Tlalpan, D.F.: Subdelegación del Ajusco.

Subdelegación del Ajusco (1985b). "Programa de Actividades." Tlalpan, D.F.: Subdelegación del Ajusco.

Subdelegación del Ajusco (1985c). "Programa de Trabajo: Lineamientos para su Ejecución." Tlalpan, D.F.: Subdelegación del Ajusco.

Subdelegación del Ajusco (1986). "Programa de Actividades." Tlalpan, D.F.: Subdelegación del Ajusco.

Subdelegación del Ajusco (1987a). "1. Diagnóstico, 2. Programa de Trabajo, 3. Avance Programático." Tlalpan, D.F.: Subdelegación del Ajusco.

Subdelegación del Ajusco (1987b). "Inventario de Procedimientos." Tlalpan, D.F.: Subdelegación del Ajusco.

Sunkel, Osvaldo (1985). *América Latina y la Crisis Económica Internacional: Ocho Tesis y una Propuesta.* Buenos Aires: Grupo Editor Latinamericano S.R.L.

Sunkel, Osvaldo and Nicolo Gligo (eds.) (1981). *Estilos de Desarrollo y Medio Ambiente en la América Latina.* México, D.F.: Fondo de Cultura Económica.

"Surge la Primera Colonia Ecológica del País" (1984). *El Nacional,* June 31.

"Surgen Fraccionadores Clandestinos con la Autorización de Poseedores de Tierra." (1991). *UnoMásUno,* February 21.

Tang, Wing-Shing (1994). "Urban Land Development under Socialism: China between 1949 and 1977." *International Journal of Urban and Regional Research.* 18 (3): 392–415.

Tarver, James D. (ed.) (1994). *Urbanization in Africa: A Handbook.* London: Greenwood.

Taylor, P. and R. Garcia Barrios (1995). "The Social Analysis of Ecological Change: From Systems to Intersecting Processes." *Social Science Information.* 34 (1): 5–30.

Third World Planning Review (1994). Special issue, "The Contribution of Cities to Sustainability." 16 (2): 1–9.

Thorup, Cathryn (1991). "The Politics of Free Trade and the Dynamics of Cross-Border Coalitions in U.S.-Mexican Relations." *Columbia Journal of World Business.* 26 (2).

Toledo, Víctor Manuel (1983). "Ecologismo y Ecología Política: La Otra Guerra Florida." *NEXOS.* 6, no. 6 (September): 15–24.

"Toman Vecinos Delegación" (1996). *Reforma,* May 23.

Toynbee, Arnold (1973). *Cities of Destiny.* New York: McGraw-Hill.

Tudela, Fernando (1991). "El Laberinto de la Complejidad: Hacia un Enfoque Sistémico del Medio Ambiente y la Gestión de los Servicios Urbanos en América Latina." In *Servicios Urbanos, Gestión Local y Medio Ambiente,* edited by Martha Schteingart and Luciano d' Andrea (pp. 41–55). México, D.F.: Colegio de México and CE.R.FE.

Tulchin, Joseph S. and Andrew I. Rudman (1991). *Economic Development and Environmental Protection in Latin America.* Boulder, Calif.: Lynne Rienner.

Tuñón Pablos, Esperanza (1992). "Women's Struggles for Empowerment in Mexico: Accomplishments, Problems, and Challenges." Translated by David and Victor Arriaga. In *Women Transforming Politics: Worldwide Strategies for Empowerment,* edited by Jill M. Bystydzienski (pp. 95–107). Bloomington: Indiana University Press.

Turner, B. L., II and K. W. Butzer (1992). "The Columbian Encounter and Land-use Change." *Environment.* 34:16–20, 37–44.

Turner, B. L. and William B. Meyer (1993). "Environmental Change: The Human Factor." In *Humans as Components of Ecosystems: The Ecology of Subtle Human Effects and Populated Areas,* edited by Mark J. McDonnell and Stewart T. A. Pickett. New York: Springer-Verlag.

Turner, B. L., II, W. C. Clark, R. W. Kates, J. F. Richards, J. T. Mathews, and W. B. Meyer (eds.) (1990). *The Earth as Transformed by Human Action.* New York: Cambridge University Press.

Turner, John F. C. (1976). *Housing by People: Towards Autonomy in Building Environments.* London: Marion Boyars.

Ulrich, W. (1983). *Critical Heuristics of Social Planning: A New Approach to Practical Philosophy.* Bern: Verlag Paul Haupt.

Ultria, Rubén D. (1981). "La Incorporación de la Dimensión Ambiental en la Planificación del Desarrollo: Una Posible Guía Metodológica." In *Estilos de Desarrollo y Medio Ambiente en la América Latina,* edited by Osvaldo Sunkel and Nicolo Gligo (pp. 471–539). México, D.F.: Fondo de Cultura Económica.

United Nations Conference on Trade and Development (UNCTAD) (1996). "Midrand Declaration and a Partnership for Growth and Development." Adopted by the UNCTAD at the Ninth Session of the United Nations Conference on Trade and Development, Midrand, South Africa April 27–May 11. (Available on the world wide web at http://www.un.org/)

UNDP (United Nations Development Program) (1992). *Human Development Report.* New York: Oxford University Press.

United Nations Environmental Programme (1973). *Proceedings of the UNEP Governing Council.* Nairobi, Kenya: UNEP.

United Nations Environmental Programme (1981). *In Defense of the Earth.* Nairobi, Kenya: UNEP.

United Nations Environmental Programme (1989). *Environmental Data Report: 1989/1990.* 2d ed. Oxford: Blackwell.

United Nations General Assembly, Preparatory Committee for the UN Conference on Human Settlements (Habitat II) (1995). *Draft Statement of Principles and Commitments and Global Plan of Action: Report of the Secretary-General.* Nairobi, Kenya: UNCHS and the Habitat II Secretariat.

Universidad Autónoma Metropolitana Unidad Azcapotzalco, División de Ciencias Sociales y Humanidades (ed.) (1986). "Movimiento Urbano Popular." *El Cotidiano,* 2 (11): 1–72.

Ureña, José and Víctor Ballinas (1989). "Se Expropiarán 727 Hectáreas del Ajusco Medio: Camacho," and "Programa para la Conservación del Agua en el D.F. y Area Conurbada, Anunció Salinas." *La Jornada,* June 29.

Urrutia, Alonzo (1992a). "Gobierno Propio para Acabar con la Excepcionalidad de la Ciudad." *La Jornada.* November 24.

Urrutia, Alonzo (1992b). "Proponen que Organizaciones Vecinales Tengan Facultades de Decision: Techan de Inorperante al Consejo Consultivo del DF." *La Jornada.* November 19.

Urrutia, Alonso (1992c). "Sostienen Dos Investigadores de la UAM: La Estructura de Gobierno del D.F. Desalienta la Participación Social." *La Jornada,* December 1, "La Capital" sec., p. 35.

Urrutia, Alonso (1995). "Los Servicios toman el Lugar de la Industria en la Economía Capitalina." *La Jornada.* December 18.

U.S. Environmental Protection Agency and SEDUE (Secretaría de Desarrollo Urbano y Ecología, México) (1992). *Integrated Environmental Plan for the Mexican-U.S. Border Area (First Stage, 1992–1994).* Washington, D.C.: GPO.

Vaillant, George C. (1972). *Aztecs of Mexico.* Baltimore: Penguin.

Valasis, Adriana D. (1996). "Buscan Llenar Vacíos de Ley Ambiental: Presentan en la ALDF Iniciativa de Ley Ecológica para la Ciudad." *Reforma.* April 20, sec. B, "Ciudad y Metrópoli," p. 6.

Vargas, Fernando (1984). *Parques Nacionales de México y Reservas Equivalentes.* México, D.F.: Instituto de Investigaciones Económicas, Universidad Nacional Autónoma de México.

Vargas Vidales, Martín (1992). "Equilibrio de Facultades entre DDF y ARDF para Modificar el Uso del Suelo en la Capital." *UnoMásUno,* August 29, "Valle de México" sec., p. 12.

Varley, Ann (1985). "La Zona Urbana Ejidal y la Urbanización de la Ciudad de México." In *Revista de Ciencias Sociales y Humanidades* 6:71–96, and in *Revista A.* 6 (15).

Velázquez Zárate, Enrique (1992). "Acerca de la Problemática del Medio Ambiente y la Contaminación en el Valle de México." *El Cotidiano,* 47: 32–42.

Venegas, Juan Manuel (1992). "Necesita el D.F. Consejos Locales de Gestoría y Planeación." *La Jornada,* November 24, "La Capital" sec., p. 35.

Vitale, Luis (1983). *Hacia una Historia del Ambiente en América Latina: De las Culturas Aborígenes a la Crisis Ecológica Actual.* México, D.F.: Nueva Sociedad.

Wade, I. (1987). *Food Self-Reliance in Third World Cities.* Food Energy Nexus Report no. 22. Paris: United Nations University.

Walker, R. B. J. (1988). *One World, Many Worlds: Struggles for a Just World Peace.* Boulder, Colo.: Lynne Rienner.

Walker, R. B. J. (1990). *One World, Many Worlds: Struggles for a Just World Peace.* Boulder, Colo.: Lynne Rienner.

Walton, John (1991). "Review Articles." *International Journal of Urban and Regional Research.* 15 (4): 629–633.

Ward, Peter (1986). *Welfare Politics in Mexico: Papering over the Cracks.* Boston: Allen & Unwin.

Ward, Peter (1989). "Political Mediation and Illegal Settlement in Mexico City." In *Housing and Land in Urban Mexico,* edited by Allen Gilbert (pp. 135–156). Monograph Series no 31. San Diego: Center for U.S.-Mexican Studies.

Ward, Peter (1990). *Mexico City: The Production and Reproduction of an Urban Environment.* London: Belhaven.

Ward, Peter (ed.) (1982). *Self-Help Housing: A Critique.* London: Mansell.

Ward, Peter M. and Chris Macoloo (1992). "Articulation Theory and Self-Help Housing Practice in the 1990s." *International Journal of Urban and Regional Research.* 16 (1): 60–80.

WCED (World Commission on Environment and Development) (1987). *Our Common Future.* New York: Oxford University Press.

Wegelin, Emiel A. (1994). "Everything You Always Wanted to Know about the Urban Management Program (But Were Afraid to Ask)." Comment on "The World Bank's 'New' Urban Management Programme: Paradigm Shift or Policy Continuity?" by Gareth A. Jones and Peter M. Ward.) *Habitat International.* 18 (4): 127–137.

Wei, Yehua (1994). "Urban Policy, Economic Policy, and the Growth of Large Cities in China." *Habitat International.* 18 (4): 53–65.

Weiner, Jonathan (1991). *The Next 100 Years: Shaping the Fate of Our Living Earth.* New York: Bantam Books.

Weintraub, Sidney (1996) "Detour on the Way to the Promised Land." *Hemisfile.* 7, no. 3 (May/June): 5–7.

Wells, Jill (1995). "Population, Settlements and the Environment: The Provision of Organic Materials for Shelter." *Habitat International.* 19 (1): 73–90.

Whitmore, T. M. (1991). A Simulation of the Sixteenth-century Population Collapse in the Basin of Mexico." *Annals of the Association of American Geographers,* 81:463–487.

Whitmore, T. M., B. L. Turner, II, D. L. Johnson, R. W. Kates, and T. R. Gott-schang (1990). In *The Earth as Transformed by Human Action,* edited by B. L. Turner, II, W. C. Clark, R. W. Kates, J. F. Richards, J. T. Matthews, and W. B. Meyer (pp. 92–119). Cambridge: Cambridge University Press.

Wilk, David (1989). "Planeación del Uso del Suelo y Medio Ambiente en la Ciudad de México." In *Una Década de Planeación Urbano-Regional en México, 1978–1988,* edited by Gustavo Garza (pp. 329–348). México, D.F.: Colegio de México.

Wilk, David (1991a). "Controles del Uso del Suelo y Contención del Crecimiento en Areas Conurbadas: Estudios del Caso de Tlalpan y Chalco en el Area Metropolitana de la Ciudad de México." Report prepared for the conference "Controles de Crecimiento en Areas Metropolitanas de México y Estados Unidos: Políticas, Instrumentos y Técnicas de Análisis," June 17–19, México D.F.

Wilk, David (1991b). "Growth Management in the Eastern Fringes of Mexico City (State of Mexico): Policies, Institutions, and Technical Bases of Planning." Preliminary Report prepared for the Lincoln Institute of Land Policy, presented at the conference "Controles de Crecimiento en Areas Metropolitanas de México y Estados Unidos: Políticas, Instrumentos y Técnicas de Análisis," June 17–19, México D.F.

Wilk, David (1991c). "Growth Management in the Southern Fringes of Mexico City (Federal District): Policies, Institutions, and Technical Bases of Planning." Preliminary Report prepared for the Lincoln Institute of Land Policy, presented at the conference "Controles de Crecimiento en Areas Metropolitanas de México y Estados Unidos: Políticas, Instrumentos y Técnicas de Análisis," June 17–19, México D.F.

Williams, Michael (1993). "An Exceptionally Powerful Biotic Factor." In *Humans as Components of Ecosystems: The Ecology of Subtle Human Effects and Populated Areas,* edited by Mark J. McDonald and Steward T. A. Pickett, (pp. 24–39) New York: Springer-Verlag.

World Bank (1991). *Environmental Assessment Sourcebook..* Vol. 1, 2, 3. Washington, D.C.: Environment Department.

World Bank (1992). *World Development Report 1992: Development and the Environment.* New York: Oxford University Press.

World Bank and International Bank for Reconstruction and Development (1994). *World Development Report 1994: Infrastructure for Development.* New York: Oxford University Press.

World Bank, UNDP, and UNCHS (n.d.) *Urban Management Program: Helping to Manage the Future of Cities.* Washington, D.C.: UMP Publications.

World Resources Institute (1990). *World Resources 1990–91.* Oxford: Oxford University Press.

Wright, Angus (1990). *The Death of Ramón González: The Modern Agricultural Dilemma.* Austin: University of Texas Press.

Yip Yat Hoong and Kwai Sim Low (eds.) (1984). *Urbanization and Ecodevelopment, with Special Reference to Kuala Lumpur.* Kuala Lumpur: Institute of Advanced Studies, University of Malaya.

Zedillo, Ernesto (1994). *Las Políticas de Bienestar.* México, D.F.: Partido Revolucionario Institucional, Comité Ejecutivo Nacional, Secretaría de Información y Propaganda.

Zedillo, Ernesto (1995). First State of the Union Address, September 1, 1995. Archived on the Mexican government's world wide web site labeled "Archivo General—Informes Presidenciales," located at http://www.presidencia.gob.mx/ar-pto2.html.

Zedillo, Ernesto (1996). Second State of the Union Address, September 1, 1996. Archived on the Mexican government's world wide web site labeled "Archivo General—Informes Presidenciales," located at http://www.presidencia.gob.mx/ar-pto2.html.

"Zedillo Outlines His Peso-Rescue Plan, but Mexican Currency's Slide Continues" (1995) *The Wall Street Journal,* January 4, sec. A, p. 3.

"Zedillo Outlines Rescue Plan, Calls for Deep Sacrifices" (1995). *Los Angeles Times,* January 4, sec. A, p. 1.

Zepeda, Pedro José and Virginia Pérez Cota (1988). "Alimentación y Energía en la Zona Metropolitana de la Ciudad de México." In *Medio Ambiente y Calidad de Vida,* edited by Sergio Puente and Jorge Legorreta (pp. 137–164). México, D.F.: DDF and Plaza & Janés.

Zimmer, Karl S. (1994). "Human Geography and the 'New Ecology': The Prospect and Promise of Integration." *Annals of the Association of American Geographers.* 84 (1): 108–125.

Index